CELL GROWTH PROCESSES: NEW RESEARCH

CELL GROWTH PROCESSES: NEW RESEARCH

DAIKI KIMURA
EDITOR

Nova Biomedical Books
New York

For permission to use material from this book please contact us:
Telephone 631-231-7269; Fax 631-231-8175
Web Site: http://www.novapublishers.com

NOTICE TO THE READER

Library of Congress Cataloging-in-Publication Data

Cell growth processes : new research / Daiki Kimura, editor.
p. ; cm.
Includes bibliographical references and index.
ISBN 978-1-60456-132-6 (hardcover)
1. Cells--Growth. 2. Cell transformation. 3. Cell proliferation. I. Kimura, Daiki.
[DNLM: 1. Cell Growth Processes. 2. Aging. 3. Cell Transformation, Neoplastic. 4. Neoplasms--etiology. 5. Neoplasms--pathology. QU 375 C39269 2007]
QH604.7.C45 2007
571.8'4--dc22 2007045983

Published by Nova Science Publishers, Inc. ✦ New York

Contents

Preface

When used in the context of reproduction of living cells the phrase "cell growth" is shorthand for the idea of "growth in cell populations by means of cell reproduction." During cell reproduction one cell (the "mother" cell) divides to produce two daughter cells.

Cell proliferation, which depends on the intimately linked processes of growth and division, is a fundamental systems-level attribute of all life forms. The precise regulation of proliferation in response to internal and external cues is critical for development, tissue renewal and evolutionary fitness, while the disregulation of cell proliferation underlies a variety of human diseases, most notably cancer and aging. Historically, breakthroughs in our understanding of cell growth and division have derived from cross-fertilization of results and ideas from researchers studying a wide range of model organisms, from yeast to humans. The basis for cell proliferation entails the control of key signaling and cell cycle regulators through transcriptional, translational, post-translational, genetic and epigenetic mechanisms. Indeed, many conceptual breakthroughs in cell regulation have derived from analyses of basic cell cycle mechanisms. This new book presents the latest research in the field.

Chapter I - Cell-material interface plays a key role in the interaction of cells with artificial materials designed for construction of body implants or tissue engineering. Physical and chemical properties of the material surface, such as its polarity, wettability, electrical charge and conductivity, roughness and morphology, rigidity or elasticity as well as the presence of various chemical functional groups, strongly influence the cell adhesion. The latter is mediated by adsorption of cell adhesion-mediating molecules (e.g. vitronectin, fibronectin, collagen, laminin, fibrinogen) from the serum of the culture medium or body fluids in appropriate spatial conformation or flexibility enabling the accessibility of specific sites on these molecules (e.g., certain amino acid sequences or saccharide-based ligands) by adhesion receptors on the cell membrane (e.g., integrins). After binding these ligands, the adhesion receptors are recruited into focal adhesion plaques, where they associate with cascades of various structural and signalling molecules, such as cytoskeletal proteins and kinases. By these cascades, the signal provided by the cell-material interaction is delivered to the cell nucleus. It influences the gene expression, and thus the further behavior of cells, manifested by their viability, synthesis of various molecules, proliferation, differentiation, functioning, immune activation, stress adaptation or cell death.

This interdisciplinary review involves:

i) physical modifications of the material surface by plasma discharge, irradiation with ions or ultraviolet light and effects of these modifications on the surface wettability, electrical conductivity and formation of chemical functional groups influencing cell adhesion,
ii) deposition of organic or inorganic layers on the material surfaces, especially those nanostructured (e.g., nanofibers of fibrin and extracellular matrix molecules, carbon nanoparticles, such as nanodiamonds, fullerenes and nanotubes, nanocomposite metal/C:H layers),
iii) functionalization of the material surface with amino acids and oligopeptidic ligands for cell adhesion receptors (e.g., GRGDSG),
iv) effects of all these modifications on the extent and strength of adhesion of vascular and bone-derived cells, cell proliferation activity, switch between cell proliferation and differentiation, expression of cell-type specific markers of differentiation and other cell functions.

Chapter II - Skeletal muscle is a complex tissue composed predominantly by contractile elements, but the intimate association between muscle fibers and the surrounding endothelial cells and fibroblasts is determinant for the maintenance of the integrity and proper function of the muscle. The cellular unit of adult skeletal muscle is the myofiber, a highly specialized syncytium sustained by hundreds of post-mitotic myonuclei within a continuous cytoplasm. This terminally differentiated cell unit has not the ability of self-renewal. Paradoxically, skeletal muscle is daily submitted to a variety of potentially harmful stimuli, thereby requiring an ever-available, renewable source of cells for muscle adaptation and repair. Since its identification, the satellite cell population located between the basal lamina and myofiber sarcolemma has been a popular candidate for the adult skeletal muscle "stem cell". Satellite cells are mitotically quiescent in adult muscle but re-enter the cell cycle following muscle injury to generate a population of myoblasts, which may subsequently fuse with the adjacent myofiber or with other myoblasts to form new myofibers.

The identification of proliferating cells in different tissues has become a useful laboratory tool in the comprehension of the mechanisms behind tissue adaptation. In this context, The authors have evaluated the time-course of the contribution of cellular proliferation within the process of muscle regeneration by submitting mice *gastrocnemius* muscle to crush injury and studying the cellular events throughout the following 20 days. The morphological data obtained point to a significant inflammatory reaction between the 2nd and 4th days after damage. The levels of cell proliferation (evaluated by BrdU incorporation) were significantly higher during the 4th and 6th days after injury, mainly due to the activation of satellite cells and interstitial non-endothelial cells. With this approach the authors intended to highlight the efficiency of repair mechanisms used in a crucial tissue as skeletal muscle to minimize the incapacitating effects of severe damage. Indeed, there has been much debate as to whether so complex and highly organized array of syncytial cells could be regenerated and, if so, how this could occur in postnatal muscle. Unexpectedly, the authors also observed that the cellular proliferation activity is also a meaningful issue when the skeletal muscle is subjected to unloading conditions. In fact, an unpredicted but prominent activation of satellite cells can be noticed in the earliest phases of the atrophic process in hindlimb suspended mice. Interstitial endothelial and non-endothelial cells also evidence signals of proliferation in a typical time-course. By focusing on the cell populations with enhanced regeneration abilities, advances in clinical therapies for muscle injuries should be obtained in a near future.

Chapter III - Despite their wide differences, most theories about cancer proposed during the past century agree that cancer is a *biological nonsense* for the organism in which it originates since cancer cells are believed to be *autonomous* (or relatively autonomous), meaning that they are not subject to the rules and regulations that control normal cell proliferation and differentiation in the organism.

In this chapter the authors have challenged this interpretation on the basis that, throughout the animal kingdom, cancer seems to arise only in organs and tissues that display lost or diminished regenerative ability. In these organs, any injury causing loss of cells or cellular function cannot be compensated by cellular division and in consequence, the original size and function cannot be restored.

The authors suggest that this situation induces a *crisis*, which, through putative danger signals resulting from retardation of tissue repair, acceleration of cell loss and functional compromise, might promote some degree of variability in the remaining live but arrested cells of the injured organ. The outcome of this situation would be the emergence, by chance, of a genetically or epigenetically modified cell variant bearing mitotic ability to respond to the reparative signal. This new variant would begin to divide and if it were poorly functional or non-functional, the organ would be numerically but not functionally restored. In consequence, it would not score the regeneration as effective and it would continue to send mitotic signals to restore the lost or diminished organ function. As a result, the new variant would grow over and over and the outcome would be a tumor.

According to this hypothesis, cancer would *not be an autonomous* entity disobeying the mechanisms controlling cell proliferation; in fact, in the injured organ which has its regenerative ability lost or diminished, a tumor cell would be the only one able to respond to the demand to proliferate surrounded by "normal" but arrested cells that cannot respond to that signal. According to this interpretation, cancer would have a profound *biological sense*: it would be the ultimate attempt to restore organ functions and structures that have been lost or altered by aging and noxious environmental agents. However, unlike normal structures, cancer would have *no physiological value*, because the usually poor or non-functional nature of its cells would make their reparative task unattainable.

The hypothesis advanced in this chapter might have significant practical implications. All conventional therapies against cancer attempt to kill all tumor cells. However, according to the authors hypothesis, the problem might not be solved even if all the tumor cells were eradicated. In effect, if the organ failure remained, new tumor cells would emerge and the tumor would reinitiate its progressive growth in response to the permanent regenerative signal of the non-restored organ. Therefore, efficient anti-cancer therapy should combine an attack against the tumor cells themselves with the correction of the organ failure, which, according to this hypothesis, would be in the core of the cancer problem.

Chapter IV - Atherosclerosis is a chronic inflammatory-proliferative disease of the arterial wall characterized by progressive accumulation of lipids and inflammatory cells and enhanced proliferation and migration of smooth muscle cells. Nitric oxide (NO) is well recognized as the anti-atherogenic regulator by inhibiting the inflammatory-proliferative processes. Endothelial dysfunction with reduced nitric oxide (NO) production and/or bioavailability is common in patients with apparent coronary atherosclerosis or patients with risk factors and may contribute to the initiation and the progression of atherosclerosis. An increase of endogenous NO synthase inhibitors such as asymmetric dimethylarginine (ADMA) are now believed to be a major causative factor for the NO deficiency. Plasma ADMA levels are elevated in patients with atherosclerosis and predict future cardiovascular events, thus being a novel biomarker for atherosclerotic vascular diseases. In this review, the authors focus current knowledge on the role of ADMA in endothelial cell growth in atherosclerosis.

Chapter V - The EGF-receptor belongs to the HER (human) or ErbB (mouse) family of receptor tyrosine kinases that includes four members (HER1-4/ErbB1-4). Activation of these receptors by their appropriate ligands will induce the formation of homodimers and/or heterodimers. Heterodimer formation between different EGF-receptor family members has been shown to provide greater signal duration, diversification and amplification, as compared to their corresponding homodimer receptor complex. This is particularly true for the HER2/ErbB2 and the HER3/ErbB3 receptors because the HER2/ErbB2 receptor has no known ligand, whereas the HER3/ErbB3 receptor has no intrinsic tyrosine kinase activity, but heterodimer formation thereby allows these EGF-receptor family members to participate in mitogen-dependent signal transduction. Studies have also shown that the most potent EGF-receptor receptor complex for stimulating cell growth and transformation is composed of HER3/ErbB3 with either HER1/ErbB1 or HER2/ErbB2. Following ligand binding and homo- or heterodimerization, the EGF-R becomes transphosphorylated by its dimer partner, and the tyrosine phosphorylated EGF-receptor dimer is now able to interact and activate several important intracellular signaling proteins involved in the activation of several mitogenic signaling pathways including the mitogen-activated protein kinase (MAPK) cascade or Ras/Raf/MEK/ERK; the phosphatidylinositol 3-kinase (PI3K)/PI3K-dependent kinase (PDK)/Akt pathway; and the cAMP-dependent protein kinase (PKA) pathway. For example, studies have shown that HER3/ErbB3 heterodimers are the most important and potent activators of the PI3K/PDK/Akt mitogenic pathway in neoplastic mammary epithelial cells. Excessive Akt signaling promotes oncongenic expression by stimulating tumor cell proliferation, transformation and metastasis, and overexpression of constitutively active Akt has been shown to reverse the proliferative block induced by treatments targeted at inhibiting EGF-receptor mitogenic signaling in neoplastic mammary epithelial cells. In addition, these distinct mitogenic signaling pathways can greatly influence the activation and intensity of the other signaling pathways through the process of cross-talk. This review will describe the role of EGF-receptor family members in the activation and modulation of various mitogenic signaling pathways in normal mammary epithelial cells, and how alterations in the expression or activation of these EGF-receptors can lead to neoplastic transformation and enhance the development, growth and malignant progression of breast cancer in women.

Chapter VII - Solid free form (SFF) technologies offer nowadays established means to manufacture 3D scaffolds for tissue engineering and regenerative medicine applications. The resulting structures are characterized by defined porosity, pore volume and shape. Their pore network can be periodically designed and is completely accessible and interconnected, which is beneficial for optimal cell nutrient supply and diffusion. For these reasons, SFF scaffolds are interesting 3D models to study not only tissue regeneration, but also cell growth processes. No systematic study has been done on the influence of different scaffold's porosities, pore volume and shape on cell growth and tissue formation. Therefore, the aim of this investigation was to evaluate whether SFF scaffolds with different pore architectures can influence cell seeding efficiency, attachment, and tissue regeneration. 3D fiber deposition (3DF) was used as a fabrication technique and cartilage regeneration was considered as a specific tissue engineering application. Results showed that scaffolds with lower porosity and pore volume supported a higher amount of attached cells both after 10 and 30 days in dynamic culture. Scaffolds with a decreased pore volume along the longitudinal direction, but equal overall porosity, supported a higher number of entrapped cells as well. This was accompanied by a correspondent higher production of extracellular matrix (ECM), as analyzed by quantitative GAG expression. No significant influence of the analyzed pore shapes could be detected in terms of cell attachment and growth, or in terms of tissue formation, as also noticed by different microscopic analysis. Hence, when 3D scaffolds with completely accessible and interconnected pore networks are used for tissue engineering and regenerative medicine, pore volume may be selectively tailored for the overall structure, or regionally within the same structure, to appropriately modulate cell growth and tissue formation for specific applications.

Chapter VIII - Glial-cell-line-derived neurotrophic factor (GDNF) was known as a survival factor for midbrain dopaminergic neurons. GDNF and related ligands, neurturin, persephin and artemin maintain several neuronal populations in the central nervous systems, including midbrain dopamine neurons and motoneurons. In addition, GDNF, neurturin and artemin support the survival and regulate the differentiation of many peripheral neurons, including sympathetic, parasympathetic, sensory and enteric neurons. The GDNF family ligands signal through a receptor complex consisting of a ligand binding component, GDNF family receptor α (GFRα) and RET receptor tyrosine kinase. A recent study showed that systemic GDNF family ligands prevents and reverses neuropathic pain and normalizes spinal nerve ligation (SNL)-induced morphological changes. It was suggested that GDNF family ligands plays a role in inducing neurite outgrowth in the early stages of neuron development. Furthermore, the mature DRG neurons showed some response to artemin, suggesting that they exhibited a developmental shift to the ligand as prenatal DRG neurons.

Upon ligand binding, GDNF in complex with GFR may interact with heparan sulphate glycosaminoglycans to activate the Met receptor tyrosine kinase through cytoplasmic Src-family kinases. GDNF family ligands also signal through the neural cell adhesion molecule NCAM. In cells lacking RET, GDNF binds with high affinity to the NCAM and GFRα complex, which activates Fyn and FAK. At the final stage of intracellular signaling, a set of genes involved in regulation of actin dynamics including Wiskott-Aldrich syndrome protein (WASP) interacting protein, cofilin and dynamin were transcriptionally regulated by artemin, suggesting a previously undefined role in the regulation of neurite growth by artemin. DNA methylation provides a mechanism for artemin-dependent genetic regulation responsible for axonal growth.

In: Cell Growth Processes: New Research
Editor: Daiki Kimura, pp. 1-3
ISBN: 978-1-60456-132-6

Expert Commentary

Cell Growth Measurement and Kinetic Parameters in Cutaneous Melanoma: Is MIB-1 Index a Useful Indicator to Predict Patients' Outcome?

Pierre Vereecken, Marianne Laporte, Michel Petein, Bernard Van Den Heule,and Michel Heenen

Departments of Dermatology, CHU-Brugmann/ Institut Jules Bordet, Hôpital Erasme, Université Libre de Bruxelles (U.L.B.), Brussels, Belgium, IPG, Gosselies, Belgium and CMP, Brussels, Belgium

Cell proliferation is one of the most complex biological processes, as it assures tissue growth by increasing the number and the size of the cells and the intercellular substance. In any tissue, there is always a fraction of the cell population that it is actively cycling; the other non-cycling cells are resting or can no more divide. This proliferative or growth fraction can be defined as the ratio of cycling to cycling plus non-cycling cells. This fraction of proliferating cells can be measured by in-vivo and in-vitro methods.

Cell kinetics and kinetic parameters have been reported to have a significant prognostic value in various tumors i.e. gastro-intestinal and breast carcinomas for which cell proliferation level have been correlated to the development of metastases, and overall survival. Kinetic parameters are not a measurement of tumor growth rate. Indeed the determination of tumor growth rate (KG) needs a time component and the calculation of cell fluxes: the birth rate (KB) and the loss rate (KL) : KG=KB-KL. Static, or time independent parameters i.e. mitotic rate and MIB-1 index are thus an easier way to estimate tumor growth.

Since the end of the 20th century melanoma has become the cancer whose incidence has increased the most amongst Caucasians doubling every ten years to reach a current rate of 10 -15 per 100,000 per annum. Epidemiological models suggest that in fact one person in 75 is likely to develop a melanoma sometime in their life. Alongside this rise in the incidence of melanoma there has been an increase in secondary mortality. The number of deaths from melanoma was observed throughout the second half of the 20th century which then leveled out due to awareness campaigns about primary and secondary detection. Therapeutic progress in this field has been somewhat limited despite the enthusiasm generated by immunotherapy clinical studies, which has encouraged the pursuit of prevention and early detection, as well as a better knowledge of tumor prognosis. The prognosis of melanoma is mainly determined by the Breslow index which translate the thickness of the lesion in mm : how thicker the lesion is, how worse is the prognosis. However, this rule cannot explain the outcome of all patients : some patients with thinner lesions, with a breslow index less than 1mm, can develop metastases, while other patients with thick melanoma can remain disease free during mainy years. New prognostic factors are thus needed to predict the outcome of melanoma patients. Previous studies have shown that cell growth and kinetic parameters can be considered as prognostic indicators in melanoma field, but because of conflicting results, this prognostic value is still questionable.

In order to clarify the problem let us go back to some definitions. KB (cells produced /cell/ hour) is the reciprocal of the cell cycle time (Tc) assuming that all cells are cycling. If a proportion of cells are non-cycling (G0 cells or differentiating cells), the knowledge of the growth fraction (Ip) is necessary. Ip= Nc/N where Nc is the number of proliferating cells and N the total number of cells. In that case, KB = Ip/Tc. The cell loss rate (cells lost/cell/ hour) is the sum of the cells lost by differentiation and by death (apoptosis or necrosis). The number of cells lost by differentiation is the reciprocal of the transit time or time taken by a cell to differentiate. The number of cells lost by apoptosis (or necrosis) is calculated from the ratio of the apoptotic index (or necrosis index) and the time taken to perform apoptosis or necrosis.

These parameters represent cell fluxes and they need a time component to be measured. One possible technique is the curve of labelled mitoses implying first the uptake in vivo by the tumour cells of specific DNA synthesis precursors such as 3H-thymidine or bromodeoxyuridine (by infusion or by intralesional injections) followed by multiple biopsies. This technique is impossible to realize routinely in human tumours like melanoma for practical and ethical reasons. An alternative method is the incubation of a tumour fragment in a culture medium containing successively two different doses of 3H-thymidine or two different DNA precursors as 3H-thymidine and bromodeoxyuridine. This double labelling technique is easy to perform but time consuming. In melanoma this technique is also a problem since primary lesions are usually excised by office-based dermatologists and sent to a department of pathology in order to obtain a reliable Breslow index.

Static (time independent) parameters are easier to obtain. They measure the proportion of cells occupying a given cell cycle fraction at one time point in the total population. The most commonly used is probably the MIB-1 index (number of cells expressing Ki-67 nuclear antigen, detected by immunochemistry using the MIB-1 antibody and representing the proportion of cycling cells or growth fraction). Ki-67 antigen is a non-histone protein that can be detected in the nuclei of cells in G1, S, G2 and M phases but not in the Go phase of the cell cycle. The monoclonal antibody MIB-1 against Ki-67 antigen is widely used for the assessment of the growth fraction in paraffin embedded specimens of human tumours. An absolute increase of the growth fraction (i.e. an expansion of the number of proliferating cells) of a tumour is a way to increase the cell production rate. This increase can be reflected by an increase of the number of cells stained by MIB-1 antibody. It must kept in mind that a lengthening of the cell cycle time or the blocking of cells in the cell cycle can lead to an identical picture.

Detection of the Proliferating Cell Nuclear Antigen (PCNA), another nuclear protein involved in DNA synthesis, has also been used to estimate tumour growth fraction. A standardized use is however difficult due to the exceptional solubility of PCNA : formalin fixation reduces the intensity of staining and the number of positive cells to approximately 25% of those I methanol-fixed material. On the other hand, PCNA plays also a role in DNA repair synthesis. For these reasons, PCNA immunostaining has been shown to display paradoxical results and is not used routinely.

Ten published studies have tried to assess the utility of MIB-1 immuno-staining analysed as a categorical or as a continuous variable in the prognosis of stage I and stage II melanoma. All, except two, concluded that MIB-1 index was a significant prognosis factor. Multivariate analysis showed that thickness was a more powerful prognostic factor than MIB-1 expression in some studies. In thin melanomas (<1mm) and in thick melanomas (>4mm), a high MIB-1 index was demonstrated to be a good predictor of systemic dissemination. One study used a proliferation-based prognostic index (PBPI) derived as tumour thickness x Ki-67 index/100 and showed that after stratification for thickness, only PBPI showed a independent prognostic significance restricted to tumours thicker than 4 mm.

Actually an increased MIB-1 staining is the reflection of either an increased growth fraction or a blocking of cycling cells before mitosis. An increased proportion of aneuploidy as it has been described in melanoma has no impact on MIB-1 staining. This can explain the rather limited prognostic value of MIB-1 index in melanoma and the need for developing other approaches.

In: Cell Growth Processes: New Research
Editor: Daiki Kimura, pp. 5-56
ISBN: 978-1-60456-132-6

Chapter I

Cell Colonization Control by Physical and Chemical Modification of Materials

L.Bačáková[1] and V. Švorčík[2]

[1]Department of Growth and Differentiation of Cell Populations, Institute of Physiology, Academy of Sciences of the Czech Republic, 142 20 Prague, Czech Republic,
[2]Department of Solid State Engineering, Institute of Chemical Technology, 166 28 Prague, Czech Republic

Abstract

Cell-material interface plays a key role in the interaction of cells with artificial materials designed for construction of body implants or tissue engineering. Physical and chemical properties of the material surface, such as its polarity, wettability, electrical charge and conductivity, roughness and morphology, rigidity or elasticity as well as the presence of various chemical functional groups, strongly influence the cell adhesion. The latter is mediated by adsorption of cell adhesion-mediating molecules (e.g. vitronectin, fibronectin, collagen, laminin, fibrinogen) from the serum of the culture medium or body fluids in appropriate spatial conformation or flexibility enabling the accessibility of specific sites on these molecules (e.g., certain amino acid sequences or saccharide-based ligands) by adhesion receptors on the cell membrane (e.g., integrins). After binding these ligands, the adhesion receptors are recruited into focal adhesion plaques, where they associate with cascades of various structural and signalling molecules, such as cytoskeletal proteins and kinases. By these cascades, the signal provided by the cell-material interaction is delivered to the cell nucleus. It influences the gene expression, and thus the further behavior of cells, manifested by their viability, synthesis of various molecules, proliferation, differentiation, functioning, immune activation, stress adaptation or cell death.

This interdisciplinary review involves:

(i) physical modifications of the material surface by plasma discharge, irradiation with ions or ultraviolet light and effects of these modifications on the surface wettability, electrical conductivity and formation of chemical functional groups influencing cell adhesion,

(ii) deposition of organic or inorganic layers on the material surfaces, especially those nanostructured (e.g., nanofibers of fibrin and extracellular matrix molecules, carbon nanoparticles, such as nanodiamonds, fullerenes and nanotubes, nanocomposite metal/C:H layers),
(iii) functionalization of the material surface with amino acids and oligopeptidic ligands for cell adhesion receptors (e.g., GRGDSG),
(iv) effect of all these modifications on the extent and strength of adhesion of vascular and bone-derived cells, cell proliferation activity, switch between cell proliferation and differentiation, expression of cell-type specific markers of differentiation and other cell functions.

1. Introduction

Artificial materials are of growing importance in the fields of medicine and biology. *Tissue Engineering,* a new and modern interdisciplinary scientific field, has been developed to design biocompatible materials in order to substitute irreversibly damaged tissues and organs.

The tissue damage or loss due to disease, inborn defect or trauma is not only a major socio-economic burden on world healthcare systems but leads to a reduced quality of life for the patient. Currently, autogenous tissue, i.e. tissue derived from the same patient, is the most widely used and effective graft material. However, this approach brings a number of related problems, such as availability of the graft material in limited quantity or damage of additional patient's tissues, prolonged surgical time, blood loss, infection risk and prolonged rehabilitation. Allogenous or xenogenous grafts, i.e. transplants derived from human or animal donors, are associated with a severe risk of immune rejection and disease transmission. Consequently, alternative strategies with potential to overcome these limitations should be generated. One of them is represented by creation of appropriate biocompatible artificial substitutes, which offers enormous promise for the repair of lost tissue function.

In addition, artificial materials have been employed in diverse diagnostic and therapeutic applications and biotechnologies, e.g., tracers for advanced imaging technologies, carriers for controlled drug and gene delivery, biosensors and growth supports for cells in a culture.

Artificial materials designed for the biomedical use should be *biocompatible*, i.e. free of adverse effects on cells and tissues, such as cytotoxicity, immunogenicity, mutagenicity and carcinogenicity. Biocompatible materials can be constructed as *bioinert*, i.e. not allowing adsorption of proteins and adhesion of cells, or as *bioactive*, i.e., promoting colonization with cells - if possible, in a controllable manner. Thus, bioactive materials for advanced tissue engineering are not only simply tolerated by cells but they actively regulate the cell adhesion and growth activity, switch between proliferation and differentiation program, secretion of various molecules by cells, the viability or programmed death of cells and other cell functions. In other words, bioactive materials act similarly as natural extracellular matrix in controlling the cell behavior.

The bioactivity of materials can be enhanced by a wide range of physical and chemical methods. This interdisciplinary review is focused on the modification of materials, particularly synthetic polymers, by irradiation with plasma, various ions and ultraviolet light, followed by functionalization of the modified material with bioactive molecules, namely amino acids and adhesion oligopeptides. The ligands for cell adhesion receptors were also tethered on degradable materials through flexible polymeric chains, and thus templates mimicking natural extracellular matrix were constructed. In addition, coating material by various soft and hard biocompatible layers, including those nanostructured, is also discussed. The main part of this review is based on our results obtained in the field of biomaterials during more than 10 years.

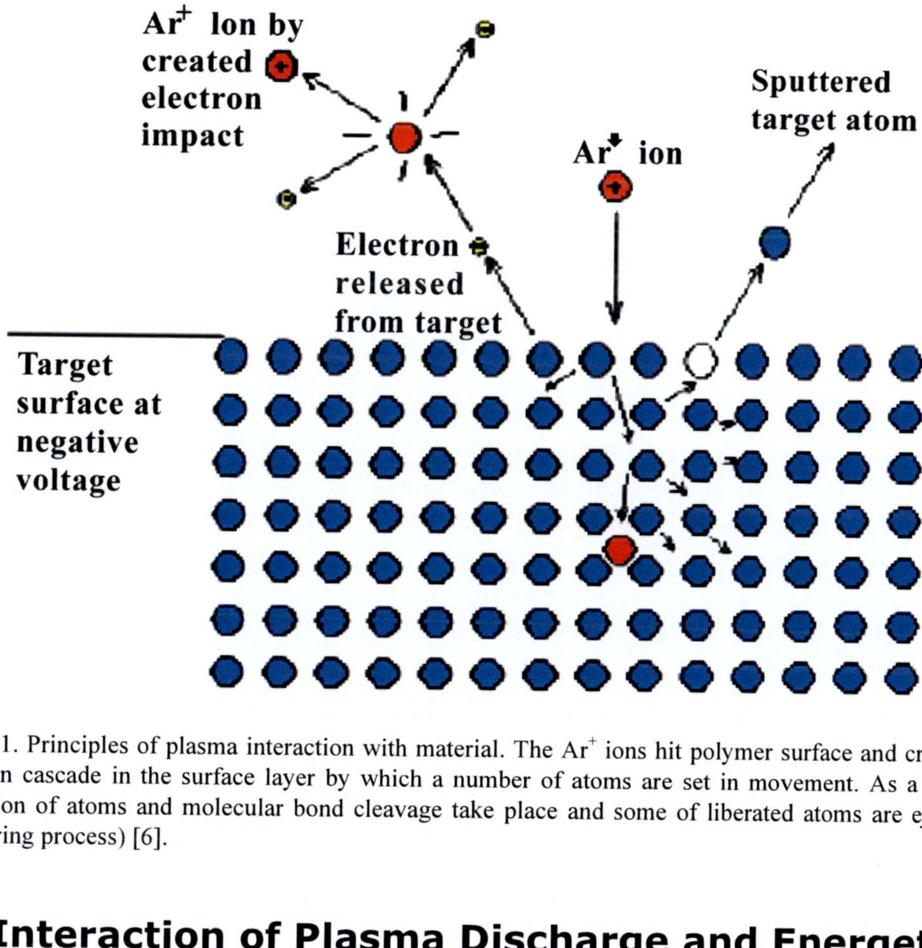

Figure 1. Principles of plasma interaction with material. The Ar^+ ions hit polymer surface and create a collision cascade in the surface layer by which a number of atoms are set in movement. As a result ionization of atoms and molecular bond cleavage take place and some of liberated atoms are ejected (sputtering process) [6].

2. Interaction of Plasma Discharge and Energetic Ions with Materials Surface

2.1. Principles of Plasma Treatment [1,2]

Plasma treatment, which is an electric gas discharge, is usually performed in a vacuum chamber in a pressure ranging from 1-100 Pa. Electric field, applied to the gas, accelerates free electrons to energies which are sufficient for ionization, fragmentation and excitation of the gas molecules (see Figure 1). In this way highly reactive gaseous products are produced which are able to react chemically with exposed surfaces. The processes of the plasma-polymer interaction and the resulting surface properties of the modified polymer are dependent on several factors such as process gas and its properties, chemical structure of polymer and last but not least the properties of the plasma discharge (density and energy of electrons and ions). The nature of the interaction of the plasma constituents with polymers is also affected by plasma reactor configuration. Depending on adjustment of all these factors different outcome of plasma-polymer interaction may result. The most important from the point of view of practical applications are:

(i) *Reactive removal of surface layer* leading to etching or cleaning of the material surface.
(ii) *Activation or more generally modification* of the polymer surface. The plasma treatment changes polymer structure and elemental composition within few moleculare layers on the polymer surface and creates new functional groups on the polymer surface. The functional groups may increase adhesion, wettability, bioactivity of polymer surface or provide more inert surface properties.
(iii) *Deposition or plasma polymerisation* is the process when a coating with thickness in sub-micrometer range is deposited onto exposed surface. Generaly the coatings exhibits highly cross-linked structure and they provide useful properties for potential applications.

The surface analytical techniques mentioned above provide wealth information on the composition and structure of polymer surface and changes resulting from modification by plasma discharge. It should be however stressed that, despite of broad spectrum of analytical techniques available, the information is not sufficient to understand all underlying processes in their complexity. Especially, it is the case of plasma treatment when the interactions of many plasma constituents with polymer surface may play a role. Existing theoretical models are restricted to some specific cases and they usually describe only some part of the process.

2.2. Principles of Ion Implantation [3,4]

Ion implantation is performed in high vacuum (see Figure 2) and, contrary to plasma treatment; there is no complex interaction of reactive gaseous species from ambient atmosphere with material surface. Instead, the polymer modification is caused solely by interaction of energetic ions with polymer matrix. The basic instrumentation consists of an ion implanter, which is commonly a small dedicated accelerator with associated ion optics, beam scanning system and target chamber with a suitable sample holder. The ions are usually accelerated to energies ranging from few keV up to several MeV. The ions are mass separated and send toward the sample. A beam scanning system ensures homogenous irradiation of the sample surface. Indespensable part is a beam monitoring system measuring the total fluence of the ions. The ions entering the polymer substrate interact with atoms which are ionized and exited.

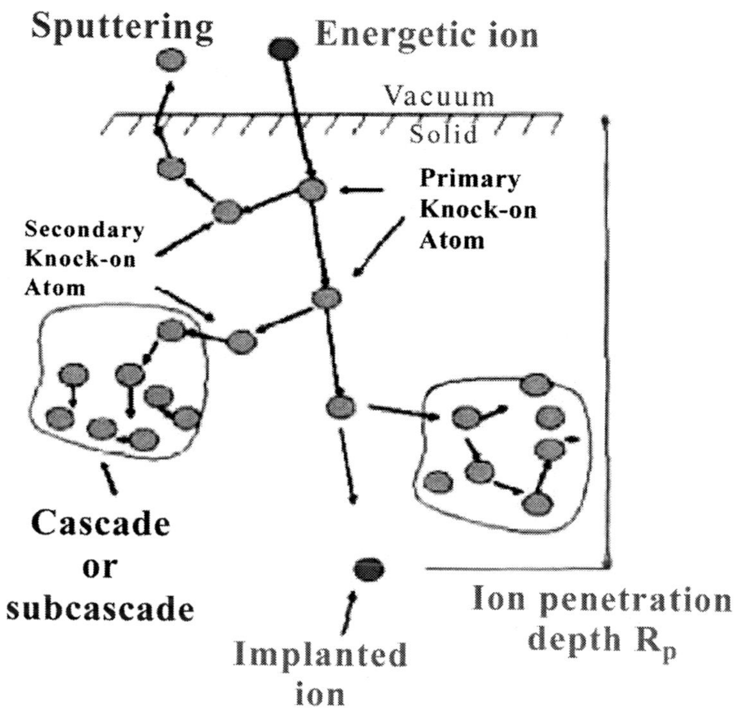

Figure 2. Principles of ion implantation. Basic processes of ion interaction with matter. The thickness of modified surface layer is determined by ion projected range R_p, measured in the direction of the sample surface normal.

The ionization and excitation may lead to chemical bond cleavage and production of highly reactive species, free radicals, ions and molecular fragments, which subsequently interact with each other and at last stable degradation products are created. This complex sequence of processes can deliberately be divided into two basic phases, the initial physical phase, in which the ion energy is dissipated to electrons and atoms, and the chemical one comprising interaction of the reactive species and production of the final stable products. Two kinds of degradation products are in principle created, low mass species created by macromolecular chain scission and two- to three dimensional structures produced by macromolecule cross-linking (mediated by free radical reactions).

Specific phenomenon is desorption of light weight, volatile products in the course of the ion implantation. Since the desorbed species are hydrogen-enriched, the polymer surface is progressively carbonized. The thickness of the polymer surface layer modified by implanted ions is determined by ion projected range which is an increasing function of the ion energy and varies, depending on the ion mass and energy and the substrate composition, from several nanometers to tens of micrometers. Typical ion fluences vary from 10^9-10^{18} cm^{-2}. It should be mentioned that the polymer modification by ion bombardment is not lateraly homogenous at least during initial phases of implantation. In fact the modification takes place only in discrete regions along the trajectories of individual ions (ion latent tracks). Once the ion fluence is increased, the sample surface is progressively covered with the ion tracks and at still higher fluences an overlapping of the ion tracks occurs, i.e. the same place is hit by more than one ion. The properties of the modified polymer could also be influenced by implanted atoms which remain embeded in the polymer matrix.

2.3. Principles of Photochemical Modification by Ultraviolet (UV) Light

UV irradiation is applied typically on surfaces on synthetic polymers, such as polytetrafluoroethylene (PTFE) or polyethylene terephtalate (PET), i.e. materials often applied in biology and medicine, e.g. for construction of clinically used vascular prostheses. Principles of photochemical modification of PTFE by ultraviolet light in ammonia or acetylene atmosphere see in Figure 3 [5]. The pristine polymers show either very low cell adhesion (PTFE) or medium cell adhesion (PET), but exposure to UV-light at wavelengths $\lambda < 200$ nm in a reactive ammonia (NH_3), acetylene (C_2H_2) and oxygen (O_2) atmosphere can result in a drastic change of the surface properties, especially in a higher wettability with the first two gases. Treatment in oxygen can also result in etching. The high wettability of treated surfaces is ascribed to polar groups incorporated into the PTFE surface by substitution of F atoms. The subtraction of F is mediated via atomic H which is generated in the dissociation reaction $NH_3 + h\nu \rightarrow NH_2 + H$ [6].

Modified PTFE surfaces show a high degree of biocompatibility with good cell adhesion and proliferation [7-11]. However, the UV-treatment results also in a loss of mechanical stability due to the scission of polymer chains, especially for light-sources with wavelengths below 193 nm [6]. Similarly to the ion implantation or plasma modification, also the UV light-irradiation is performed on both sides of a polymer foils in order to avoid the material torsion.

(a) surface reactions **(b) reactions in atmosphere** **(c) reactions in polymer**

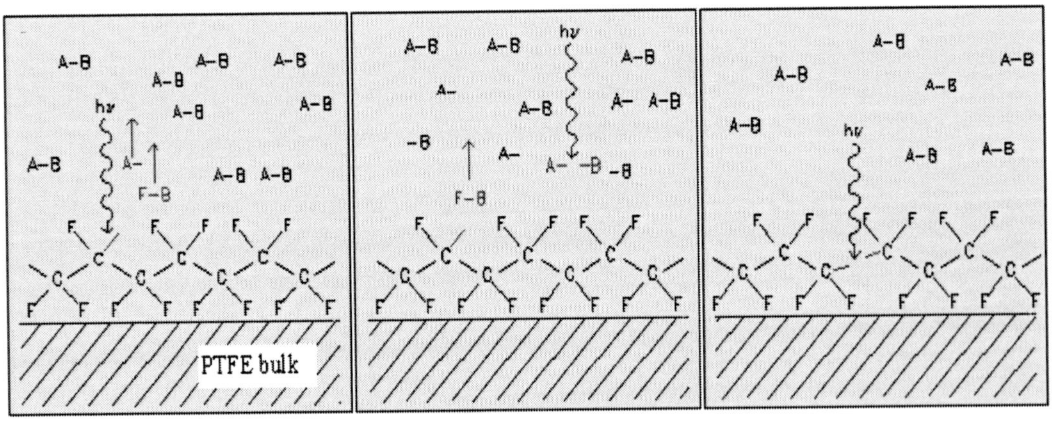

Figure 3. Principles of photochemical modification of polymer (e.g. PTFE) by ultraviolet (UV) light in ammonia or acetylene atmosphere (A-B). Basic processes of photochemical modification of polymer by UV light (hv) in atmosphere are: (a) surface reactions, (b) reactions in atmosphere and (c) reactions in polymer. [5] .

3. Physical and Chemical Characterization of the Modified Material Surface

Several techniques can be used for characterization of the physical and chemical properties of pristine polymers and polymers modified by various chemical and physical treatments as well. The techniques can deliberately be divided in two groups, those determining the polymer properties of application importance (mechanical strength and hardness, surface morphology, wettability, electrical conductivity, dielectric properties and bioactivity) and those suitable for the characterization of the polymer structure and composition on microscopic scale, which in turn affect the macroscopic properties of polymers mentioned before. The latter techniques include, besides others, different sorts of optical and electron microscopies, X-ray photoelectron spectroscopy (XPS≡ESCA), X-ray and neutron diffraction, infrared and Raman spectroscopy (FTIR), secondary ions mass spectrometry (SIMS), atomic force microscopy (AFM) and several techniques based on elastic scattering of charged particles (RBS, ERDA, MEIS). In what follows selected techniques are briefly described with respect to their use for polymer characterization. Most of the diagnostic methods listed below are notoriously known and their principles are described in standard, easily available surveys [1]. For few techniques references to specialized publications are given.

Goniometry, measurement of surface wettability. Measurement of surface polarity and wettability (contact angle goniometry) is based on measurement of contact angle between a drop of a liquid and polymer surface. The contact angle is a direct measure of the polymer surface energy and consequently of the polymer wettability.

Ion beam analytical techniques [12]. Analytical techniques based on the use of energetic charged particles (protons, alpha particles, heavy ions) such as Rutherford Back Scattering (RBS), Elastic Recoil Detection Analysis (ERDA) and Nuclear Reaction Analysis (NRM) are very powerfull because of broad spectrum of elements, which can be determined simultaneously with high sensitivity and excellent depth resolution down to few nanometers. The energy spectra of scattered or recoiled particles bring information on the sample composition and the depth structure of the sample as well. For all these methods the analysed depths are limited to several micrometers by the range of primary particles and typical detection limits vary from 10^{11}-10^{15} at.cm^{-2}. So that the time during which the samples are exposed to particle irradiation should be as short as possible and the radiation damages caused by analyzing beam should be taken into account when evaluating the experimental data. These methods are insensitive to chemical state of elements in the sample. General disadvantage of these methods is high cost of basic equipment, accelerator and data acquisition technique.

Neutron depth profiling technique (NDP) [13]. NDP is a special method for depth profiling of few light elements, namely He, Li, B and N in any solid material. The method makes use of specific nuclear reactions of these elements with thermal neutrons. The samples are placed in the neutron beam from nuclear reactor and the charged products of the neutron induced reactions (protons or alpha particles) are registered using a standard multichannel spectrometer. From the measured energy spectra the depth profiles of above mentioned elements can be deduced by a simple computational procedure.

In *X-ray Phtotelectron Specroscopy* (XPS) the sample is irradiated by monoenegetic soft X-rays which induce photoionization of the atoms and photoeletrons emitted from the sample surface layers, several nanometers thick, are analyzed by an electrostatic spectrometer. Photoelectron energy spectrum consists of peaks corresponding to electron energy states of the sample. By the XPS method it is possible to determine elemental composition of near surface layer of the sample and identify chemical state of the elements. Chemical state identification is based on measured energy shifts related to electron density in the atoms of interest. Apparent disadvantage of XPS metod in some instances is too shallow sampling depth and rather high cost of instrumentation.

InfraRed Spectroscopy (IR). Infrared spectroscopy is an effective method for characterization of polymers as to chemical structure. IR spectra of the sample examined are obtained by two basic types of IR spectrometers: dispersive or Fourier transform (FTIR) ones. Infrared spectra are usually presented as a dependence of absorption (in percent transmission) on wave length or wave number. Any molecule or its part may be viewed as mechanical system exhibiting certain natural frequency at which it can be exited. The absorption bands in IR spectra correspond to these frequencies, which are characteristic for different chemical structures.

Raman spectroscopy. Raman spectroscopy is an analytical technique that yields information about the molecular structure of sample materials based on observation of scattered light spectra. Small fraction of light scattered by molecules exhibits a frequency shift from the incident light which is related to the frequency of molecule vibrational states. It is similar to FTIR spectroscopy, but it has a few advantages, for example Raman can be used to study solids, liquids, powders, gels, slurries and aqueous solutions. Raman spectroscopy is based on detection of scattered light, i.e. the Raman effect. The radiation may be scattered elastically, that is without any change in its wavelength and this is known as Rayleigh scattering. Conversely the radiation may be scattered inelastically resulting in the Raman effect.

Both *Raman and infrared spectroscopy* provide qualitative and quantitative information about chemical species through the interaction of radiation with molecular vibrations. Raman spectroscopy complements infrared spectroscopy, particularly for the study of non-polar bonds and certain functional groups. It is often used as an additional technique for elucidating the molecular structure and symmetry of a compound. Raman spectroscopy also provides facile access to the low frequency region (less than 400 cm^{-1} Raman shift), an area that is more difficult for infrared spectroscopy.

UV-Vis-NIR spectroscopy. UV-Vis spectroscopy probes the electronic transitions of molecules as they absorb light in the UV and visible regions. Any species with an extended system of alternating double and single bonds will absorb UV light and anything with with color absorbs visible light, making UV-Vis spectroscopy applicable to a wide range of materials. While UV-Vis and UV-Vis-NIR spectroscopy are well established analytical techniques being effectively applied across a broad spectrum of well-known industries, high performance UV-Vis spectroscopy has expanded into new applications, especially in the life sciences. Yet even in the traditional application areas such as materials characterization, optics, coatings, glass, pharmaceutical and color control, UV-Vis spectroscopy has become more sophisticated, providing higher quality results faster and easier than ever.In the middle range of high performance instruments, you should expect UV-Vis coverage from 190 to 900 nm, and if NIR is needed, the range should extend from at least 190 to 3000 nm.

Measurement of *electrical sheet resistance* (two- or four-point method). The simplest technique for measurement of sheet resistance of polymer samples is two-point method, prerequisite of which is creation of two ohnmic contact on the sample surface. The metalic contacs are commonly produced by sputtering or vaccum evaporation of suitable metal through a mask onto material surface. The conductivity measurement proceeds in standard manner.

Atomic Force Microscopy (AFM). The AFM provides quantitative morpholographic images of solid surfaces at a nanometer scale. It uses a sharp tip scanned over the sample surface to sense attractive or repulsive forces between the tip and the surface. The microscopic image of the surface is obtained as a surface, representing the locus of points of constant force between the tip and the sample.

The *Scanning Electron Microscope* (SEM) is a standard imaging technique based on electron back-scattering from the sample surface. It analyses the surfaces of solid objects, producing images with the resolution which is about order of magnitude better than that of optical microscopy (typically 10 nm). The SEM avoids the problem of thin samples (TEM) but the SEM observation requires the deposition of a thin conductive metal film on the sample surface to prevent sample charging.

Transmission Electron Microscopy (TEM) is a standard laboratory technique. TEM is indispensable tool for high resolution observation of very fine structures on material surface. The resolution of TEM is abou one order of magnitude better than that of SEM. It corresponds to 1 nm. There exist also high resolution transmission microscopes (HRTEM) with a resolution down to 0.1 nm, capable to resolve individual atomic lattice planes. Samples must be stable enough to withstand the electron beam impact during their examination. This can be a problem for polymers.

X-Ray Diffraction (XRD) is well known technique based on scattering of X-rays. It can be applied to semicrystalline or crystaline polymers and structural changes induced by modification. Occasionally observed shifts in the X-ray peak positions might indicate a distortion of the crystal structure due to increasing strain. The width of the X-ray reflection peaks yields information on the size of the crystals under investigation.

Electron Paramagnetic Resonance (EPR) is a technique used for detection and identification of unpaired free electrons (radicals) exhibiting spin magnetic moment. The method makes use of electron paramagnetic resonance corresponding to transition of electrons between states with different spin orientation initiated by absorption of rf electromagnetic signal in the sample. It is indispensable for detection of free radicals produced by modification of polymers.

Surface mechanical properties (nanoindentation). Nanoindentation technique has an important role in the characterisation of mechanical properties of materials on submicron scale. The ultra-low load indentation systems (nanoindentors) carry out continuous records of load and displacement as an indentor is pushed into a surface. Indentation curve can be used to determine mechanical properties even when the indents are too small to be imaged conveniently. Analyse of this data gives a number of useful parameters and information for complete assessment of elastic, plastic and fracture properties of surface at the scale of real asperity contacts. The hardness, elastic modulus and other properties are evaluated from the indentation curve. The indentation methods can de used for the measurement of adhesion between the coating and the substrate when the indents create cracks and delamination on the coating or on the interface on polished section. The scratch test characterizes the surface mechanical properties of thin films and coatings, e.g. adhesion, fracture and deformation.

4. Evaluation of Cell-Material Interaction

4.1. Cell Culture as an Advantageous Evaluation System for Cell-Material Interaction

This review is focused on the evaluation of the interaction between the artificial material and cells in cell culture conditions. In comparison with *in vivo* experiments, where both material and contacting cells are subjected to multiple and complex influences of the whole organism, the cell culture represents a relatively simple and well-defined system. This system enable screening of a wide range of materials and their surface modification in order to select the most appropriate variants for testing on laboratory animals and then for clinical studies. Therefore, the use of cell culture considerably saves these animals, and thus fulfills the ethical requirements for advanced scientific research.

Use of Cell Lines

In the first stage, the studies on cell-material interaction can be advantageously carried out on commercially available cell lines, including those derived from human patients undergoing surgery or from fetal tissues, such as human osteoblast-like MG-63 cells (European Collection of Cell Cultures, Salisbury, UK), obtained during osteosarcoma surgery in a 13-year-old boy, vascular smooth muscle cells of the line T/G HA-VSMC, derived from the human aorta, or human umbilical vein endothelial cells of the line HUVEC. The cell lines represent a relatively homogeneous and stable cell population, currently used for reproducible studies of the biocompatibility of various artificial materials, as well as for investigations of humoral factors influencing growth and differentiation of various cells. For these studies, it is necessary to select lines retaining markers characteristic for the identity, differentiation and maturation of a certain cell type, e.g., activity of alkaline phosphatase or production of osteocalcin typical for osteoblastic differentiation [14-17], presence of contractile proteins (alpha-actin, SM1 and SM2 myosins) in vascular smooth muscle cells or von Willebrand factor in endothelial cells [18-23].

Primocultures and Low-Passaged Cultures from Laboratory Animals

In the second stage of the *in vitro* studies on cell-material interaction, the selected results obtained on cell lines are verified on primary or low-passaged cultures obtained from the tissues of laboratory animals. For example, osteogenic cells can be isolated from the calvarias and long bones of newborn rats, e.g. by sequential collagenase digestion or the explantation method [24-27]. Similarly, also vascular smooth muscle cells can be relatively easily isolated by the explantation method or enzymatical digestion from the rat aorta [18,19,21,23,28]. For isolation of vascular endothelial cells, vessels from larger animals, at least from rabbits or minipigs, are necessary in order to facilitate the manipulation with the tissue and to obtain a satisfactory cell yield. In order to avoid a potential contamination with smooth muscle cells, which are very vital and prone to overgrow the endothelial cells, the best approach seem to be to fill the vessel with collagenase diluted in cell culture media and incubate the vessel in a cell incubator without any scraping of endothelial cell layer. The released endothelial cells are then collected by centrifugation ad seeded into small amount of culture media of appropriate chemical composition, e.g. a higher (20%) content of fetal bovine serum, presence of glutamine, pyruvate, non-essential amino acids and other specific supplements [29-31]. The artificial materials material can be also tested with stem cells, e.g. those obtained from bone marrow, in order to set appropriate physical and chemical properties of the developed artificial material, which would promote the differentiation of stem cells into a desired cell type [32].

Primocultures and Low-Passaged Cultures from Human Patients

In the third stage of the cell-material interaction studies, the selected results can be further verified in primary or low-passaged cultures obtained from human patients undergoing surgery or biopsy. Such cells are also commercially available from specialized companies (e.g., Provitro GmbH, Berlin, Germany; http://www.provitro.de), thus scientific workers from non-clinical institutions have possibility to avoid potential difficulties associated with obtaining the ethical approval, which is usually required for the experimental use of human cells and tissues.

Static and Dynamic Cell Culture Systems

The cultivation of cells obtained from all mentioned sources on the investigated materials usually starts in a conventional static cell culture system. The material samples are inserted in plastic dishes or bottles, immersed with cell culture media, containing defined concentrations of suspended cells, and incubated at 37°C in a humidified air atmosphere with 5% of CO_2 [33]. However, the relatively simple and highly reproducible manipulation with cells and materials in the static cell culture system is compromised by a relatively low supply of oxygen and nutrients to cells, as well as relatively slow waste removal, especially if the investigated artificial material is constructed in a so-called three-dimensional form, i.e. as scaffolds enabling ingrowth of cells inside the material. For these advanced tissue engineering constructs, a more appropriate environment is represented by dynamic cell culture systems, e.g. perfusion systems (http://www.provitro.de) or rotary bioreactors (http://www.cellon.lu). These systems enable the flow or penetration of fresh cell culture media through the material. In addition, these systems can provide defined mechanical stimulation of cells, so that they better resemble the physiological environment of cells *in situ* conditions and accelerate the maturation and establishment of specialized function of a given cell type. The dynamic cell culture systems are neccessary especially for the construction of bioarteficial vascular prostheses, i.e. synthetic grafts with reconstructed tunica intima containing endothelial cells or even tunica media with smooth muscle cells [30,34-36], as well as for the creation of bioarteficial bone tissue on the basis of three-dimensional porous scaffolds [37].

4.2. Specific Markers of Cell Behavior in Cell-Material System *in Vitro*

During years of studies on cell-material interaction, we have developed a complex test of basic and more specialized markers of cell adhesion, growth, differentiation, viability/cellular damage, stress adaptation, immune activation as well as potential mutagenicity and cancerogenicity of the material. These markers can be investigated by a wide range of microscopical and biochemical methods, such as conventional optical, confocal and scanning electron microscopy, immunofluorescence and immunoperoxidase staining, enzyme-linked immunosorbent assay (ELISA), flow-cytometry, electrophoresis, immunoblotting, immunoprecipitation, microarray techniques or cell transfection, depeding on the facilities in a given scientific laboratory.

Table 1. Examples of specific amino acid sequences in extracellular matrix molecules, which represent ligands for cell adhesion receptors and can be used in tissue engineering.

Amino acid sequence	Typical for a certain ECM molecule	Preferred by a certain cell adhesion receptor	Preferred by a certain cell type
RGD*	Vitronectin Fibronectin	Integrin $\alpha_v\beta_3$# Integrin $\alpha_5\beta_1$	Without significant preference
DGEA	Collagen	Integrin $\alpha_2\beta_1$	Without significant preference
KQAGDV	Vitronectin, fibronectin	Integrin $\alpha_v\beta_3$#	Vascular smooth muscle cells
VAPG	Elastin	Non-integrin receptor	Vascular smooth muscle cells
REDV	Fibronectin	Integrin $\alpha_4\beta_1$	Vascular endothelial cells
YIGSR, IKVAV	Laminin	Integrins $\alpha_6\beta_1, \alpha_7\beta_1$	Neurons
KRSR	Part of the heparin binding domain	Non-integrin receptor	Osteoblasts

*The possibility of binding RGD to different receptors is probably caused by a different spatial conformation of RGD due to the vicinity of different amino acids or existence of different synergistic sequences on various extracellular matrix (ECM) molecules. # At the same time, one type of integrin can bind more ligands [38-40].

Indicators of Cell Adhesion

Protein adsorption. The cell adhesion on artificial materials is mediated by adsorption of extracellular matrix (ECM) molecules or molecules participating in tissue regeneration. These molecules, represented e.g. by vitronectin, fibronectin, collagen, laminin, fibrinogen and fibrin, can be spontaneously adsorbed to the material surface from the serum supplement of the culture media *in vitro* or blood and other body fluids *in vivo*. In addition, many cell types are able to synthetize and secrete these molecules, and to deposit them on the material surface. Even after trypsinization of cells (in order to detach them from the original cell culture flask or dish and seed them on the tested material), the cells usually contain a considerable amount of these molecules on their membrane. After adsorption of cell adhesion-mediating molecules, the cells bind active sites on these molecules, represented by a wide range of specific amino acid sequences, such as Arg-Gly-Asp (RGD), Asp-Gly-Glu-Ala (DGEA), Lys-Gln-Ala-Gly-Asp-Val (KQAGDV), Val-Ala-Pro-Gly (VAPG), Arg-Glu-Asp-Val (REDV), Lys-Arg-Ser-Arg (KRSR), Ile-Lys-Val-Ala-Val (IKVAV) or Tyr-Ile-Gly-Ser-Arg (YIGSR) (Table 1, Figure 4). These sequences are usually typical for a certain ECM molecule. Also they are usually preferred by a certain cell adhesion receptor as well as by a certain cell type (Table 1) [38-41], which can be advantageously used in tissue engineering, as decribed bellow. In addition, these adhesion motifs cooperate with other synergistic amino acid sequences, e.g. PHSRN, which help to maintain appropriate spatial conformation of these ligands as well as integrin receptors [40-42].

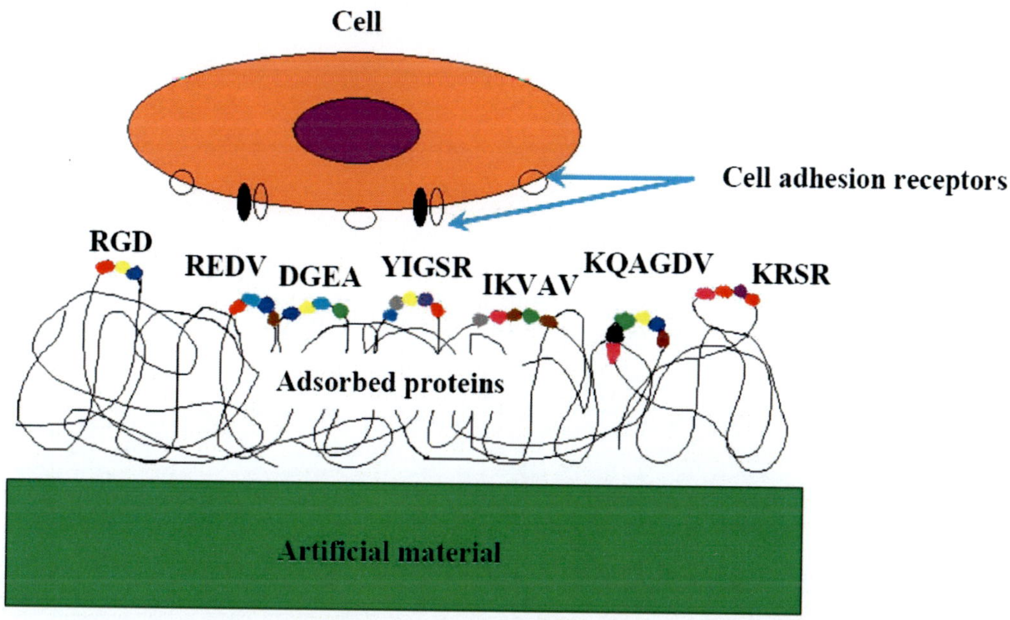

Figure. 4. Principle of the cell adhesion to artificial materials. In cell culture media or body fluids, the material is spontaneously adsorbed with cell adhesion-mediating extracellular matrix proteins (e.g., vitronectin, fibronectin). The cells then adhere to specific amino acid sequences of these proteins by their adhesion receptors of integrin or non-integrin type [38-41].

As evident from the text mentioned above, for appropriate cell adhesion, the cell adhesion-mediating molecules should be adsorbed not only in a sufficient amount and homogeneity, but also in an appropriate spatial conformation enabling a good accessibility of active sites on these molecules by adhesion receptors on the cell membrane (Figures 4-6). If the primary spatial conformation of the adsorbed proteins is not suitable for their binding to cell adhesion receptors, the cells can actively reorganize it, if the proteins are adsorbed in a flexible form, which usually happens on wettable surfaces. On the contrary, on highly hydrophobic surfaces (water drop contact angle about 100°), the cell adhesion mediating molecules are usually too rigid, and thus the cell adhesion is often insufficient, although the absolute amount of the adsorbed molecules can be even higher than on wettable surfaces. In addition, highly hydrophobic surfaces are known to promote the preferential adsorption of albumin, which is non-adhesive for cells [18,19,40,41,44-48].

The type and amount of proteins adsorbed to the biomaterial surface can be determined by immunocytochemical studies using specific antibodies. Also commercially available proteins conjugated with fluorescence markers can be used for quantification of protein adsorption as well as for the evaluation of the homogeneity of the protein distribution on the material surface (Figures 5,6). Hovewer, the studies on the spatial conformation of the adsorbed ECM molecules are more difficult and are still in their beginning. This problem can be partially solved by some physicochemical methods originally developed for the investigation of artificial materials, e.g. observation of the surface morphology of the adsorbed protein layer by scanning electron microscopy (SEM) or atomic force microscopy (AFM), studies of the protein layer using surface plasmon resonance (SPR), mass spectrometry (e.g., MALDI-TOF), determination of the chemical composition of the protein layer by various spectroscopic methods, like Fourier transform infrared spectroscopy (FTIR), X-ray photoelectron spectroscopy (XPS), Auger spectroscopy (AES) or Raman spectroscopy, or by indirect immunocytochemical studies using antibodies against specific regions on ECM molecules or certain types of adhesion receptors preferentially expressed on adhering cells in response to the composition of the surface of the adsorbed protein layer [16,31,44-46,48-51].

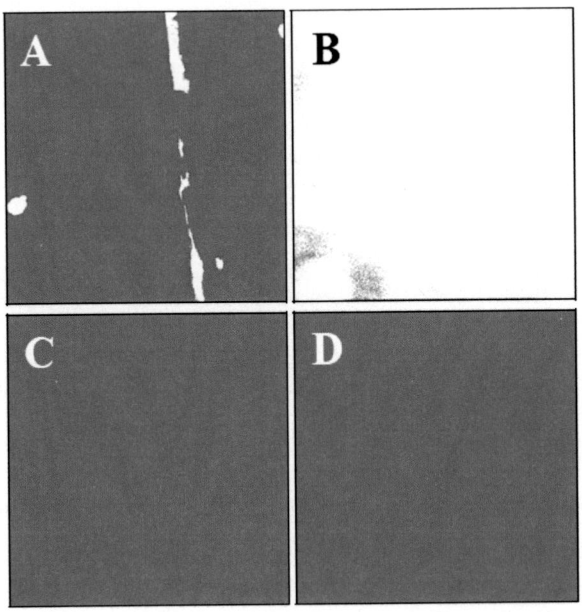

Figure 5. Fluorescence of collagen IV conjugated with Oregon Green 488 (Molecular Probes, Eugene, OR, USA.) spontaneously adsorbed to unmodified polyethylene foils (A) or polyethylene modified with 1014 O+ ions/cm^2 at the energy of 30 keV (B). Collagen was diluted in phosphate-buffered saline to the concentration of 0.02 mg/ml (10 μg/cm^2) and incubated with the foils for 24 h at room temperature. For autofluorescence control, umnodified (C) and ion-irradiated foils (D) were not exposed to the collagen solution. Microscope Nikon, CCD camera, Image Pro Plus 3.0 software [43].

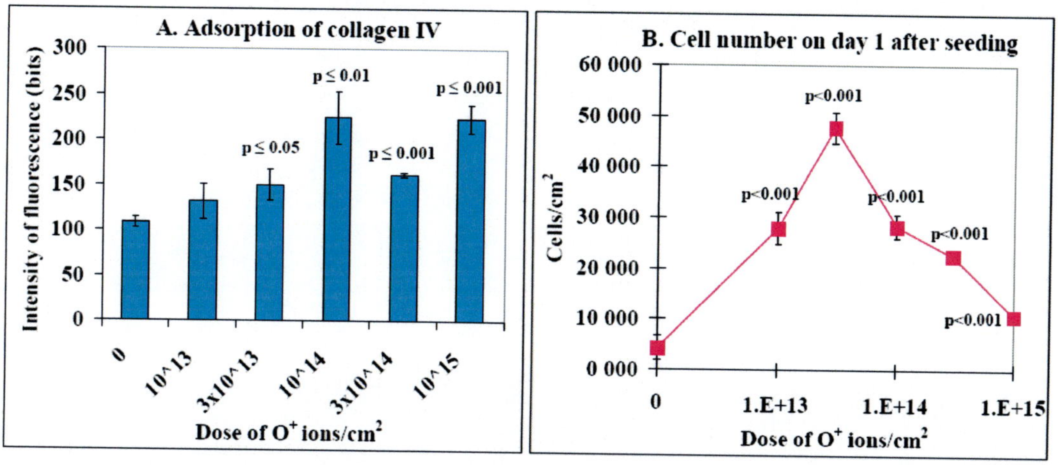

Figure 6. Semiquantitative evaluation of spontaneous adsorption of collagen IV conjugated with a fluorescence marker Oregon Green 488 (A) to polyethylene foils irradiated with O^+ ions (energy 30 keV, dose from 10^{13} to 10^{15} ions/cm^2), and its correlation with the number of rat aortic smooth muscle cells initially adhering to the polymer on day 1 after seeding (B). Collagen was diluted in phosphate-buffered saline to the concentration of 0.02 mg/ml (10 μg/cm^2) and incubated with the foils for 24 h at room temperature. Mean±SEM from 4-9 experiments, Student t-test for unpaired data [43].

Expression of cell adhesion receptors. Proteins adsorbed on the biomaterial surface are bound by adhesion receptors on the cell membrane. The most known and most systematically described adhesion receptors are integrins. Integrins are heterodimeric transmembrane glycoproteins consisting from one alpha and one beta chains. Their function is dependent on calcium, which binds on the alpha subunit. About 16 different subunits alpha and 8 subunits beta were described. However, these numbers are not definitive, because integrin receptors are still under intensive research and new subunits are being revealed. Various combinations of alpha and beta chains result in constitution of receptors with preferential affinity to certain ECM molecules. In addition, certain integrins are present preferentially on certain cell types. Thus, these cell types adhere preferentially to certain amino acid sequences and ECM proteins (Table 1) [38-40,52,53]. Recent investigations have shown that other receptors than integrins can also take place in cell-matrix adhesion, e.g. proteoglycan-based receptors. These non-integrin receptors bind not only aminoacid sequences on ECM molecules but also saccharide-based parts of these molecules. For example, heparan sulphate proteoglycan on osteoblasts recognizes a bone-specific oligopeptide KRSR, or an asialoglycoprotein receptor on hepatocytes binds a galactose ligand. Also the VAPG motif on elastin is recognized by a non-integrin receptor, represented by a multifunctional 67-KDa peripheral membrane protein [38-40,52,53].

Assembly of focal adhesion plaques. It was well described in integrin receptors that after ligand binding, these receptors are recruited into distinct dot-like or streak-like nano- or microdomains on cell membrane, called "focal adhesion sites", "focal adhesion plaques" or simply "focal adhesions" (Figure 7). In these regions, the integrins communicate with many specific structural and signaling molecules. The former are represented by proteins of membrane-associated cytoskeleton, also called "focal adhesion proteins", such as talin, α-actinin, filamin, paxillin or vinculin. These proteins act as linkers between the integrin receptors and the cytoplasmatic actin cytoskeleton, which is associated with nuclear membrane, membranes of cellular organelles as well as with various enzymes, and thus influences intracellular processes important for cell behavior, including transport and secretion of various molecules, endocytosis, decision between cell proliferation and differentiation or apoptosis (Table 2). The signaling molecules are represented e.g. by cytohesin-1, focal adhesion kinase (FAK), integrin-linked kinase (ILK), mitogen-activated protein kinases (MAPK), extracellular signal-regulated kinase (ERK), β_3-endonexin, integrin cytoplasmic-domain-associated protein-1 (ICAP-1), receptor for activated protein kinase C (Rack-1), calcium- and integrin-binding protein (CIB) or small GTPases.

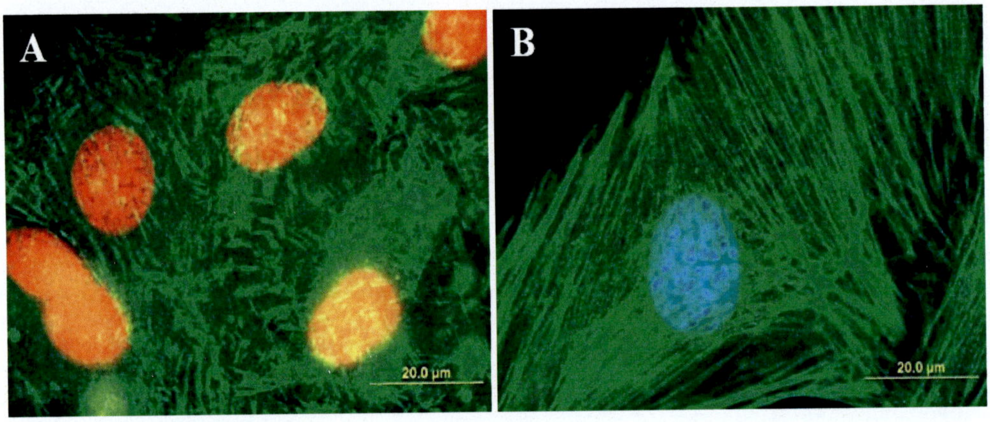

Figure 7. Talin-containing streak-like focal adhesion plaques (A) and beta-actin cytoskeleton (B) in vascular smooth muscle cells in cultures derived from the rat aorta. Immunofluorescence staining, the nuclei couterstained with propidium iodide (A) or Hoechst 33342 (B). Microscope Olympus IX 50, digital camera DP 70, obj. 100, bar=20 µm.

Table 2. Correlation between the extent of cell adhesion and the subsequent cell behavior in a conventional two-dimensional cell culture system [40].

Cell behavior/ Cell spreading	Viability	Migration	Proliferation	Differentiation
Small 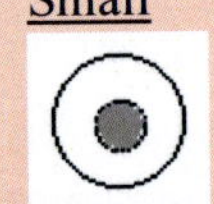	↓	↓	↓	↓
Medium 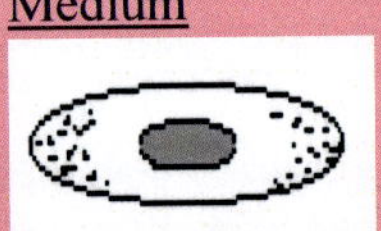	↑	↑	↑	↓
High 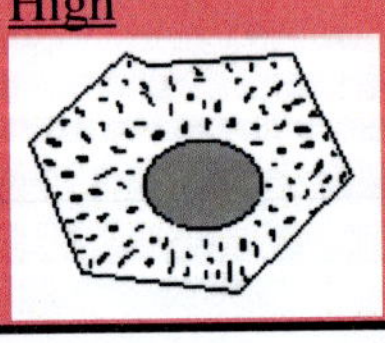	↑	↓	↓	↑

Both structural and signaling molecules play a decisive role for further behavior of cells after contacting a biomaterial, i.e., switching between cell proliferation and differentiation, survival or apoptosis and other functions [38-40,52,53].

Number, spreading and shape of adhering cells. The engagement of cell adhesion receptors and formation of focal adhesion plaques are determining factors for the number of initially attached cells as well as the size and shape of their spreading area. The number and spreading area can be used as relatively simple but accurate and important criteria for evaluation of initial cell adhesion as well as the cell behavior in late culture intervals. The cell number can be determined by counting cells directly on the material in a microscope. On transparent materials, living cells in native cultures can be evaluated. On non-transparent materials, both living and fixed cells can be visualized by various fluorescence methods including transfection of cells with fluorescence proteins or scanning electron microscopy [10,14,16,18,19,21,23,41,54]. The microscopic methods are also suitable for measurement of the size of cell adhesion area, i.e. the cell area projected on the material surface, which can be measured on digital pictures using an image analysis software [16,31,41,54]. For cell counting, the cells can be also detached from the material surface by protelytic enzymes (usually trypsin) and/or ethylene-diamino-tetraacetic acid (EDTA), a potent calcium chelator. The cells in suspension then can be counted manually in a hemocytometer or automatically in various cell counters or cell viability analysers, i.e. instruments in which cells are stained with trypan blue. This stain penetrates the membrane of dead cells, whereas the live cells remain unstained. In addition, the resistance of cells to the detachment by proteolytic enzymes can serve as an important indicator of the strenght of cell adhesion [18,19,21,23].

The cell number is often estimated indirectly by evaluation of the cell metabolic activity using tests of activity various enzymes, e.g. esterases or mitochondrial enzymes (MTT, MTS or XTT tests), or mesurement of DNA content in cell lysates [11,37,43,55,56]. The indirect methods are suitable paticularly for cells growing inside three-dimensional materials, not accessible for direct cell counting.

Indicators of Cell Growth

Growth of cells comprises their *hyperplasia (proliferation)*, i.e. increase in cell number, and *hypertrophy*, manifested by the increase in cell volume and protein content, which is often associated with the increase in the DNA content and chromosome number, resulting in polyploidia. Polyploidia is a well known feature e.g. in vascular smooth mucle cells, where it is considered as an adaptation of cells to increased mechanical or hormonal stimulation [21].

The cell proliferation can be evaluated by counting cells on the tested material in several time intervals, followed by the constructing growth curves and calculating cell population doubling time. Doubling time (DT) can be calculated according the the following formula:

$$DT = (t-t_o)\log 2/\log Nt-\log Nt_o,$$

where t_o and t represent earlier and later time intervals after seeding, respectively, and Nt_o and Nt represent the number of cells at these intervals [7,16,18,19,21,23].

Important markers of cell proliferation are also represented by various parameters of the cell cycle, such as the duration of the whole cell cycle and its phases, distribution of cells within the phases of the cell cycle, distribution of DNA content, incorporation of bromodeoxyuridine (BrdU) into newly synthesized DNA (i.e., a parameter showing the S-phase fraction, i.e., the percentage of cells actually synthesizing DNA), expression of proliferation antigens, such as PCNA or Ki 67 (the presence of Ki-67 antigen reveals the growth fraction, i.e., the percentage of cells capable of entering the mitotic cycle). The cell cycle parameters can be advantageously evaluated by immunocytochemical and flow-cytophotometric approaches [18,21,23,57]; an older but accurate approach is represented by ^{3}H-thymidine autoradiography [58,59]. Also the presence and activity of various enzymes taking part in the cell cycle, such as cyclin-dependent kinase 2 (CDK2) and its endogenous activators and inhibitors can serve as precise indicators of the cell proliferation capacity. CDK2 is a key stimulator of cell cycle progression, acting in G1/S, S-phase and G2/M boundary of the cell cycle. Its activity is supported by both cyclin A and E and inhibited by proteins p21 and p27 [60-62].

The cell volume can be calculated from cell diameters measured either manually in cells in suspension on microscopic slides or automatically in a flow cytometer. Also certain types of cell counters provide automated information on cell diameter [14,18,19,21,23].

The protein content is measured by biochemical methods, such as methods of Lowry or Bradford [14,18,19,21,23], now usually by commercially available kits. The DNA content can be measured by flow cytometry, and the chromosomes can be counted directly in cells after their arrest in the metaphase of the mitosis by Colcemide [21].

Indicators of Cell Differentiation

Recently it appears that the cell diferentiation can be essentially influenced not only by the presence of various humoral factors in the surrounding environment, but also by physical and chemical properties of the artificial growth support. For example, human mesenchymal stem cells in cultures on soft polyacrylamide matrices that mimicked brain differentiated towards neurons, whereas stiffer matrices mimicking muscle were myogenic, and comparatively rigid matrices that mimicked collagenous bone proved osteogenic [32]. As indicators of cell differentiation, the presence, content and spatial distribution of molecules specific or typical for certain cell types are usually followed. For example, typical markers of VSMC differentiation are represented by a higher content of contractile proteins alpha-actin and SM-myosins, and their assembly into filaments (Figure 8). Other markers of VSMC differentiation are desmin, caldesmon, calponin and meta-vinculin [18-23,40,41]. Mature vascular endothelial cells are characterized by a high content of von Willebrand factor stored in Weibel-Palade bodies, presence of CD 31 and CD 34 antigens, as well as the formation of thick beta-actin cables [19,31,34,63]. As markers of osteogenic differentiation, high activity of alkaline phosphatase, production of collagen I and non-collagenous calcium-binding ECM glycoproteins osteocalcin and osteopontin are currently evaluated [14-17,24-27,37,47].

Indicators of Cell Viability, Stress Adaptation and Death

Good adhesion, spreading, assembly of focal adhesion plaques and cytoskeleton as well as a high proliferation and differentiation activity mentioned above are also signs of good cell viability and non-cytotoxicity of the tested material. In addition, the above mentioned tests of the activity of mitochondrial enzymes, such as MTT, MTS or XTT, often used for evaluation of cell growth, can be considered as tests of cell viability and metabolical activity rather than tests of cell proliferation.

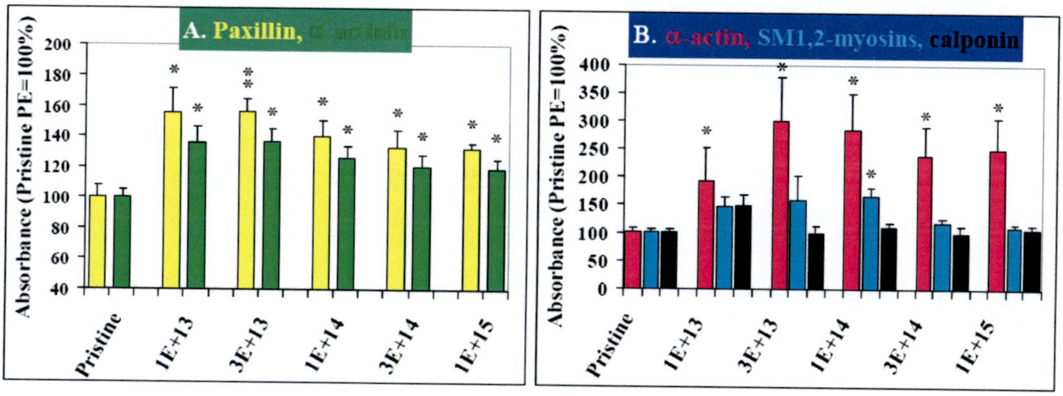

Figure 8. Concentration of markers of adhesion (A) and differentiation (B) in rat aortic smooth muscle cells in cultures on polyethylene (PE) modified by irradiation with O^+ ions (energy 30 keV, doses from 10^{13} to 10^{15} ions/cm^2). Measured by enzymatic immunosorbent assay (ELISA) per mg of protein, absorbances expressed in % of the values obtained on pristine non-modified PE. Mean ± SEM from 4 experiments, Student t-test for unpaired data, *: $p<0.05$ **: $p<0.01$ compared to the values on pristine PE.

Sometimes the cells can be highly metabolically acive without a considerable proliferation [10,37,43,55]. Long-term tolerance of the newly developed layers by cells can be assessed by cultivating cells on these materials for a period of several months [18].

Stress adaptation of cell growing on artificial materials can be estimated by the presence ad content of heat-shock proteins, such as HSP 60 and HSP 70, and other chaperone molecules [64]. For detection of dead cells, staining of dead cells with trypan blue, propidium iodide and chromatin dye Hoechst 33258 can be applied (the latter visualizes chromatin clumps in cell nuclei typical for apoptosis; [65]). These dyes penetrate through damaged cell membrane, whereas viable cells with intact membrane remain unstained. Simple fluorescence staining is often a principle of various commercially available kits developed for quick evaluation of potential material cytotoxicity. For example, in a commercially available LIVE/DEAD viability/cytotoxicity kit (Invitrogen), the the cells are incubated in two probes, i.e. calcein AM producing green fluorescence and detecting the esterase activity in living cells, and ethidium homodimer-1 emitting red fluorescence and signalizing the membrane damage in dead cells. For deeper studies on cell death (necrosis and apoptosis), a wide range of more sophisticated approaches are available, e.g., studies on annexin V, caspase activity, p53 expression or DNA fragmentation [66,67]. However, the studies on cell death represent a large independent scientific field exceeding the frame of this review.

Indicators of Potential Mutagenicity, Carcinogenicity and Immunogenicity of the Material

Some new materials perspective for advanced biomedical technologies, especially carbon nanoparticles like fullerenes, are potentially mutagenic, carcinogenic and immunogenic [16,65]. Therefore, standard tests of the morphological transformation of Syrian hamster embryonic cells in cultures on these materials (described in detail by [68,69]) can be performed. Immune activation of bone and vascular cells on the materials can be estimated by increased concentration of immunoglobulin and selectin adhesion molecules (ICAM-1, VCAM-1, ELAM-1), which bind cells of the immune system [15,16,18,19,23], as well as by the production of cytokines, such as tumor necrosis factor alpha or interleukins beta [55].

5. Reactions between Plasmas and Polymer Surface

Surface Chemistry

The characterization of the surface chemistry of the modified polymer is one step in understanding the mechanism for cells adhesion. The next crucial step is to determine the nature and extent of chemical interactions between the overlayer of interest and the modified polymer surface. This step presents a challenge because currently there are no techniques available with the sensitivity to characterize chemical interactions for an atomically thin buried interface. Several approaches have been used to analyze buried interfaces. Ion sputter depth profiling (typically done with Ar ions) in conjunction with XPS can be used to evaluate a buried interface for overlayers >10 nm [2].

The parameters of treatment were chosen since these led to the most pronounced changes of polymer surface in our previous experiments [70-74]. It was observed elsewhere that plasma treatment of polymer macromolecules results in their cleavage, ablation, alterations of chemical structure and thus affects surface properties e.g. solubility [75]. The chemical structure of modified polyethylene (PE) was characterized by FTIR and XPS spectroscopy. Exposition to discharge leads to cleavage of polymeric chains and C–H bonds followed by generation of free radicals which easily oxidize [10,76]. By FTIR spectroscopy the presence of new oxidized structures within whole specimen volume can be detected. IR spectra in the 1710–1745 cm^{-1} interval [71,77] from PE, exposed to Ar plasma, show the presence of the oxidized structures (carbonyl, carboxyl, ester).

The chemical structure of plasma-treated PE was examined using XPS, the oxygen concentration profile was determined from RBS measurement and the concentration of free radicals was determined by EPR technique [78]. The -CH a $-CH_2$ (of pristine PE), oxidized (-C=O, -COO, -COC-), $-NH_2$, and also -C=C- groups (typical for aromates) were reported present at Ar plasma-treated high-density PE (HDPE) [70].

Depth profile and total content of oxygen in PE was determined by RBS method. The profiles of plasma-treated sample and that subsequently etched in water are compared in Figure 9. The mean oxygen contents in the surface layer accesible by RBS, ca 140 nm thick, are (39 ± 5) and $(28\pm3).10^{15}$ cm^{-2} for plasma-treated and water etched samples respectively. Both concentration profiles exhibit a maximum at the depth of ca 20 nm, further the concentration decreases slowly, falling to negligible concentrations at 60 nm. Similar trend was observed in previous report dealing with PE exposed to Ar plasma at the power of 1.7 W [70].

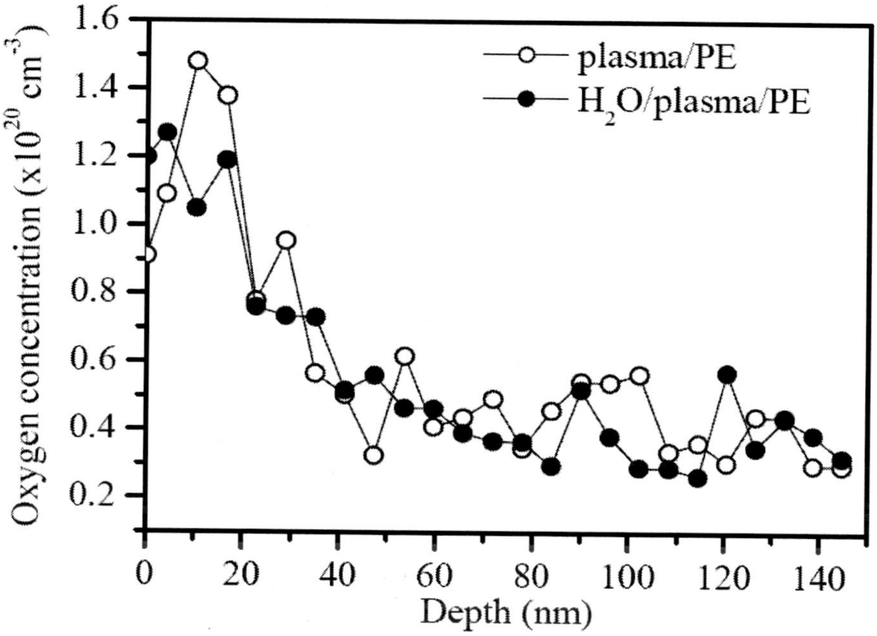

Figure 9. Concentration depth profiles of oxygen obtained from RBS measurement of plasma-treated PE and that of plasma-treated and water etched for 24 hours [78].

Due to plasma treatment the radicals (R, free spin) are generated on polymer chain. Not only C-H but also C-C bonds are likely to brake, the later leading to fragmentation of polymer chain. The radical concentration determined by EPR method is in Table 3. The concentration of free radicals R decreases during the aging to about quarter of the initial after 80 days. The decrease of R is a result of radical recombination [h]. Detailed comment of EPR observation ("aging" of radicals) is described too [79].

Table 3.C (1s) a O (1s) concentrations (at. %) determined from XPS measurements and radical densities (R, 10^{18} g^{-1}) determined by EPR method. The concentrations and densities were determined on the pristine PE, plasma-treated PE (1 or 24 hours after the treatment) and plasma-treated and water etched PE (see left hand side of the table). Before the measurement the samples were stored under standard laboratory conditions [78].

PE modification	C	O	Radical density
pristine PE	100	0	-
1 h after plasma treatment	69.2	30.8	-
24 h after plasma treatment	75.7	24.3	2.77
plasma treatment and 24 h in water	63.4	36.6	2.31

The oxygen concentration in the first 10 monolayers of modified PE was followed independently by XPS (see Table 3) [80]. The oxygen concentration was determined from survey spectra shown in Figure 10. It is confirmed that plasma treatment leads to oxidation of the surface layer and the oxygen concentration is 31 at.% 1 hour after plasma treatment. One can see from Table 3 that with increasing aging time the oxygen concentration decreases due to rearrangement of the treated polymer chains and their fragments [74,81]. The profile and area of individual peak in XPS spectra enable to determine chemical group concentration in PE surface (see Table 4). Model deconvoluted C(1s) peak is shown in Figure 11. In Table 4, the concentration of detected groups -CH_2-, -C-O-, -C=O and -O-C=O is presented [78,82]. One can see that during sample aging the concentration of -C=O group increases and -O-C=O decreases. Only a trace concentration of nitrogen was detected.

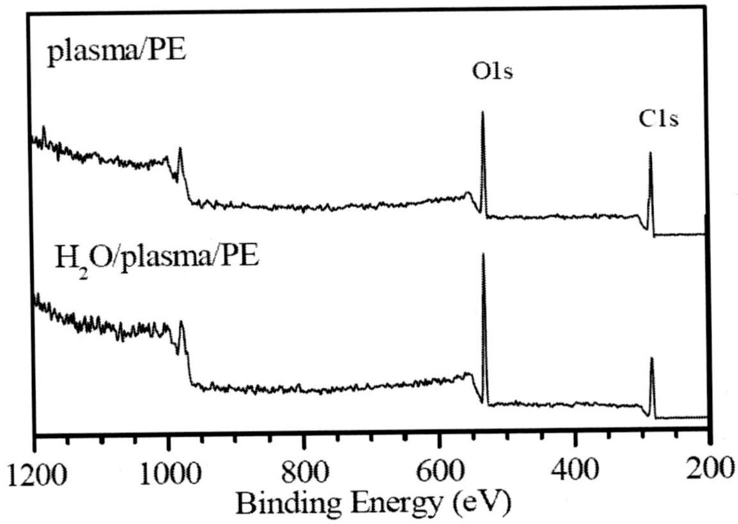

Figure 10. The XPS survey spectra of PE modified by plasma (plasma/PE) or modified by plasma and water etched (H_2O/plasma/PE) measured 24 h after the plasma treatment. Before the measurement the samples were stored under standard laboratory conditions. [78].

Table 4. Result of evaluation of XPS C(1s) peak for the PE samples 1 and 24 hours after the plasma treatment and plasma treated PE sample subsequently water etched for 24 hours. Concentrations of selected groups are in at. % [78]

Chemical group	1 h after plasma	24 h after plasma	PE modification plasma and 24 h in water
-CH_2-	65.8	66.5	47.7
-C-O-	26.1	26.8	39.3
-C=O	1.6	5.3	12.4
-O-C=O	6.5	1.4	0.6

Above described results prove that plasma treatment of PE leads to cleveage of molecular chain and production of polar groups. Since these changes may enhance interaction with polar solvents, the water etching of modified PE was examined; the calculated thickness of the surface layer removed from plasma-modified PE (240 s, 8.3 W) by 24 hour water etching is 20.6±4.6 nm [78].

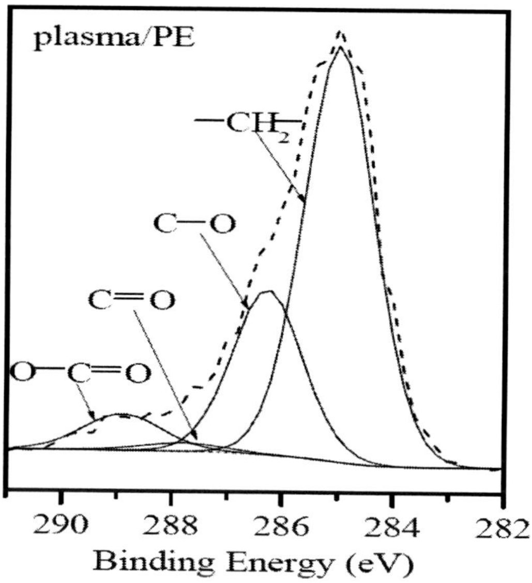

Figure 11. C(1s) XPS spectrum of PE measured 1 h after the plasma treatment (240 s, 8.3 W). The dashed line represents the measured intensities. [78].

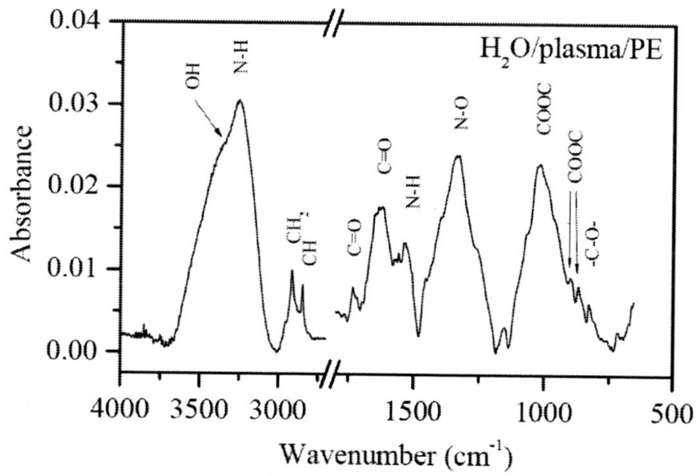

Figure 12. FTIR spectrum of the layer removed from plasma-treated PE by dissolving in deionized water (details of the sample preparation are given. [78]).

Chemical composition of the dissolved material was determined by FTIR spectroscopy. FTIR spectrum of the sample obtained as a deposit after evaporation of water is shown in Figure 12. The etched material contains -CH and -CH_2 groups of original PE molecular chain. Furthermore the nitrogen (-NH, -N-O) and oxygen containing (-C=O, -COOC-, C-O-) groups were detected. These were introduced either by the residual reactor atmosphere or during the post treatment storage. The polar groups are responsible for enhanced water solubility of the modified PE [78].

Ablation and Depth of Modification

In general, the depth of surface modification mainly depends on the power of the plasma discharge and treatment time [83]. For plasma-treated polymer samples, the depth of the surface modification is typically of several tens nm and has been studied by surface-sensitive techniques such as angle-resolved XPS [n] or RBS [70,78]. In [70,78] the concentration depth profile of oxygen, incorporated in the polymer surface layer (high- and low-density PE, HDPE and LDPE) as a result of the plasma treatment, was determined (Figure 13) [70]. With increasing exposure time to plasma discharge oxygen amount in polymer surface layer also increases in both, LDPE and HDPE. The deeper penetration of oxygen up to about 120 nm is observed in HDPE plasma-treated for 25 s. After treatment for 250 s, however, most of oxygen is incorporated in surface layer only about 50 nm thick. The evolution of the oxygen concentration depth profiles in LDPE is similar.

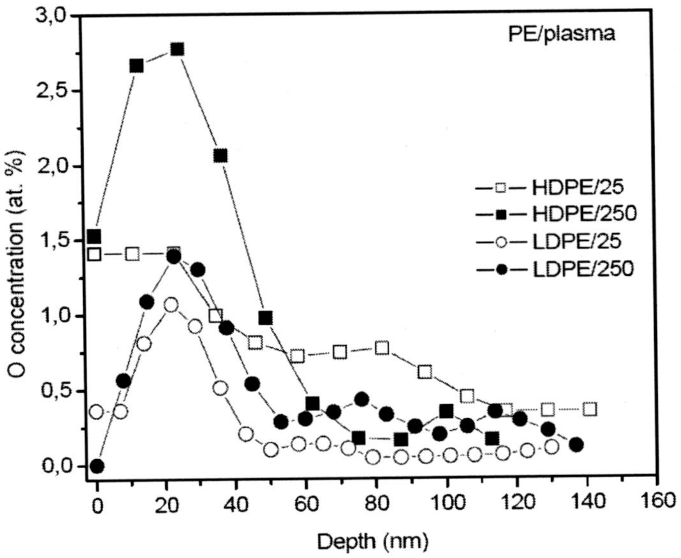

Figure 13. Concentration depth profile of oxygen incorporated in HDPE and LDPE modified in the Ar plasma at the discharge power of 1.7 W. The numbers are the times of plasma treatment in seconds. The profiles were determined by RBS technique [70].

It was observed elsewhere that plasma treatment of polymer macromolecules results in their cleavage, ablation, alterations of chemical structure and thus affects surface properties e.g. solubility [75]. The effects of the treatment in Ar plasma on the surface properties of PE were examined in [72]. The parameters of the plasma discharge (240 s, 8.3 W power) were chosen, on the basis of our previous experiments [70-74,78], to guarantee the most pronounced changes of polymer surface. Mean thickness of the ablated PE layer was calculated from the weight difference measured by gravimetry. By Ar plasma (8.3 W, 240 s) 30±8 nm thick surface layer is ablated and thickness of the surface layer removed from plasma-modified PE by 24 hour water etching is 21±5 nm [78]. The surface morphology of PE was examined by AFM method (Figure 14, [78]). Obviously the exposition to plasma discharge leads to dramatic topology change. As ablation result a lamellar structure on the sample surface appears [70]. It can be concluded, with regard to the earlier data [84], that PE amorphous phase is ablated faster than the crystalline one. Surface morphology of this sample after subsequent water-etching is shown in Figure 14.

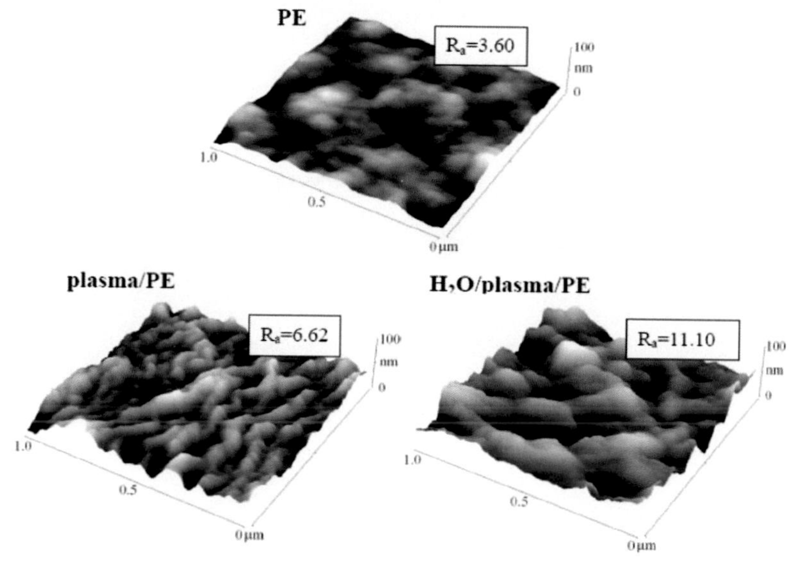

Figure 14. AFM images of pristine, plasma-modified and plasma-modified and water-etched PE. R_a is the surface roughness in nm [78].

It is seen that water etching removes tiny, sharp formations from the sample surface and the etched surface exhibits lower diversity but higher surface roughness R_a. In analogy with earlier reports [75,85] obtained on LDPE and polypropylene (PP) plasma-modified samples, it can be concluded that the observed changes correspond to removal of low-molecular oxidized segments from plasma-modified polymer surface [72].

Surface Properties

Surface Morphology

In most of published studies of polymer modification by plasma treatment, great attention is devoted to the roughness and surface morphology of modified polymers e.g. reviews [1,2]. Exposure to plasma discharge results in significant changes of surface morphology of high- and low-density polyethylene (HDPE and LDPE) which are seen from SEM images of pristine and plasma-treated polymers [70]. While on the surface of plasma-modified HDPE a lamellar structure of macromolecular chains is visible, on the surface of modified LDPE formations are created which may reflect spherulitic arrangement of macromolecular chains [70]. AFM examination of the same samples was accomplished to verify the SEM observation. The AFM results for selected samples are shown in Figure 15. Surface roughness is an increasing function of the scan dimension and plasma exposure [70]. Analogous results were reported in [86]. Plasma treatment results in a mild increase of the surface roughness for LDPE but in much stronger increase for HDPE.

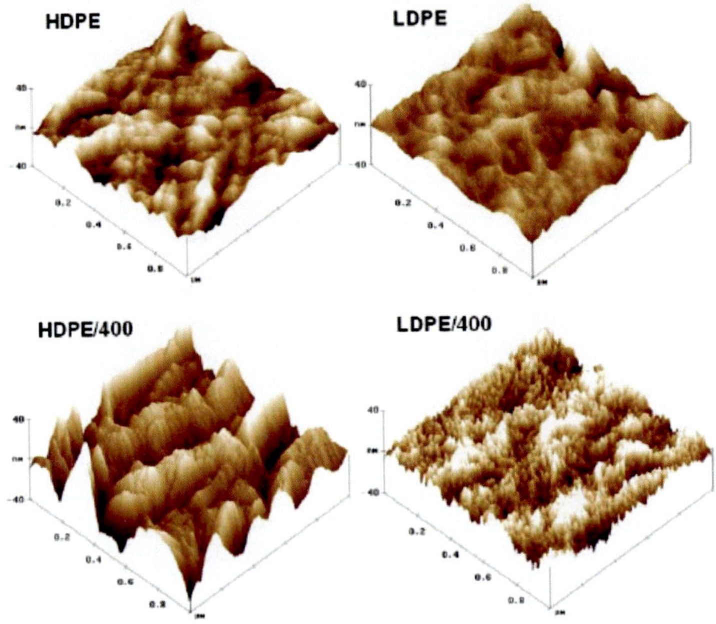

Figure 15. Contact mode AFM height images of PE foils before (HDPE and LDPE) and after 400 s modification in Ar plasma (HDPE/400, LDPE/400). [70].

Surface Wettability and Aging Effects (Stability)

Long-term stability of a modified polymer surface is important for their potential practical applications. Adverse aging effects may be significant if the polymer is not stored in a controlled environment or coated immediately after treatment. The stability of a modified surface depends on several factors including chemical structure of the modified polymer which in turn depends on the polymer initial structure and composition, plasma treatment procedure (working gas, discharge power) and the environment in which the polymer is stored. It has been observed by many researchers that plasma-modified polymer surfaces are highly susceptible to aging effects [1,2,83,87-92].

It is known that the value of the contact angle (i.e. polarity and wettability) is mostly affected by the chemical structure and morphology of the polymer surface layer [93,94]. In Figure 16, the dependence of the contact angle on the exposure time to Ar plasma at 8.3 W discharge power is shown [71]. With increasing exposure time the contact angle decreases. It was found that the enhanced wettability (low contact angle) of the plasma-treated PE is not permanent and it declines with the time elapsed from the plasma treatment. With increasing time from the plasma exposure the contact angle increases and the increase is more pronounced for the specimens exposed to plasma for longer time. The increase is so strong that after 120 h of ageing the contact angle of the PE sample, plasma-treated for 240 s, becomes larger than that of pristine PE. It was found that the contact angle gradual increase practically stops after 336 h, i.e. 14 days after exposition to plasma. Relaxation of the surface hydrophilicity is obviously connected with a rearrangement of degraded macromolecules on the surface of modified PE [84]. An inward diffusion of low mass oxidized fragments and an orientation of polar groups toward specimen bulk are also supposed [89].

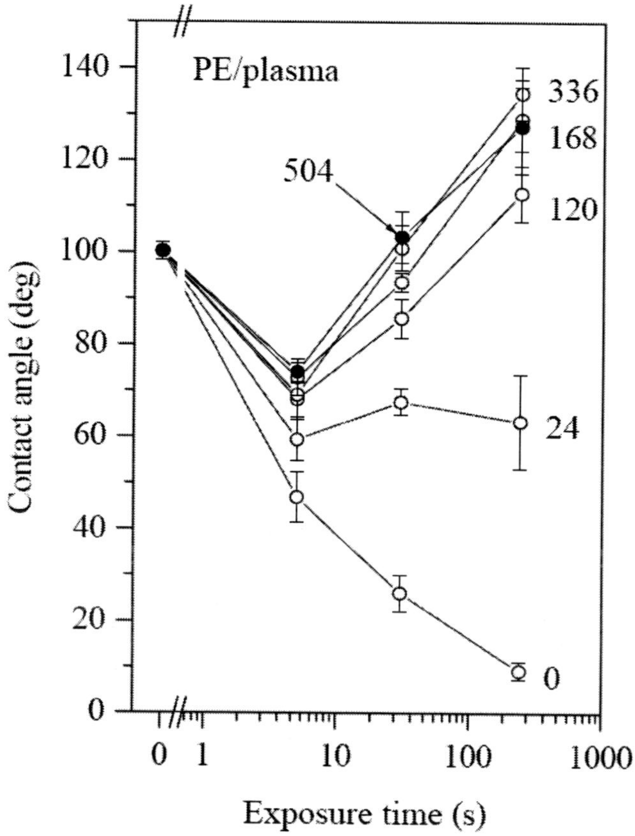

Figure 16. Dependence of the PE contact angle on the exposure time to 8.3 W plasma discharge. The contact angle was measured at different times elapsed from the plasma treatment. The numbers are the times in hours elapsed from the plasma treatment [71].

Using angle resolved XPS (ARXPS) technique, the chemical structure of the first ca 10 monolayers (a surface layer about 5 nm thick) of the sample surface can be analyzed [74]. For plasma-treated samples, the oxygen concentration was followed as a function of the aging time of the samples. The oxygen concentrations were determined from ARXPS spectra using standard procedure [82]. In the first monolayer of pristine PET the oxygen concentration is below the value expected from the PET stoichiometry, the deficit could indicate the preferential orientation of the oxygen containing groups inwards resulting from the process of fabrication of oriented PET foil. Immediately after the plasma treatment the oxygen concentration increases, but then it decreases with increasing aging time. These results are in agreement with the results of the contact angle measurements [74], the evolution of which is explained by the rearrangement of degraded macromolecules and molecular fragments.

The oxygen concentration in the first 10 monolayers of plasma-modified PE was followed independently by XPS (see Table 3) [78]. The oxygen concentration was determined from survey spectra shown in Figure 10. It is confirmed that plasma treatment leads to oxidation of the surface layer and the oxygen concentration is 31 at.% 1 hour after plasma treatment. One can see from Table 3 that with increasing aging time the oxygen concentration decreases due to rearrangement of the treated polymer chains and their fragments [78,81]. The profile and area of individual peak in XPS spectra enable to determine chemical group concentration in PE surface (see Table 4). Model deconvoluted C(1s) peak is shown in Figure 11. In Table 4 concentration of detected groups -CH_2-, -C-O-, -C=O and -O-C=O is presented [82]. One can see that during sample aging the concentration of -C=O group increases and -O-C=O decreases. Only trace concentration of nitrogen was detected.

Mechanical Properties

Mechanical properties of polymer surface layer are affected by changes in chemical structure due to exposure to plasma discharge [95]. Same mechanical properties of pristine PE and PE modified in Ar plasma discharge were studied using NanoIndenter XP MTS device [71]. The scratch results for pristine PE and PE plasma-modified for time 240 s were chosen. No difference between scratch characteristic of pristine and modified PE is observed. Also there are no signs of plastic deformation and only mild mechanical damage of surface layer is seen. It may be therefore concluded that the deformation of the PE surface layer is mostly elastic. Elastic modulus and hardness as a function of the exposure time for pristine and plasma-modified PE are shown in Figure 17. It is seen that the modulus (Figure 17A) increases as a result of the plasma treatment. For 240 s exposition the modulus achieves its maximum. Dependence of the PE hardness on the exposure time and displacement into surface is shown in Figure 17B. Plasma treatment for short times does not change, within experimental errors, the PE hardness. Significant hardness increase observed on the PE sample treated for 240 s may be due to polymer crosslinking initiated by plasma discharge [95].

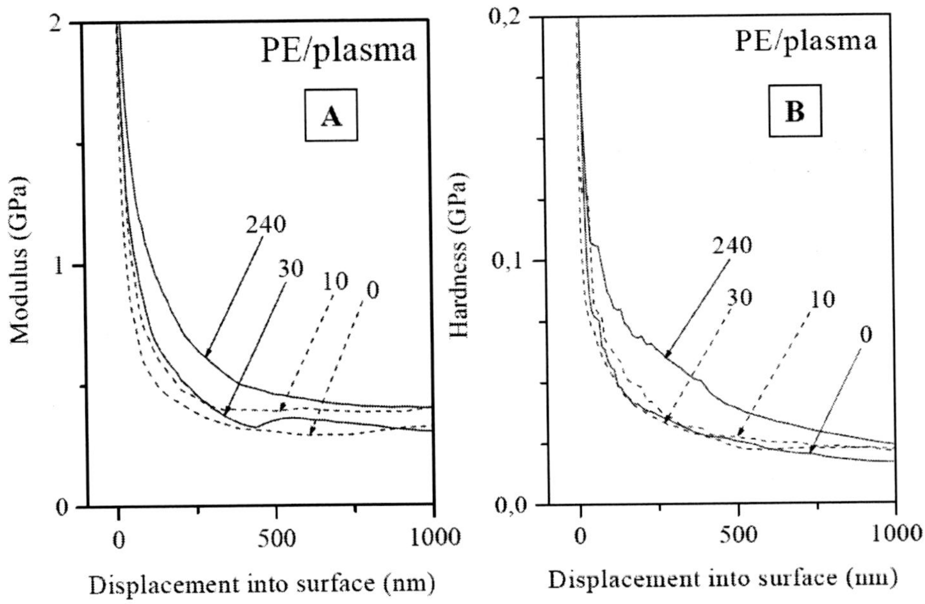

Figure 17. Dependence of the elastic modulus (A) and hardness (B) on the displacement into surface measured on pristine PE and PE exposed to Ar plasma for 10, 30 and 240 s [71].

6. Modification of Polymers by Ion Implantation

Compositional and Structural Changes of Polymers under Low-to Medium-Energy Ion Implantation

The ion implantation is a well-established technique for modification of properties of solids. It has been used for modification of polymers since its invention on the beginning of 60-ties [3,4,96]. Huge number of experimental results has been obtained on interaction of low-medium energy ions with polymers on a microscopic level and on structural and compositional changes of polymers resulting from ion irradiation. The effect of ion irradiation on polymer structure and composition is a result of complex physical and chemical processes initiated by energy deposited by ion in polymer matrix. Since the energy of bond dissociation in polymers does not exceeds 10 eV, the energy deposited by an ion leads to multiple cleavage of chemical bonds along the ion trajectory and formation of first generation of degradation products (molecular fragments, free radicals, ions and free electrons) exhibiting high chemical reactivity. Then the stable degradation products are formed by interaction of the reactive species with each other and with components of the polymer matrix as well. Typical degradation products are low mass molecular fragments and cross linked two- or three- dimensional structures. Relative fractions of these products depend on the polymer initial structure and composition and on the energy deposited by the ions [97-101]. Some of volatile degradation products (mostly light weight, hydrogen rich ones) may diffuse to the polymer surface and escape. This process leads to gradual carbonization of the polymer surface layer and creation of excessive free volumes [102-107]. The modified polymer surface may be attacked by aggressive species from ambient atmosphere (e.g. oxygen) and new structures may be formed on the polymer surface.

The implantation of low-to-medium energy ions changes only surface layer of polymers, with the thickness in nm-μm range, and preserves favorable bulk properties of polymers. By the ion implantation such surface properties as chemical structure [108-111], wettability [109,110,112], electrical conductivity [109,113-116], tribological properties [112,117,118] and bioactivity [18,111-113,119] can be changed in a manner which can be controlled by a proper choice of ion mass, energy and fluence.

Desorption of Volatile Degradation Products

Upon ion irradiation the polymer macromolecules are ionized and excited and a broad spectrum of degradation products of different mass and chemical activity are created. Low mass products, possessing higher mobility, may migrate to the sample surface and escape into evacuated implanter target chamber. The degassing rate can be determined by residual gas analysis. It was found that the main molecules and fragments liberated during ion irradiation of common polymers are H_2, CH_4, C_2H_2 CH_4, CO, CO_2, CF, CF_2, CF_3. In "pure" hydrocarbons (PE, PP) hydrogen molecules may form about 90 % of released degradation products [102]. In polymers containing oxygen (e.g. PMMA, PI) beside hydrogen also carbon oxides are released in significant quantities (up to 50% of released products). Desorption of hydrogen molecules and hydrogen-rich fragments leads to gradual declining of hydrogen concentration in the polymer layer modified by the ion irradiation. Hydrogen depletion can be studied by any hydrogen sensitive depth profiling method (e.g. ERDA). In [120,121] the hydrogen and oxygen depth profiles were measured using standard ERDA and RBS techniques on different polymers (PE, PS, PI and PET) implanted with He, C, N, F, Ne, Ar and Xe ions at the energies from 20-200 keV to ion fluences varying from 10^{12}-10^{17} cm^{-2}. The evolution of hydrogen depth profile with increasing ion fluence is illustrated in Figure 18. It was found that during initial stages of the ion implantation the hydrogen and oxygen contents in the irradiated surface layer decrease slowly but above some threshold fluence (typically 10^{14}-10^{15} cm^{-2}), which is a decreasing function of the ion stopping power, rather rapid decreae of H and O content is observed. At highest ion fluences the H and O contents achieves a saturated value which vary from 50-70 % of their original value in pristine polymer. Several semiempirical models have been suggested describing hydrogen desoption during ion implantation (see e.g. [122]).

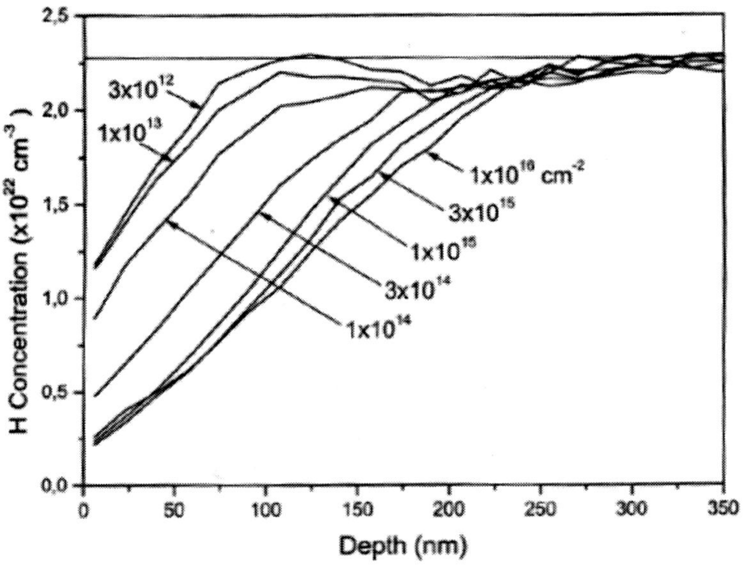

Figure 18. Depth profiles of hydrogen from polyimide (PI) implanted with 100 keV Kr ions to the fluences indicated. The horizontal line shows the hydrogen concentration in pristine PI. The profiles were determined from ERDA measurement performed with 2 MeV alpha particles [122]. The depth resolution was about 50 nm. It is seen that significant H depletion starts at the fluence of about 1x10^{14} cm^{-2}.

Transport Processes in Ion-Irradiated Polymers

By ion irradiation a large number of highly reactive degradation products are produced during whole implantation process. Desorption of volatile degradation products leads to creation of excessive free volume which may facilitate penetration of gaseous species from residual atmosphere in the ion implanter. Reactive agents such as oxygen may react with degradation products and form new oxidized structures, which can be seen in RBS spectra of implanted samples. These structures may affect significantly several properties of the ion implanted polymers. The depth profiles of the incorporated oxygen were studied on polymers implanted with B, A, F, As, Sb and I ions [123-126]. In Figure 19 [125] the depth profiles of oxygen incorporated in PE during implantation of 150 keV Sb ions are shown as a function of the ion fluence. It is seen that oxygen penetrates through the implanted surface layer and is trapped within. With increasing ion fluence the oxygen content also increases up to some critical fluence which varies from 10^{14} to 10^{15} cm^{-2} for heavy and light ions respectively. At the critical fluence the oxygen content achieves a maximum. For still higher fluences the oxygen content declines probably due to destroying of the oxidized structures created in initial phases of the ion irradiation. At the maximum the total amount of trapped oxygen can be quite large. It was found that in polyethylene implanted with 150 keV F ions ($5x10^{15}$ cm^{-2}) oxygen concentration in ion implanted layer achieves 25% of that of carbon.

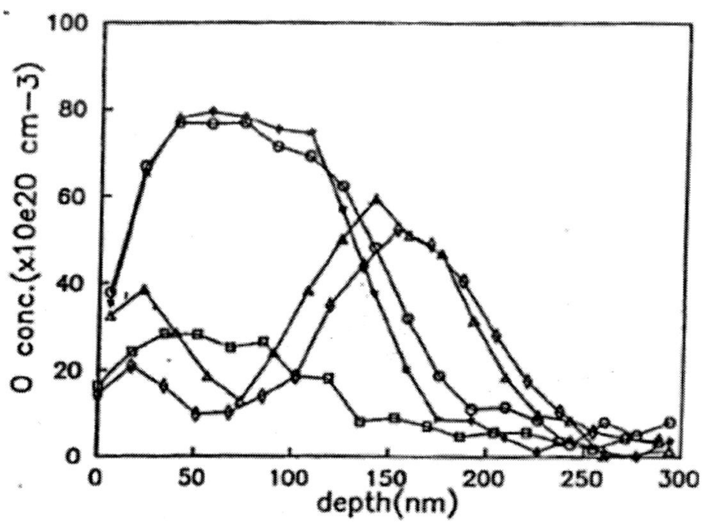

Figure 19. Depth profiles of oxygen incorporated into the surface layer of PE implanted with 150 keV Sb ions to the following fluences: $2x10^{13}$ (□), $1x10^{14}$ (o), $2x10^{14}$ (+), $1x10^{15}$ (Δ) and $2x10^{15}$ cm^{-2} (♦) [125].

As can be seen in Figure 19 the oxygen depth profile is not homogenous and for low ion fluences it follows roughly the profile of ion electronic energy loss. For highest fluences the oxygen content decreases and its depth profile acquires bimodal form, with two concentration maxima, one near the sample surface and second, more pronounced, near the ion range. The bimodal depth profile gives evidence of deep structural changes induced by high fluence implantation.

Surface Morphology of Ion-Implanted Polymers

Ion-implanted polymers exhibit altered surface morphology, wettability, chemical affinity, bioactivity and electrical conductance. The observed changes are due to deep structural and compositional changes of polymer surface layer taking place during the ion irradiation. Surface morphology of pristine and modified polymers may be changed by different processes mentioned before, such as desorption of volatile degradation products through sample surface or preferential sputtering of amorphous and crystalline phases of polymers. The morphology can best be examined by AFM technique. Surface morphology of polyimide (PI) implanted with 90 keV N ions to the fluences from $1x10^{14}$–$2x10^{17}$ cm^{-2} was investigated by standard AFM [127]. It was found that the pristine PI exhibits rather diverse, bumpy surface with formations with typical dimensions of 200 nm and 2 nm in lateral and vertical directions respectively. The irradiation to 10^{14} cm^{-2} leads to dramatic smoothing of the polymers surface but for increasing fluences the surface roughness increases again. The maximum roughness is observed at the fluence of $1x10^{17}$ cm^{-2} but for the fluence $2x10^{17}$ cm^{-2} the roughness decreases again probably due to progressive compaction of the polymer surface (see Figure 20).

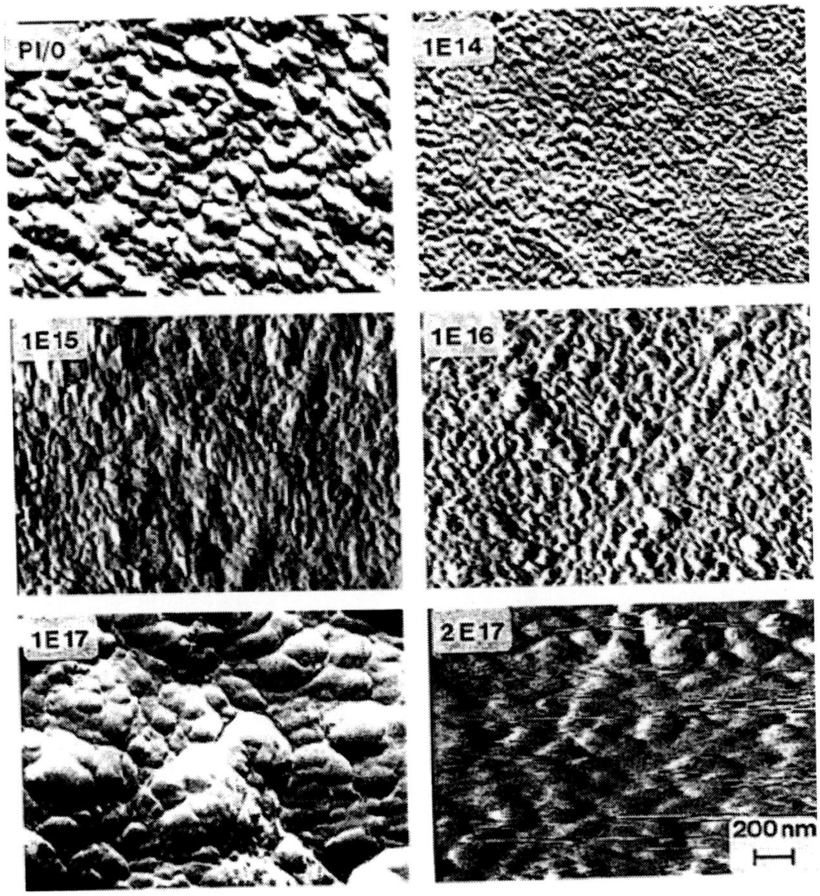

Figure 20. AFM deflection images of pristine polyimide (PI) and PI implanted with 90 keV N ions to fluences indicated [127].

Occurence of a graphitic phase on PI irradiated to the fluences $1x10^{16}$ and $1x10^{17}$ cm^{-2} was proved by XRD measurement. Similar surface evolution was observed in [118] on PI implanted with 100 keV C and N ions to the fluences from $3x10^{15}$-$1x10^{17}$ cm^{-2}. In [128] PI was implanted with 40 and 80 keV Ar ions at high ion flux and to the fluences from10^{14}-10^{17} cm^{-2}. At lowest fluence small craters at the ion impact sites were observed by AFM. With increasing ion fluence bubles occur on the polymer surface and the surface roughness increases significantly. The bulb appearance is explained by effect of escaping Ar gas through the sample surface. Similar AFM examination was accomplished on PE implanted with 63 keV Ar ions and 155 keV Xe ions to the flunces from $1x10^{13}$-$3x10^{15}$ cm^{-2} [117]. The irradiation to lowest fluence does not result in any significant change in the PE surface morphology but at higher fluences rather dramatic changes are observed. New structures appear on the PE surface and the surface roughness increases in comparison with that of pristine PE, the increase being larger for Ar ions. Great differences were observed between the topografic effects created by two different ions. The Ar ions produce rather fine but sharp surface structures while the structures produced by heavier Xe ions are broader and shallower. It appears that AFM technique is very useful for investigation of polymer surface morphology and its changes under ion irradiation. Specific phenomenon, i.e. ablation under ion irradiation was studied on PTFE, the polymer which is extremely radiation sensitive [129]. The PTFE samples were irradiated with 300 keV Ar ions to fluences from $1x10^{14}$-$1x10^{16}$ cm^{-2}. The thickness of ablated layer, determined by optical microscopy, was found to be an increasing function of the ion fluence; for the fluence of $1x10^{16}$ cm^{-2} about 14 μm of PTFE was removed by ion irradiation. XRD measurement shows gradual loos of PTFE crystalline phase with incresing ion fluence.

New Functional Groups Created by Ion Irradiation

Destruction of initial structure of polymers and the occurence of new structures formed by ion irradiation can be followed using FTIR, UV-Vis and XPS spectroscopy. Structural changes of PTFE implanted with Ar ions to the fluences from $1x10^{14}$-$1x10^{16}$ cm^{-2} were examined by FTIR method [129]. It was found that new C=C and $C=CF_2$ bonds are produced together with C=O group, the formation of which indicate significant oxidation of the irradiated PTFE. The concentration of these new structures is an increasing function of the ion fluence. On the other hand, the irradiation to fluences above $3x10^{15}$ cm^{-2} results in a significant decrease of concentration of C-C bonds and CF_2 group. The decrease indicates splitting of C-C bonds and release of CF_2 fragments. At the same time the increase of absorbance in UV-Vis spectra with increasing ion fluence points to production and accumulation of conjugated double bonds in irradiated PTFE. In [118] gradual destruction of imide group (-CO-N-CO-) in PI irradiated with C and N ions to increasing ion fluences was observed from FTIR spectra. The destruction is accompanied by significant oxygen depletion in polymer surface layer which is observed in RBS spectra. Similar results were obtained in [130], where PET, polyetheretherketone (PEEK) and PI were irradiated with 150 keV Ar and 1.76 MeV He ions to the fluences from 10^{12}-10^{15} cm^{-2}. The compositional and structural changes were examined using RBS and ERDA techniques, UV-Vis spectroscopy and positron annihilation spectroscopy (PAS). Significant absorbance increase in UV-Vis spectra was observed on PEEK and PET irradiated with He ions and PET irradiated with Ar ions. The absorbance increases with increasing ion fluence. Also decline in hydrogen and oxygen concentration in the surface layer about 250 nm thick was observed on PET sample irradiated with Ar ions to the fluence of $1x10^{15}$ cm^{-2}. No such effect was initiated by He ions, probably due to their much lower stopping power.

7. Photochemical Modification of Polymers by UV Light-Irradiation

Similarly as the ion implantation or plasma irradiation described above, also the irradiation with UV light leaded to splitting chemical bonds in the irradiated polymer and creation of radicals. The radicals then interacted mutually, and in the presence of oxygen and ammonia atmosphere, new chemical functional groups, such as C-H, C-O-, C=O and $C\text{-}NH_2$ as well as conjugated double bonds (C=C) and different free radicals were created. In PTFE, the newly formed groups often replaced F-atoms. These groups increased the surface polarity and wettability, which was positively correlated with the exposure time and well apparent on the originally highly hydrophobic PTFE surface. The UV light-irradiation of PET in acetylene atmosphere leaded to the formation of hydrogenated amorphous carbon (a-C:H) [7-11].

8. Cell Colonization on Materials Modified by Plasma Discharge, Ion Implantation or UV Light Irradiation

The changes in physicochemical properties of the material surface after irradiation with plasma, ions or UV light, such as formation of new chemical functional groups, increased surface polarity, wettability, electrical charge and conductivity as well as the formation of nano-sized irregularities on the material surface, are generally associated with the improved adhesion of various types of cells, including vascular endothelial and smooth muscle cells, osteoblasts, chondrocytes, fibroblasts or cells of the nervous system [7-11,18,19,40,132-134]. As the main underlying mechanism, improved adsorption of cell adhesion-mediating molecules to the material surface can be considered (Figure 4). On physicochemically modified surfaces, these molecules are adsorbed more homogeneously. In addition, their amount, spectrum and probably also the spatial conformation are appropriate for a good accessibility of active sites (i.e., specific amino acid sequences or saccharide-based ligands, see above) on these molecules by adhesion receptors on the cell membrane (Figures 5,6). The importance of spatial conformation of adsorbed molecules for the cell attachment is suggested on the Figure 6. It is apparent that the intensity of fluorescence, i.e. an indicator of the amount of collagen IV adsorbed on ion-implanted polyethylene samples, has an increasing tendency which is proportional to the increasing ion dose, while the number of initially adhered vascular smooth muscle cells reached its maximum at the medium ion doses and then strongly decreased. Thus, not only the absolute amount of the adsorbed proteins is important for cell adhesion, but also some other factors are needed - most probably the appropriate spatial conformation of the adsorbed molecules. In addition, it is known that on wettable surfaces, like those modified by ion-, plasma- or UV light-irradiation, the adsorbed cell adhesion-mediating molecules are flexible. The flexibility enables their active spatial reorganization, resorption and resythesis by cells, which further contributes to the better accessibility of the active sites by the cell adhesion receptors [18,19,40,44-46]. On the other hand, a sufficient rigidity of the adhesion substrate is also required for a good cell adhesion and spreading. If the material is too elastic, compliant, flexible and irreversibly deformable, it does not allow the anchorage of cells, even if the ligands for adhesion receptors are present in satisfactory amounts and accessibility, and are bound by these receptors. Such type of substrate cannot resist the cell tractional forces generated by the assembling cytoskeleton. For example, when collagen was adsorbed or covalently bound on glass or hard polyacrylamide gel (PAG), the vascular smooth muscle cells were normally spread with multiple streak-like focal adhesion sites and rich actin cytoskeleton, and were viable. However, if soft PAG or pure collagen gel was used, the cells remained round (without signs of spreading and assembly of adhesion apparatus), and underwent apoptosis [40,54].

Not only the cell number, but also the cell spreading area was generally enlarged on ion-, plasma- or UV light-irradiated polymers (Figure 21). The cells adhered by nore numerous and larger focal adhesion plaques and contained a well developed actin cytoskeleton. As revealed by ELISA, also the concentrations of focal adhesion proteins vinculin, talin, paxilin and alpha-actinin were usually increased in comparison with cells growing on non-nodified polymers (Figure 8).

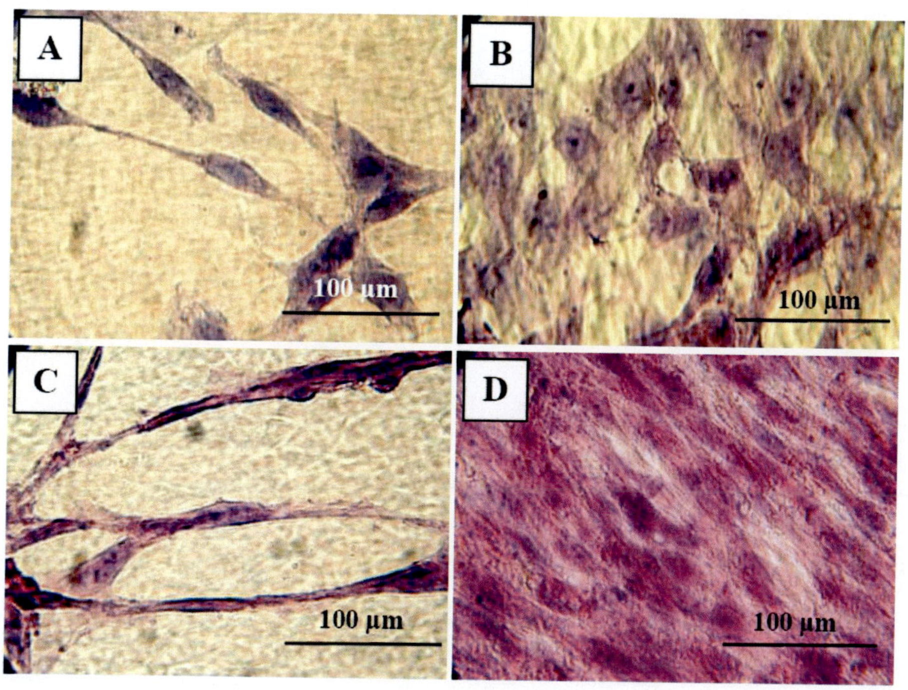

Figure 21. Morphology of rat aortic smooth muscle cells on day 1 (A, B) and 3 (C, D) after seeding on pristine unmodified PE foils (A, C) or foils irradiated with $3x10^{13}$ ions C+ cm-2 (B) or $3x10^{14}$ ions O^+ cm^{-2} (D); energy 150 keV. Stained with hematoxylin and eosin, microscope Opton Axioplan, Leica, Germany. Bar=100 μm.

The cells were more resistant to the proteolytic enzyme-mediated as well as spontaneous detachment. They were able to persist on the modified polymer surfaces in confluent, viable and metabolically active state for several months [7-11,18,19,40,132-134].

Also the proliferation activity of cells was often increased on the ion-, plasma- or UV light-modified surfaces. However, in some cases, when the initial adhesion and spreading of cell was very high, the proliferation activity was similar or even lower than in cell on non-modified surfaces. It is well known that the cell behavior, such as migration, proliferation, differentiation, appropriate functioning or long-term viability, is regulated by the extent of initial cell adhesion. The size and shape of cell spreading area, as well as the number, size, shape and distribution of focal adhesion plaques are decisive for further migratory, proliferative and differentiation behavior of anchorage-dependent cells (Table 2). If this extent is very small (i.e. attachment of round cells without formation of focal contacts and spreading), the cells usually do not survive. At the intermediate adhesion strength, the cells are most active in migration and proliferation, and if the cell-material contact area is very large with multiple well-developed focal adhesion sites associated with a rich cytoskeleton, the cells tend to skip the proliferation phase and enter sooner the differentiation program (Table 2, Figure 8). The latter cell behavior is advantageous for construction of bioarteficial vascular prostheses, where a soon maturity of both endothelial and smooth muscle cells is required in order to prevent thrombogenicity and immunogenicity of the cell-material constructs and excessive proliferation of vascular smooth muscle cells, often leading to restenosis and even full obliteration of vascular prostheses [18,19,31,40,41,134].

9. Patterned Surfaces and Grafting Bioactive Molecules

Physicochemical modifications of the polymer surfaces by ion-, plasma- or UV light-irradiation can be advantageously applied for the construction of micropatterned surfaces used for regionally selective adhesion and growth of cells. Directed cell adhesion and growth is needed for some specific tissue engineering applications, as well as construction of microarrays used in advanced geomics and proteomics. Typically, the polymers are irradiated through metallic mask with openings of various sizes and shapes. In addition, if the openings are relatively small, allowing the adhesion and spreading of one individual cell, the size and shape of these openings can control the size and shape of cell adhesion area, and thus the following cell behavior. In our studies, the most selective cell adhesion and growth of cells on modified spots was obtained using UV light irradiation in ammonia atmosphere (Figure 22) [10,11].

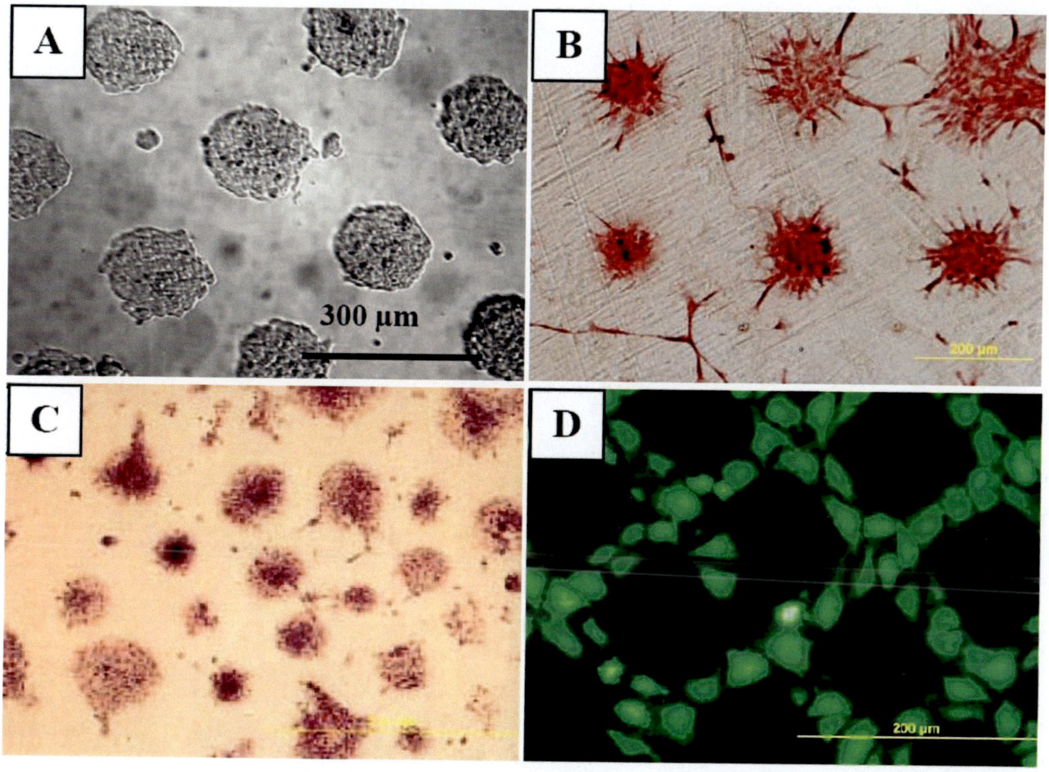

Figure 22. Human embryonic kidney cells (A), rat vascular smooth muscle cells (B, C) and human osteoblast-like MG 63 cells (D) in cultures on micropatterned surfaces. A, B: PTFE irradiated with UV light produced by a Xe_2*-excimer lamp for 30 min in an ammonia atmosphere through a mask with holes 100 µm in diameter and center-to-center distance 300 µm; C: PE irradiated with Ar^+ ions (energy 150 keV, ion dose 10^{13} ions/cm^2) through a mask with holes 100 µm in diameter and center-to-center distance 200 µm; fullerenes C_{60} deposited through a mask with rectangular holes with an average size of 128±3 µm per 98±8 µm on glass coverslips. Day 7 after seeding. A: native cells in an inverted phase-contrast microscope; B, C: cells stained with hematoxylin and eosin, Olympus microscope IX 50; D: cells stained with fluorescence-based LIVE/DEAD viability/cytotoxicity kit (Invitrogen), Olympus microscope IX 50. Bars: 300 µm (A), 200 µm (B, D), 1mm (C) [10,11].

The presence of free radicals, double bonds and various chemical functional groups on ion-, plasma- or UV light-modified surfaces may be also utilized for chemical bonding of other substances, particularly various bioactive molecules, such as amino acids, oligopeptidic ligands for cell adhesion receptors or active sites of various enzymes and hormones. When polymers modified by UV light in ammonia atmosphere or by Ar plasma were grafted with amino acids, such as glycine, alanine or leucine, the cell adhesion and growth were further enhanced (Figure 23) [8,9].

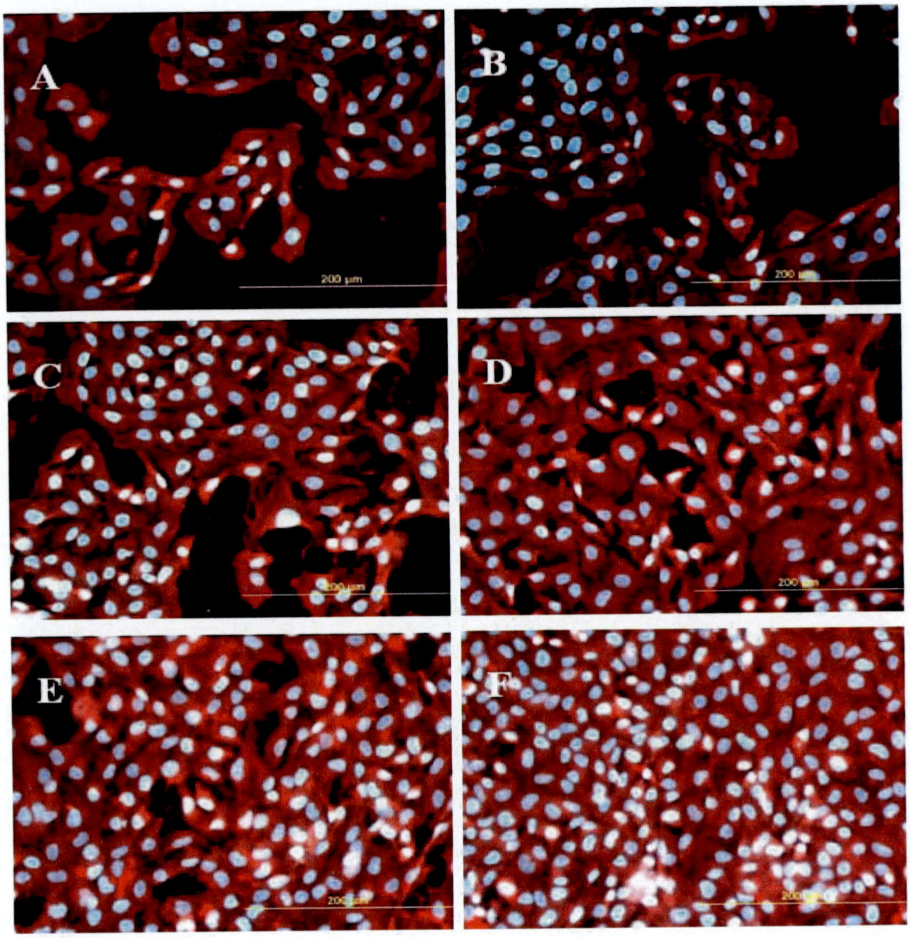

Figure 23. Morphology of rat aortic smooth muscle cells 48 h after seeding on high-density (A, C, E) and low-density (B, D, F) polyethylene. A, B: pristine polymers, C, D: plasma-irradiated polymers, E, F polymers irradiated with plasma and subsequently grafted with glycine. Cell membrane stained with Texas Red C_2-maleimide, the nuclei counterstained with Hoechst #33342. Olympus IX 50 microscope, digital camera DP 70. Bar=200 µm.

Although the single amino acids cannot bind the adhesion receptors on cells, they provide other chemical functional groups which may improve the adsorption of cell adhesion-mediating molecules on the material surface. Similar effect was observed on ion-implanted polyethylene grafted with the Arg-Gly-Asp (RGD) sequence [135]. Although this tripeptide represents a minimum ligand for integrin adhesion receptors, it was not probably bound directly to the receptors, because it was attached to the material surface relatively tightly and rigidly. In order to facilitate binding to adhesion receptors, the oligopeptides must be tethered on the material surface through a flexible chain ensuring good accessibility of these ligands to cell adhesion receptors, together with an appropriate spatial conformation of these ligands (i.e., circular of loop-like shape). A specific binding of a RGD-containing oligopeptide, i.e. GRGDSG, was achieved on polylactide-based surfaces, where the oligopeptides were tethered through polyethylene oxide (PEO) chains of an appropriate length and concentration (Figure 24) [41].

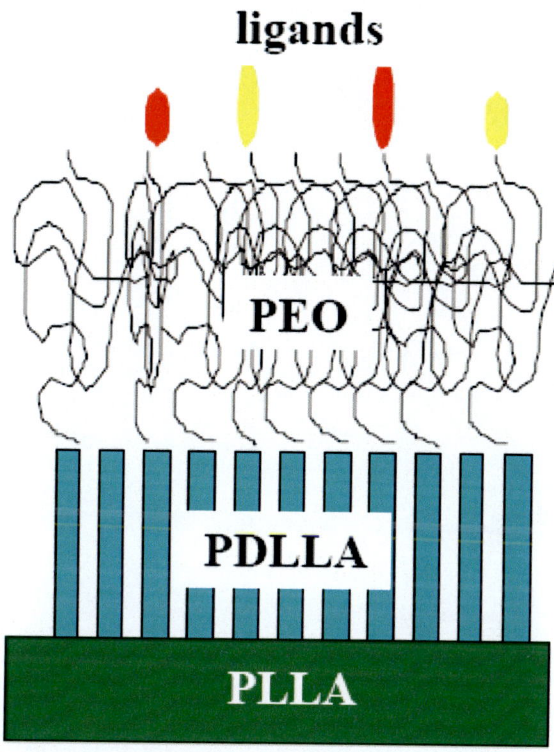

Figure 24. Scheme of a synthetic biodegradable polymer for tissue engineering, endowed with ligands for cell adhesion receptors. PLLA: poly-L-lactide, representing the bulk material of a newly constructed bioartificial tissue, e.g. vascular wall. PDLLA: poly-DL-lactide chains for copolymerization with polyethylene oxide (PEO) in order to prepare a bioinert surface. Due to its extreme hydrophilia, PEO do not allow spontaneous protein adsorption and aberrant cell adhesion. In addition, flexible PEO chains ensure the accessibility of ligands for cell adhesion receptors. These ligands could be represented by various amino acid sequences, such as RGD-containing oligopeptides, as well as their cooperating sequences like PHSRN [41].

These chains not only ensured a good accessibility of GRGDSG by integrin receptors but provided the material surface with a bioinert background preventing aberrant protein adsorption and cell adhesion, which is due to the extreme hydrophilia of PEO. As a result, the cell adhesion and growth could be fully controlled by the type, concentration and spatial distribution of oligopeptidic and other adhesion ligands on the material surface [38-41]. The effects of adhesion oligopeptides tethered on the material surface can be enhanced in serum-free medium, because the absence of serum minimizes the potential adsorption of cell adhesion mediating ECM molecules on the regions with oligopeptides. Vascular smooth muscle cells cultured on the GRGDSG-grafted surfaces in a chemically-defined serum-free medium in our experiments were able to synthesize DNA and proliferate (Figure 25) [41].

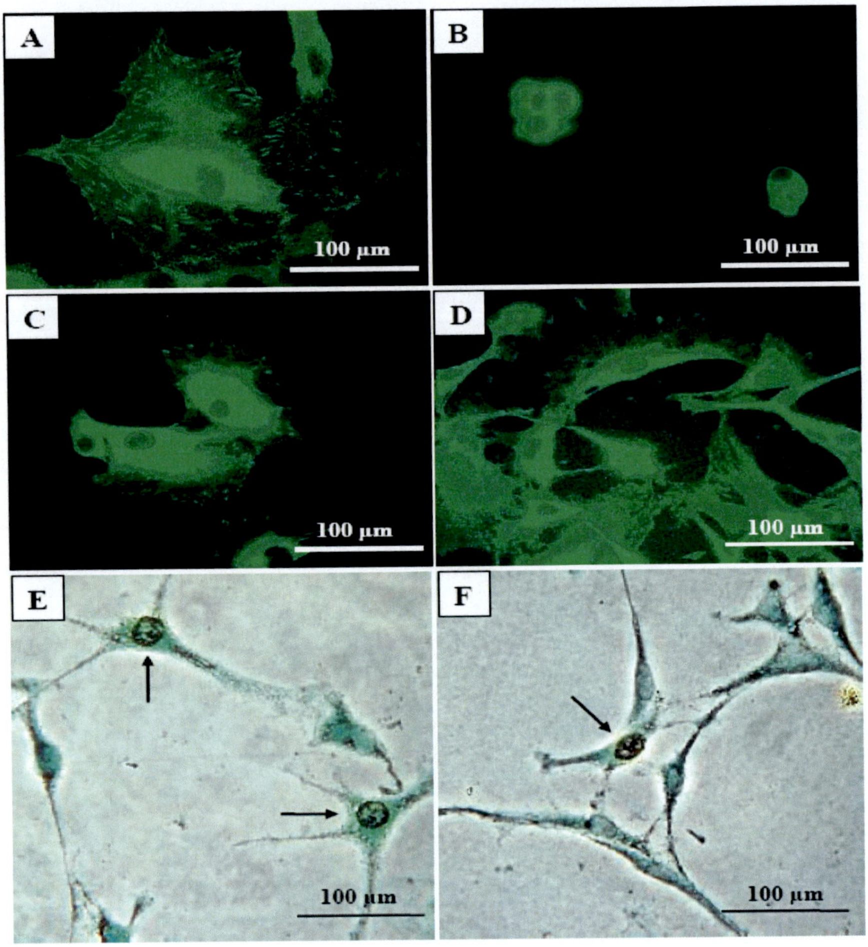

Figure 25. A-D: Immunofluorescence staining of vinculin in vascular smooth muscle cells on day 3 after seeding on polymeric surfaces (medium supplemented with 10% fetal bovine serum). A: poly(DL-lactic acid), PDLLA; B: block copolymer of poly(DL-lactic acid) and poly (ethylene oxide) (PEO), PDLLA-*b*-PEO; C, E: PDLLA-*b*-PEO with 5% GRGDSG-PEO-*b*-PDLLA; D, F: PDLLA-*b*-PEO with 20% GRGDSG-PEO-*b*-PDLLA. E, F: Immunoperoxidase staining of bromodeoxyuridine (arrows) incorporated into DNA newly synthesized in vascular smooth muscle cells cultured for 3 days in serum-free medium on PDLLA-*b*-PEO with 5% (E) or 20% (F) GRGDSG-PEO-*b*-PDLLA. Cells counterstained with light green. Bar=100 μm [41].

10. Modification of Materials by Deposition of Organic or Inorganic Layers

The bioactivity of various materials designed for implantation into the human body could be also enhanced by coating with various soft or hard, degradable or durable layers.

The soft degradable layers are represented by various extracellular matrix molecules and proteins involved in tissue regeneration, such as collagen, laminin, fibronectin, fibrin and their combinations [31,48,49]. Controlled assemblies of these molecules are suitable e.g. for modification of the inner surface of vascular prostheses in order to improve their endothelialization (Figure 26).

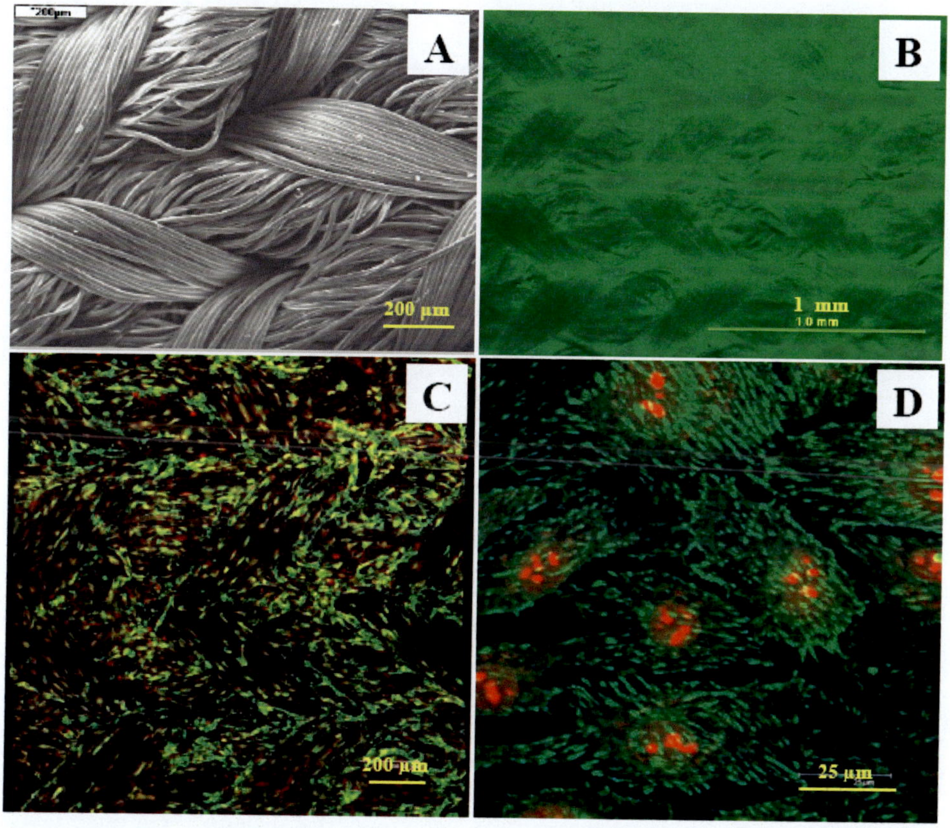

Figure 26. Reconstruction of the tunica intima on the inner surface of a clinically used polyethylene terephtalate vascular prosthesis. A: non-modified inner surface of the prosthesis, B: immobilization of defined assemblies of protein molecules (e.g., collagen+laminin or collagen+fibrin) on the inner surface of the graft, C: immunofluorescence of von Willebrand factor, a marker of the identity a differentiation of vascular endothelial cells, in human saphenous vein endothelial cells in cultures on the inner surface of a prosthesis coated with collagen and laminin, D: detail of a layer of endothelial cells growing on a layer of collagen and fibrin. Note well developed talin-containing focal adhesion plaques. A, B: conventional optical microscope, C, D: confocal microscope Leica DM 2500 [30,31].

These molecules represent a natural source of ligands for cell adhesion receptors in an appropriate amount, spectrum, spatial conformation as well as the presence of cooperating sequences. On the other hand, allogenous or xenogenous ECM molecules can be immunogenic or associated with a risk of pathogen transfer. These problems could be avoided by isolation of autologous fibrin from the patient's blood in a sufficient quantity. Thin protein layers could be also used for coating of the inner surfaces of three-dimensional scaffolds. In other words, the walls of the pores inside the material could be coated with a thin layer of cell adhesion-mediating molecules without filling the entire volume of pores, which could improve the ingrowth of cells (Figure 27).

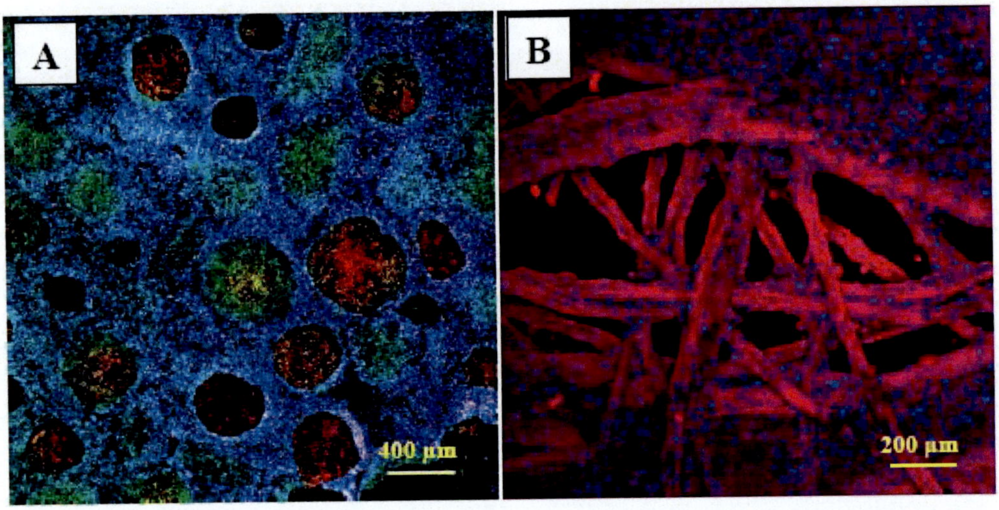

Figure 27. Human osteoblast-like MG 63 cells in cultures on porous (A) or fibrous (B) poly(L-lactide-co-glycolide) scaffolds. A: A summarizing picture of horizontal optical sections. The depth of cell ingrowth into the pores (average pore diameter of 400-600 mm) is indicated by spectral colors (blue: 0–60 mm, green: 80–160 mm, yellow: 180–220 mm, orange: 240–300 mm, red: 320–400 mm, violet: 420–480 mm). Day 14 after seeding, cells stained with propidium iodide. B: cells grown for 4 days in static culture followed by 2 days in dynamic perfusion cell culture system. Cell membrane stained with Texas Red C_2-maleimide and the nuclei counterstained with Hoechst #33342. Leica TCS SP2 confocal microscope, objective 5x (A) or 10x (B) [37].

The hard layers are represented with a wide range of coating based on carbon, metals, ceramics, various oxides (TiO_2, SiO_2) and nitrides (ZrN) and their combinations [14,16,24,47,136-143]. These layers can streghten the material surface, especially in materials designed for hard tissue surgery, such as construction of bone and joint replacements and bone fixation. In addition, these layers can smooth various irregularities on the material surface, which can hamper the adhesion, spreading and growth of cells [14-16,136,138,140,143]. For a more pronounced surface smoothening, the deposition of layers can be combined with grinding and polishing of the material surface (Figure 28). In addition to the surface smoothening, the beneficial effects of the deposited layers on the cell adhesion are often mediated by the presence of oxygen-containing groups and appropriate wettability, and especially by the nanostructure of some of these layers, i.e. the presence of particles less than 100 nm. The nanostructured surfaces resemble, at least to a certain degree, the architecture of physiological adhesion substrates, such as extracellular matrix, which is composed from nanoscale proteins, and in the case of bone, also hydroxyapatite and other inorganic nanocrystals [16,17,24-27]. From this point of view, carbon nanoparticles, such as fullerenes, nanotubes and nanodiamonds, may serve as important novel building blocks for creating artificial bioinspired nanostructured surfaces for bone tissue engineering.

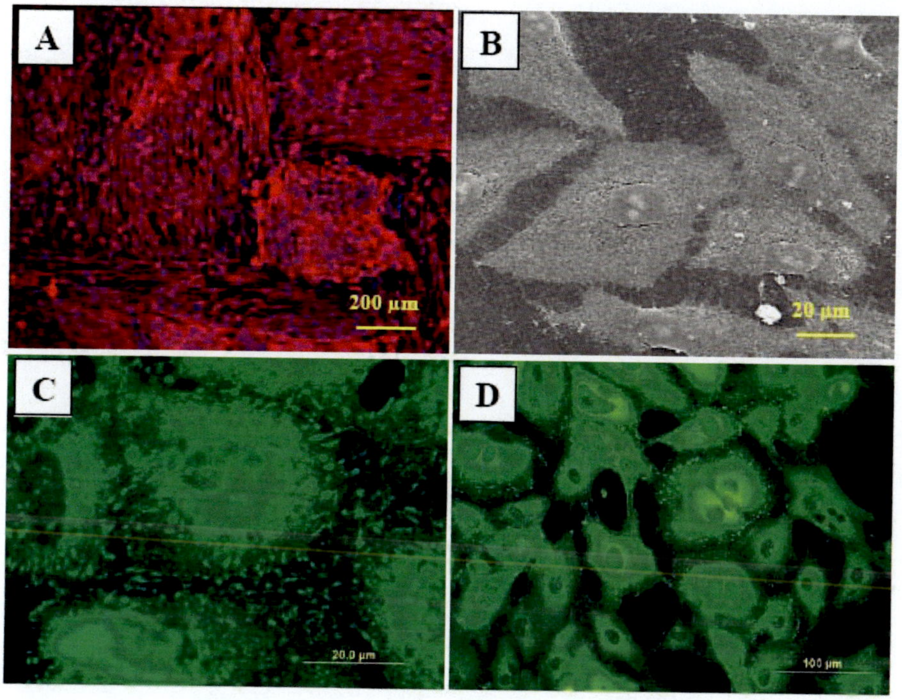

Figure 28. Human osteoblast-like MG 63 cells (A, B) and bovine pulmonary artery endothelial CPAE cells (C, D) in cultures on carbon fibre-reinforced carbon composites (CFRC) modified by deposition of zirconium nitride (A), grinding with metallographic paper, polishing with diamond paste and coating with pyrolytic graphite (B) and magnetron sputtering of nanocomposite hard hydrocarbon polymers containing 20 at.% of Ti (C) or 3 at.% of Ag (D). Confocal microscope, cells stained with Texas Red C_2-maleimide and Hoechst #33342 (A), cells in scanning electron microscope (B) and immunofluorescence staining against talin (C) or vinculin (D) in an inverted Olympus IX 50 microscope [15,47,136,137,143].

Fullerenes C_{60} were found to promote proliferation and differentiation of chondrocytes and to enhance the production of specific large proteoglycan ECM molecules, typical for cartilage, in rat embryonic limb bud cells in culture [144]. In a line of human epidermoid carcinoma cells, fullerenes protected cells from anoikis, i.e., apoptosis due to adhesion deprivation, by a mechanism supporting the formation of focal adhesion plaques, assembly of the actin cytoskeleton and cell spreading [65]. Enhanced attachment and spreading of platelets was also found on a polyurethane surface grafted with fullerene C_{60} molecules [145]. Fullerene C_{60} layers formed on polystyrene culture dishes by the evaporation of methanol from colloidal C_{60}-methanol suspensions, promoted adhesion, proliferation and assembly of an actin cytoskeleton in several lines of normal and malignant breast epithelial cells [146]. Similarly, non-cytotoxicity of nanodiamonds (2 to 10 nm in size) and their supportive effect on cell adhesion and growth has been demonstrated on a variety cell types, including neuroblastoma cells, macrophages, and keratinocytes [147].

Recently, a very interesting finding has been reported that nanostructured surfaces can promote preferential adhesion and growth of osteoblasts over other "competitive" cell types, mainly fibroblasts (which can cause fibrous encapsulation and loosening of bone implants). It is considered that the underlying mechanism is a higher adsorption of vitronectin, an ECM protein preferred by osteoblasts, to the nanostructured surfaces [24-27]. In future, carbon and other nanoparticles could be integrated into so-called "three-dimensional" matrices for constructing bone replacements, i.e. materials with a porous or scaffold-like architecture hierarchically organized at both micro- and nano-size level according principles observed in natural tissues and organs. For example, carbon nanoparticles could decorate the walls of micropores in the material and thus create a nanopattern promoting the ingrowth of osteogenic cells inside the scaffolds (Figure 27) [17,148]. In addition, carbon naoparticles, especially, nanotubes can reinforce synthetic polymers designed for bone tissue engineering. Nanotubes can resemble nanofibres of collagen and other ECM proteins of the bone, and together with other carbon nanoparticles, also hydroxyapatite and other inorganic crystals in the bone [25,26]. Furthermore, carbon nanotubes can be used effectively for electrical stimulation of osteoblasts, which has been reported to promote their proliferation, differentiation, production of mineralized bone matrix and thus healing of the damaged bone [25].

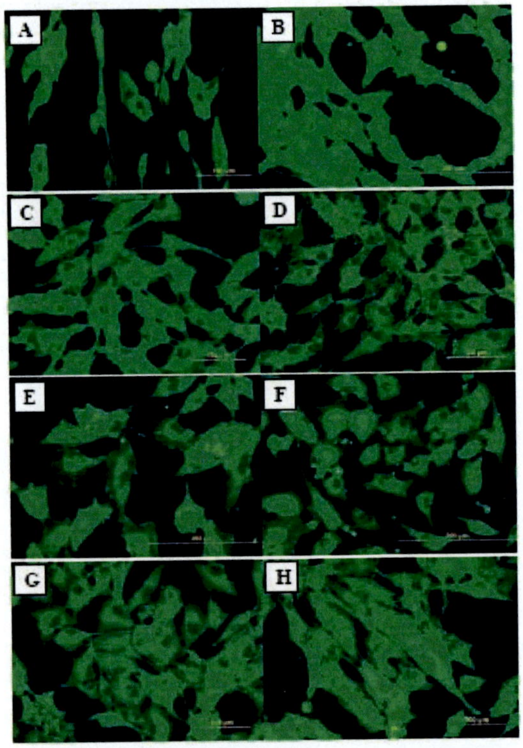

Figure 29. Human osteoblast-like MG 63 cells in cultures on material surfaces modified with carbon nanoparticles. A: fullerene C_{60} layers deposited on carbon fibre-reinforced carbon composites (CFRC), B: fullerene C_{60} layers deposited on microscopic glass coverslips, C: terpolymer of polytetrafluoroethylene, polyvinyldifluoride and polypropylene, mixed with 4% of single-wall carbon nanohorns, D: the same terpolymer with high crystalline electric arc multi-wall nanotubes, E: diamond layer with hierarchically organized micro- and nanostructure deposited on a Si substrate, F: nanocrystalline diamond layer on a Si substrate. Standard control cell culture substrates were represented by a PS culture dish (G) and microscopic glass coverslip (H). Immunofluorescence staining on day 2 (A) or 3 (B-H) after seeding, Olympus epifluorescence microscope IX 50, digital camera DP 70, obj. 20x, bar: 100 μm (A, C, D, G, H) or 200 μm (B, E, F) [16].

Promising results on carbon nanoparticles were obtained also in our earlier experiments. Fullerenes C_{60}, deposited on carbon fibre-reinforced carbon composites or microscopic glass coverslips, as well as nanocrystalline diamond on silicon substrates, promoted adhesion, growth, viability, mitochondrial activity and differentiation of human osteoblast-like MG 63 cells, usually in an extent similar as on standard cell culture substrates represented by polystyrene cell culture dishes or mcroscopic glass coverslips. On a terpolymer of polytetrafluoroethylene, polyvinyldifluoride and polypropylene, i.e. a material promising for construction of bone implants, which was mixed with single- or multiwalled carbon nanotubes, the colonization with MG 63 cells was even significantly better than on the unmodified terpolymer (Figure 29). At the same time, the presence of nanotubes did not increase the immune activation of cells, measured by the concetration of ICAM-1 per mg of protein [16].

The interactions of fullerenes, carbon nanotubes and nanodiamonds with biological systems can be further enhanced by functionalization of these molecules with various atoms, chemical groups or molecules (e.g., hydrogen, halogens, aldehydic, ketonic, alcoholic, carboxylic acid/ester moieties, amine groups, amino acids), which usually renders them soluble in water, and could stimulate the adsorption of cell adhesion-mediating ECM molecules [148-151]. The carbon nanoparticles could be also functionalized by ligands for cell adhesion receptors, such as RGD-containing oligopeptides or KRSR preferred by bone cells.

An interesting property of fullerenes and nanotubes, especially after derivatization, is the tendency to form self-assembling surface and bulk morphologies, such as patterned fullerene lipid films containing C_{60} [152], self-assembling chemisorbed amino octane thiol based-monolayers functionalized with fullerene C_{60} [153], self-assembling nanotube structures [151] or large scale stripe and circular patterns in transition metal/C_{60} composites [154]. These self-assembling layers are often organized in micro- and nano-dimensional patterns, and might serve as templates for spatially controlled adhesion and directed growth of cells. The microstructured surfaces can be also created by the deposition of carbon nanoparticles, e.g. fullerenes, through a metallic mask (Figure 22 D).

11. Conclusion

The bioactivity of artificial materials designed for biomedical use can be enhanced by various physico-chemical modifications, such as irradiation by plasma, ions or UV light in reactive atmospheres. These modifications lead to the formation of new structures, such as chemical functional groups, radicals and double bonds, which improve spontaneous adsorption of cell adhesion-mediating molecules from the culture media or body fluids. In addition, these structures can be utilized for functionalization of the materials with various biomolecules, e.g. oligopeptidic ligands for cell adhesion receptors. Another important tool for construction of bioactive materials is their coating with various soft or hard layers, especially defined assemblies of cell adhesion-mediating molecules or inorganic nanostructured films composed e.g. from ceramic, metallic or carbon nanoparticles. Thus, physical and chemical modifications of the material surface are of a great perspective for the future development of the biomaterial science and tissue engineering.

Acknowledgements

The work was supported by the Ministry of Education of the Czech Republic (Research program No. LC 06041), Academy of Sciences of the CR (Research programs No. KAN400480701, KAN101120701, A 5011301 and 1QS500110564) and the Grant Agency of the CR (Grants No. 204/06/0225 and 101/06/0226). The autors thank Mrs. O. Kesselová and Mrs. I. Zajanová for their technical assistance.

References

[1] Chan, C. M.; Ko, T. M.; Hiraoka, H. *Surf. Sci. Rep.* 1996, *4*, 1-54.

[2] Grace, J. M.; Gerenser, L. J. *J. Disp.Sci. Technol.* 2003, *24*, 305-341.

[3] Fink, D. Ed.; Fundamentals of Ion-Irradiated Polymers; *Series in Material Science* 63; Springer-Verlag: Berlin Heidelberg, 2004.

[4] Fink, D. Ed.; Transport processes in Ion-irradiated Polymers; *Series in Material Science* 65; Springer-Verlag: Berlin Heidelberg, 2004

[5] http://www.johannesheitz.at/mech.gif, 21.8.2007

[6] Gumpenberger, T.; Heitz, J.; Bäuerle, D.; Rosenmayer, T. *Appl. Phys. A* 2005, *80*, 27-33.

[7] Švorčík, V.; Walachová, K.; Heitz, J.; Gumpenberger, T.; Bačáková, L. *J. Mater. Sci. Lett.* 2001, *20*, 1941-1942.

[8] Heitz, J.; Švorčík, V.; Bačáková, L.; Ročková, K.; Ratajová, E.; Gumpenberger, T.; Bäuerle, D.; Dvořánková, B.; Kahr, H.; Graz, I.; Romanin, C. *J. Biomed. Mater. Res.* 2003, *67A*, 130-137.

[9] Švorčík, V.; Ročková, K.; Ratajová, E.; Heitz, J.; Huber, N.; Bäuerle, D.; Bačáková, L.; Dvořánková, B.; Hnatowicz, V. *Nucl. Instr. Meth. B* 2004, *217*, 307-313.

[10] Mikulíková, R.; Moritz, S.; Gumpenberger, T.; Olbrich, M.; Romanin, C.; Bačáková L.; Švorčík, V.; Heitz, J. *Biomaterials* 2005, *26/27*, 5572-5580.

[11] Pařízek, M.; Bačáková, L.; Lisá, V.; Kubová, O.; Švorčík, V.; Heitz, J. *Eng. Biomater.* 2006, *9*, 7-10.

[12] Tesmer, J. R. In Book Handbook of modern ion beam materials analysis; Nastasi, M. Ed.; Material Research Society: Pittsburg, 1995.

[13] Fink, D. In Book Report HMI-B 539, Berlin, 1996.

[14] Bačáková, L.; Starý, V.; Kofroňová, O.; Lisá, V. *J. Biomed. Mater. Res.* 2001, *54*, 567-578.

[15] Bačáková, L.; Starý, V.; Glogar, P.; Lisá, V. *Eng. Biomater.* 2003, *6*, 8-9.

[16] Bačáková, L.; Grausová, L.; Vacík, J.; Fraczek, A.; Blazewicz, S.; Kromka, A.; Vanecek, M.; Švorčík, V. *Diamond Relat. Mater.* 2007, *16*, 2133–2140.

[17] Tan, J.; Saltzman, W.M. *Biomaterials* 2004, *25*, 3593-3601.

[18] Bačáková, L.; Mareš, V.; Bottone, M. G.; Pellicciari, C.; Lisá V.; Švorčík, V. *J. Biomed. Mater. Res.* 2000, *49*, 369-379.

[19] Bačáková, L.; Mareš, V.; Lisá, V.; Švorčík, V. *Biomaterials* 2000, *21*, 1173-1179.

[20] Graves, D. C.; Yablonka-Reuveni, Z. *J. Histochem. Cytochem.* 2000, *48*, 1173-1193.

[21] Bačáková, L.; Pellicciari, C.; Bottone, M. G.; Lisá, V.; Mareš, V. *Histol. Histopathol.* 2001, *16*, 675-684.

[22] Mitani, Y.; Ueda, M.; Komatsu, R.; Maruyama, K.; Nagai, R.; Matsumura, M.; Sakurai, M. *Eur. Respir. J.* 2001, *17*, 316-20.

[23] Bačáková, L.; Lisá, V.; Kubínová, L.; Wilhelm, J.; Novotná, J.; Eckhart, A.; Herget, J. *Virchow's Archiv* 2002, *440*, 50-62.

[24] Webster, T. J.; Ergun, C.; Doremus R. H.; Siegel, R. W.; Bizios, R. *J. Biomed. Mater. Res.* 2000, *51*, 475-483.

[25] Supronowicz, P. R.; Ajayan, P. M.; Ullmann, K. R.; Arulanandam, B. P.; Metzger, D. W.; Bizios, R. *J. Biomed. Mater. Res.* 2002, *59*, 499-506.

[26] Price, R. L.; Ellison, K.; Haberstroh, K. M.; Webster, T. J. *J. Biomed. Mater. Res. A* 2004, *70*, 129-138.

[27] Webster, T. J.; Smith, T. A. *J. Biomed. Mater. Res. A* 2005, *74*, 677-686.

[28] Loukotová, J.; Bačáková, L.; Zicha, J.; Kuneš, J. *Physiol. Res.* 1998, *47*, 501-505.

[29] Baudin, B.; Bruneel, A.; Bosselut, N.; Vaubourdolle, M. *Nat. Protoc.* 2007, *2*, 481-485.

[30] Chlupáč, J.; Filová, E.; Riedel, T.; Brynda, E.; Remy-Zolghadri, M.; Bareille, R.; Fernandez, P.; Daculsi, R.; Bordenave, L.; Bačáková, L. *Eng. Biomater.* 2006, *9*, 10-13.

[31] Filová, E.; Brynda, E.; Riedel, T.; Bačáková, L.; Chlupáč, J.; Lisá, V.; Houska, M.; Dyr, J. E. *J. Biomed. Mater. Res. A.* 2007, in press.

[32] Engler, A. J.; Sen, S.; Sweeney, H. L.; Discher, D. E. *Cell.* 2006, *126*, 677-689.

[33] Freshney R.I.: In Book Culture of Animal Cells. Willey-Liss, Inc.: New York, 2000.

[34] Fernandez, P.; Bourget, C.; Bareille, R.; Daculsi, R.; Bordenave, L. *Tissue Eng.* 2007, *13*, 1607-1614.

[35] Mall, J. W.; Philipp, A. W.; Rademacher, A.; Paulitschke, M.; Buttemeyer, R. *Nephrol. Dial. Transplant.* 2004, *19*, 61-67.

[36] Buttemeyer, R.; Mall, J. W.; Paulitschke, M.; Rademacher, A.; Philipp, A. W. *Eur. J. Vasc. Endovasc. Surg.* 2003, *26*, 156-160.

[37] Pamula, E.; Bačáková, L.; Filová, E.; Buczynska, J.; Dobrzynski, P.; Nosková, L.; Grausová, L. *J. Mater. Sci. Mater. Med.* 2007, in press.

[38] Gobin, A. S.; West, J. L. *J. Biomed. Mater. Res. A* 2003, *67*, 255-259.

[39] Shin, H.; Jo, S.; Mikos, A. G. *Biomaterials* 2003, *24*, 4353-4364.

[40] Bačáková, L.; Filová, E.; Rypáček, F.; Švorčík, V.; Starý, V. *Physiol. Res.* 2004, *53*, S35-S45.

[41] Bačáková, L.; Filová, E.; Kubies, D.; Machová, L.; Proks, V.; Malinova, V.; Lisá, V.; Rypáček, F. *J. Mater. Sci. Mater. Med.* 2007, *18*, 1317-1323.

[42] Ochsenhirt, S. E.; Kokkoli, E.; McCarthy, J. B.; Tirrell, M. *Biomaterials* 2006, *27*, 3863-3874.

[43] Bačáková, L.; Walachová, K.; Švorčík, V.; Hnatowicz, V. *J. Biomater. Sci. B* 2001, *12*, 817-834.

[44] Tzoneva, R.; Groth, T.; Altankov, G.; Paul, D. *J. Mater. Sci. Mater. Med.* 2002, *13*, 1235-1244.

[45] Altankov, G.; Thom, V.; Groth, T.; Jankova, K.; Jonsson, G.; Ulbricht, M. *J. Biomed. Mater. Res.* 2000, *52*, 219-230.

[46] Arima, Y.; Iwata, H. *Biomaterials* 2007, *28*, 3074-3082.

[47] Starý, V.; Glogar, P.; Bačáková, L.; Hnilica, F.; Chmelík, V.; Kořínek, Z.; Gregor, J.; Mareš, V.; Lisá, V. *Acta Montana. AB* 2003, *11*, 19-36.

[48] Sperling, C.; Houska, M.; Brynda, E.; Streller, U.; Werner, C. *J. Biomed. Mater. Res. A* 2006, *76*, 681-689.

[49] Brynda, E.; Pacherník, J.; Houska, M.; Pientka, Z.; Dvorak P.: *Langmuir* 2005, *21*, 7877-7883.

[50] Scott, E. A.; Elbert, D. L. *Biomaterials* 2007, *28*, 3904-3917.

[51] Steiner, G.; Tunc, S.; Maitz, M.; Salzer, R. *Anal. Chem.* 2007, *79*, 1311-1316.

[52] Moiseeva, E. P. *Cardiovasc. Res.* 2001, *52*, 372–386.

[53] Aplin, A. E. *FEBS Lett.* 2003, *534*, 11-14.

[54] Engler, A.; Bačáková, L.; Newman, C.; Hategan, A.; Griffin, M.; Discher, D. *Biophys. J.* 2004, *86*, 617-628.

[55] Pesakova, V.; Klezl, Z.; Balik, K.; Adam, M. *J. Mater. Sci. Mater. Med.* 2000, *11*, 793-798.

[56] Walboomers, X. F.; Habraken, W. J.; Feddes, B.; Winter, L. C.; Bumgardner, J. D.; Jansen, J. A. *J. Biomed. Mater. Res. A.* 2004, *69*, 131-139.

[57] Bačáková, L.; Mareš, V.; Lisá, V.; Bottone, M. G.; Pellicciari, C.; Kocourek, F. *In Vitro Cell. Dev. Biol.* 1997, *33*, 410-413.

[58] Bačáková, L.; Mareš, V. *Physiol. Res.* 1995, *44*, 389-398.

[59] Bačáková, L.; Kuneš, J. *Physiol. Res.* 1995, *44*, 127-130.

[60] Arellano, M.; Moreno, S. *Int. J. Biochem. Cell Biol.* 1997, *29*, 559-573.

[61] Brooks, E. E.; Gray, N. S.; Joly, A.; Kerwar, S. S.; Lum, R.; Mackman, R. L.; Norman, T. C.; Rosete, J.; Rowe, M.; Schow, S. R.; Schultz, P. G.; Wang, X.; Wick, M. M.; Shiffman, D. *J. Biol. Chem.* 1997, *272*, 29207-29211.

[62] Preclíková, H.; Bryja, V.; Pacherník, J.; Krejčí, P.; Dvořák, P.; Hampl, A. *Cell Growth Differ.* 2002, *13*, 421-430.

[63] Podestá, F.; Roth, T.; Ferrara, F.; Cagliero, E.; Lorenzi, M. *Diabetologia* 1997, *40*, 879-886.

[64] Kato, S.; Akagi, T.; Sugimura, K.; Kishida, A.; Akashi, M. *Biomaterials* 1998, *19*, 821-827.

[65] Straface, E.; Natalini, B.; Monti, D.; Franceschi, C.; Schettini, G.; Bisaglia, M.; Fumelli, C.; Pincelli, C.; Pellicciari, R.; Malorni, W. *FEBS Lett.* 1999, *454*, 335-340.

[66] [66] Krajčí, D.; Mareš, V.; Lisá, V.; Bottone, M. G.; Pellicciari, C. *Histochem. Cell. Biol.* 2006, *125*, 183-191.

[67] Garner, E.; Martinon, F.; Tschopp, J.; Beard, P.; Raj, K. *Cancer Res.* 2007, *67*, 7631-7637.

[68] Mauthe, R. J.; Gibson, D. P.; Bunch, R. T.; Custer, L. *Toxicol. Pathol.* 2001, *29*, 138-146.

[69] Oshiro, Y.; Balwierz, P. S.; Morris, D. L.; Alden, C. L.; Bunch, R. T. *In Vitro Mol. Toxicol.* 2001, *14*, 121-127.

[70] Švorčík, V.; Kolářová, K.; Slepička, P.; Macková, A.; Novotná, M.; Hnatowicz, V. *Polym. Degr. Stab.* 2006, *91*, 1219-1225.

[71] Švorčík, V.; Kotál, V.; Slepička, P.; Bláhová, O.; Špírková, M.; Sajdl, P.; Hnatowicz, V. *Nucl Instrum. Meth. B* 2006, *244*, 365-372.

[72] Švorčík, V.; Kotál, V.; Slepička, P.; Bláhová, O.; Šutta, P.; Hnatowicz, V. *Polym. Eng. Sci.* 2006, *46*, 1326-1332.

[73] Švorčík, V.; Siegel, J.; Slepička, P.; Kotál, V.; Špírková, M. *Surf. Interf. Anal.* 2007, *39*, 79-85.

[74] Kotál, V.; Švorčík, V.; Slepička, P.; Bláhová, O.; Šutta, P.; Hnatowicz, V. *Plasma Proc Polym.* 2007, *4*, 69-76.

[75] Guimond S.; Wertheimer M. R. *J. Appl. Polym. Sci.* 2004, *94*, 1291-1303.

[76] Švorčík, V.; Hnatowicz, V.; Stopka, P.; Ochsner, R.; Ryssel, H. *Rad. Phys. Chem.* 2001, *60*, 89-93.

[77] Švorčík, V.; Prošková, K.; Hnatowicz, V.; Rybka, V. *J. Appl. Polym. Sci.* 2000, *75*, 1144-1148.

[78] Švorčík, V.; Kotál, V.; Siegel, J.; Sajdl, P.; Macková, A.; Hnatowicz, V. *Polym. Degr. Stab.*, 2007, *92*, 1645-1649.

[79] Kotál, V.; Siegel, J.; Stopka, P.; Sajdl, P.; Švorčík, V. *Mater. Sci. Forum,* accepted.

[80] Chu, P. K.; Chen, J. Y.; Wang, L. P.; Huang, N. *Mater. Sci. Eng. R* 2002, *36*, 143-206.

[81] Kim, K. S.; Ryu, Ch. M.; Park, Ch. S.; Sur, G. S.; Park, Ch. E. *Polymer* 2003, *44*, 6287-6285.

[82] Barr, T. L. In Book Modern ESCA. CRC Press Int.: Boca Raton, 1994.

[83] Siow, K. S.; Britcher, L.; Kumar, S.; Griesser, H. J. *Plasma Process. Polym.* 2006, *3*, 392-418.

[84] Cahn, R.W. In Book Concise Encyclopedia of Materials Characterization. Pergamon Press Ltd.: London; 1993.

[85] Ones, V.; Strobel, M.; Prokosch, M. J. *Plasma Proc. Polym.* 2005, *2*, 547-553.

[86] Tahara, M.; Cuong N. K.; Nakashima Y. *Surf. Coat. Technol.* 2003, *174-175*, 826-830.

[87] Gerenser, L. J.; *J. Adhesion Sci. Technol.* 1987, *1*, 303-318.

[88] Pochan, J. M.; Gerenser, L. J.; Elman, J. F. *Polymer* 1986, *27*, 1058-1062.

[89] Gerenser, L. J. *J. Adhesion Sci. Technol.* 1993, *7*, 1019-1040.

[90] Foerch, R.; Izawa, J.; Spears, G. A. *J. Adhesion Sci. Technol.* 1991, *5*, 549-564.

[91] Occhiello, E.; Morra, M.; Morini, G.; Garbassi, F.; Humphrey, P. J. *Appl. Polym. Sci.* 1991, *42*, 551-559.

[92] Strobel, M.; Lyons, C. S.; Strobel, J. M.; Kapaun, R. S. *J. Adhesion Sci. Technol.* 1992, *6*, 429-443.

[93] Ročková-Hlaváčková, K.; Švorčík, V.; Bačáková, L.; Dvořánková, B.; Hnatowicz, V. *Nucl. Instr. Meth. B* 2004, *225*, 275-282.

[94] Švorčík, V; Tomášová, P; Hnatowicz, V; Ochsner, R; Ryssel, H. *Nucl. Instr. Meth. B* 2004, *215*, 366-370.

[95] Lehocký, M.; Drnovská, H.; Lapčíková, B.; Barros-Timmons, A. M.; Trindade, T.; Cembala, M.; Lapčík Jr., L. *Colloid. Surf. A* 2003, *222*, 125-131.

[96] Macková, A.; Švorčík, V.; Strýhal, Z.; Pavlík, J. *Surf. Interf. Anal.* 2006, *38*, 335-338.
[97] Mazzoldi, P. and Arnold, G.W. Ed.; Ion Beam Modification of Insulators; Elsevier: Amsterdam, 1987.
[98] Marletta, G. *Nucl. Instr. Meth. B* 1990, *46*, 295-305.
[99] Calgano, L.; Foti, G.; *Nucl. Instr. Meth. B* 1991, *59-60*, 1153-1158.
[100] Calgano, L.; Compgnini, G.; Foti, G.; *Nucl. Instr. Meth. B* 1992, *65*, 413-422.
[101] Davenas, J.; Stevenson, I.; Clete, N.; Cambon, S. J.; Gardette, L.; Rivaton, A.; Vignoud, L. *Nucl. Instr. Meth. B* 2002, *191*, 653-661.
[102] Milinchuk, V. K.; Klinschpont, E. R.; Tulikov, V. I. In Book Osnovy radiacionoj stojkosti organiceskich materialov; Energoatomizdat: Moscow, 1994.
[103] Lewis, M. B.; Lee, E. H. *Nucl. Instr. Meth. B* 1992, *69*, 341-346.
[104] Davenas, J.; Thevenard, P. *Nucl. Instr. Meth. B* 2003, *208*, 170-175.
[105] Avasthi, D. K.; Singh, J. P.; Biswas, A.; Bose, S. K. *Nucl. Instr. Meth. B* 1998, *146*, 504-508.
[106] Steckenreiter, T.; Balanzat, E.; Fuess, H.; Trautmann, C. *Nucl. Instr. Meth. B* 1999, *151*, 161-168.
[107] Murphy, J. J.; Patel, M.; Skinner, A. R.; Horn, I. M.; Powell, S. J.; Smith, P. F.; *Nucl. Instr. Meth. B* 2004, *215*, 423-435.
[108] Švorčík, V.; Rybka, V.; Jankovskij, O.; Hnatowicz, V. *J. Appl. Polym. Sci.* 1996, *61*, 1097-1100.
[109] Švorčík, V.; Rybka, V.; Miček, I.; Hnatowicz, V.; Kvítek, J. *J. Mater. Res.* 1995, *10*, 468-472.
[110] Švorčík, V.; Rybka, V.; Hnatowicz, V.; Bačáková, L.; Lisá, V. *J. Mater. Chem.* 1995, *5*, 27-30.
[111] Bačáková, L.; Švorčík, V.; Rybka, V.; Miček, I.; Hnatowicz, V. *Biomaterials* 1996, *17*, 1121-1126.
[112] Švorčík, V.; Rybka, V.; Hnatowicz, V.; Smetana, K. *J. Mater. Sci. Mat. Med.* 1997, *8*, 435-440.
[113] Bačáková, L.; Walachová, K.; Švorčík, V.; Hnatowicz, V. *J. Biomater. Sci. B* 2001, *12*, 817-834.
[114] Švorčík, V.; Walachová, K.; Prošková, K.; Dvořánková, B.; Vogtová, D.; Ryssel, H. *J. Mater. Sci. Mater. Med.* 2000, *11*, 655-660.
[115] Švorčík, V.; Rybka, V.; Hnatowicz, V.; Kvítek, J. *Mater. Lett.* 1994, *19*, 329-332.
[116] Švorčík, V.; Rybka, V.; Jankovskij, O.; Hnatowicz, V. *J. Mater. Res.* 1994, *9*, 643-647.
[117] Švorčík, V.; Arenholz, E.; Hnatowicz, V.; Rybka, V.; Ochsner, R.; Ryssel, H. *Nucl. Instr. Meth. B* 1998, *142*, 349-354.
[118] Švorčík, V.; Prošková, K.; Hnatowicz, V.; Kluge, A. *Polym. Degr. Stab.* 1999, *65*, 131-135.
[119] Švorčík, V.; Rybka, V.; Hnatowicz, V.; Wang, L.; Angert, N. *Polym. Deg. Stab.* 1997, *55*, 115-121.
[120] Hnatowicz, V.; Vacík, J.; Švorčík, V.; Rybka, V.; Ryssel, H.; Kluge, A. *Rad. Eff. Def. Sol.* 2001, *155*, 195-202.
[121] Hnatowicz, V.; Peřina, V.; Macková, A. Švorčík, V.; Rybka, V.; Fink, D.; Heitz, J. *Nucl.Instr. Meth. B* 2001, *175-177*, 437-441.

[122] [Hnatowicz, V. *Nucl. Instr. Meth. B* 2004, *215*, 162-168.
[123] Hnatowicz, V.; Kvitek, J.; Švorčík, V.; Rybka, V. *Eur. Polym. J.* 1993, *29*, 1255-1258.
[124] Hnatowicz, V.; Kvitek, J.; Švorčík, V.; Rybka, V. *Appl. Phys. A* 1994, *58*, 349-352.
[125] Hnatowicz, V.; Kvitek, J.; Švorčík, V.; Rybka, V.; Popok, V. *Czech. J. Phys.* 1994, *44*, 621-627.
[126] Hnatowicz, V.; Havránek, V.; Kvítek, J.; Peřina, V.; Švorčík, V.; Rybka, V. *Jpn. J. Appl. Phys.* 1993, *32*, 1810-1813.
[127] Švorčík, V.; Arenholz, E.; Rybka, V.; Hnatowicz, V. *Nucl. Instr. Meth. B* 1997, *122*, 663-667.
[128] Popok, V. N.; Azarko, I. I.; Khaibullin, R. I.; Stepanov, A. L.; Hnatowicz, V.; Macková, A.; Prasalovich, S.V. *Appl. Phys. A* 2004, *78*, 1067-1072.
[129] Švorčík, V.; Miček, I.; Rybka, V.; Palmetshofer, L.; Hnatowicz, V. *J. Appl. Polym. Sci.* 1998, *69*, 1257-1261.
[130] Macková, A.; Havránek, V.; Švorčík, V.; Djourelov, N. Suzuki, T. *Nucl. Instr. Meth. B* 2005, *240*, 245-249.
[131] Chan, C. M. In Book Polymer Surface Modification and Characterization; Hanser: New York, 1994.
[132] Pignataro, B.; Conte, E.; Scandurra, A.; Marletta, G. *Biomaterials* 1997, *18*, 1461-1470.
[133] Tsai, W. B.; Wei, T. C.; Lin, M. C.; Wang, J. Y.; Chen, C. H. *J. Biomater. Sci. Polym. Ed.* 2005, *16*, 699-714.
[134] Marletta, G.; Ciapetti, G.; Satriano, C.; Perut, F.; Salerno, M.; Baldini, N. *Biomaterials* 2006, *28*, 1132-1140.
[135] Ročková-Hlaváčková, K.; Švorčík, V.; Bačáková, L.; Dvořánková, B.; Heitz, J.; Hnatowicz, V. *Nucl. Instr. Meth.B* 2004, *225*, 275-282.
[136] Starý, V.; Bačáková, L.; Horník, J.; Chmelík, V. *Thin Solid Films* 2003, *433*, 191-198.
[137] Bačáková, L.; Nosková, L.; Koshelyev, H.; Biederman, H. *Eng. Biomater.* 2004, *7*, 18-20.
[138] Bačáková, L.; Grausová, L.; Vacík, J.; Jungová, I. *Eng. Biomater.* 2005, *8*, 3-6.
[139] Lappalainen, R.; Santavirta, S. S. *Clin. Orthop. Relat. Res.* 2005, *430*, 72-79.
[140] Bačáková, L.; Kabátová, J.; Lisá, V.; Starý, V.; Fencl, J. *Eng. Biomater.* 2006, *9*, 1-3.
[141] Grössner-Schreiber, B.; Herzog, M.; Hedderich, J.; Dück, A.; Hannig, M.; Griepentrog, M. *Clin. Oral. Implants Res.* 2006, *17*, 736-745.
[142] Areva, S.; Aaritalo, V.; Tuusa, S.; Jokinen, M.; Linden, M.; Peltola, T. *J. Mater. Sci. Mater. Med.* 2007, *18*, 1633-1642.
[143] Paul, J.; Bačáková, L.; Douděrová, M.; Starý, V.; Vyskočil, V.; Lisá, V.; Slosarczyk, A.; Zima, A. *Eng. Biomater.* 2007, *10 (67-68)*, 5-8.
[144] Tsuchiya, T.; Yamakoshi, Y. N.; Miyata, N. *Biochem. Biophys. Res. Commun.* 1995, *206*, 885-894.
[145] Lin, J. C.; Wu, C. H. *Biomaterials* 1999, *20*, 1613-1620.
[146] Levi, N.; Hantgan, R. R.; Lively, M. O.; Carroll, D. L.; Prasad, G. L. *J. Nanobiotechnol.* 2006, *4*, 14.
[147] Schrand, A. M.; Huang, H.; Carlson, C.; Schlager, J. J.; Osawa, E.; Hussain, S. M.; Dai, L. *J. Phys. Chem.* 2007, *111*, 2-7.

[148] Woo, K. M.; Chen, V. J.; Ma, P. X. *J. Biomed. Mater. Res.* 2003, *67A*, 531-537.

[149] Guldi, D. M.; Prato, M. *Acc. Chem. Res.* 2000, *33*, 695-703.

[150] Banerjee, S.; Kahn, M. G.; Wong, S. S. *Chem. Eur. J.* 2003, *9*, 1898-1908.

[151] Pantarotto, D.; Partidos, C. D.; Graff, R.; Hoebeke, J.; Briand, J. P.; Prato, M.; Bianco, A. *J. Am. Chem. Soc.* 2003, *125*, 6160-6164.

[152] Wilson S.R. Fullerenes: Chemistry, Physics, and Technology. Kadish K.M. and Ruoff R.S. (Eds.); John Wiley and Sons: New York, 2000.

[153] Sahoo, R. R.; Patnaik A. *J. Colloid. Interface Sci.* 2003, *268*, 43-49.

[154] Vacík, J.; Naramoto, H.; Yamamoto, S.; Narumi, K.; Miyashita, K. *J. Chem. Phys.* 2001, *114*, 9115-9119.

In: Cell Growth Processes: New Research
Editor: Daiki Kimura, pp. 57-88
ISBN: 978-1-60456-132-6

Chapter II

The Contribution of Cellular Proliferation in the Skeletal Muscle Regenerative Response

Rita Ferreira[1,2,3]***, Maria João Neuparth***[1,3]***, Rui Vitorino***[1,2]***, Hans-Joachim Appell***[1,4]***, Francisco Amado***[2]***, and José Alberto Duarte***[1]

[1]CIAFEL, Faculty of Sport, University of Porto, Portugal;
[2]Chemistry Department, University of Aveiro, Portugal;
[3]ESSVA Health School, IPSN, CESPU CRL, Portugal;
[4]Department of Physiology and Anatomy, German Sports University, Germany

Abstract

Skeletal muscle is a complex tissue composed predominantly by contractile elements, but the intimate association between muscle fibers and the surrounding endothelial cells and fibroblasts is determinant for the maintenance of the integrity and proper function of the muscle. The cellular unit of adult skeletal muscle is the myofiber, a highly specialized syncytium sustained by hundreds of post-mitotic myonuclei within a continuous cytoplasm. This terminally differentiated cell unit has not the ability of self-renewal. Paradoxically, skeletal muscle is daily submitted to a variety of potentially harmful stimuli, thereby requiring an ever-available, renewable source of cells for muscle adaptation and repair. Since its identification, the satellite cell population located between the basal lamina and myofiber sarcolemma has been a popular candidate for the adult skeletal muscle "stem cell". Satellite cells are mitotically quiescent in adult muscle but re-enter the cell cycle following muscle injury to generate a population of myoblasts, which may subsequently fuse with the adjacent myofiber or with other myoblasts to form new myofibers.

The identification of proliferating cells in different tissues has become a useful laboratory tool in the comprehension of the mechanisms behind tissue adaptation. In this context, we have evaluated the time-course of the contribution of cellular proliferation within the process of muscle regeneration by submitting mice *gastrocnemius* muscle to crush injury and studying the cellular events throughout the following 20 days. The morphological data obtained point to a significant inflammatory reaction between the 2nd and 4th days after damage. The levels of cell proliferation (evaluated by BrdU incorporation) were significantly higher during the 4th and 6th days after injury, mainly due to the activation of satellite cells and interstitial non-endothelial cells. With this approach we intended to highlight the efficiency of repair mechanisms used in a crucial tissue as skeletal muscle to minimize the incapacitating effects of severe damage. Indeed, there has been much debate as to whether so complex and highly organized array of syncytial cells could be regenerated and, if so, how this could occur in postnatal muscle. Unexpectedly, we also observed that the cellular proliferation activity is also a meaningful issue when the skeletal muscle is subjected to unloading conditions. In fact, an unpredicted but prominent activation of satellite cells can be noticed in the earliest phases of the atrophic process in hindlimb suspended mice. Interstitial endothelial and non-endothelial cells also evidence signals of proliferation in a typical time-course. By focusing on the cell populations with enhanced regeneration abilities, advances in clinical therapies for muscle injuries should be obtained in a near future.

Introduction

The precise voluntary movement in vertebrates is achieved by the subtle and coordinated control of skeletal muscles, which are designed to do intermittent and unidirectional work against load or gravity, transmitting the force through tendinous attachments [37, 171]. Skeletal muscle, by virtue of its size, makes up as much as 50% of body mass in some athletic species, and is a key metabolic organ system involved in conserving the energy balance of the individual [31, 79, 96, 149]. The cellular unit of adult skeletal muscle, the myofiber, is wrapped in connective tissue through which the blood vessels and nerves courses [26, 48, 149].

The connective tissue framework creates a supportive skeleton for the myofibers, joining the contraction of individual fibers together into a combined effort and, thus, converting the contraction of the individual myofibers into efficient locomotion via myotendinous junctions at their ends, where myofibers attach to the skeleton by tendons [48, 106]. As well as being rich in connective tissue, skeletal muscles are highly vascularized to provide essential nutrients for muscle function [48]. Thus, besides its neural and endocrine control, the functional properties of skeletal muscle also depend on the maintenance of a complex framework involving different cellular players such as myofibers, blood vessels, and the extracellular connective tissue matrix.

Cellular Constituents of Skeletal Muscle

Myofibers

The main cellular unit of adult skeletal muscle is the myofiber, a giant and highly specialized syncytium sustained by hundreds of post-mitotic myonuclei, located preferentially at the fiber periphery [7, 26, 40, 52, 139, 169]. The reasons for having so many nuclei within a continuous cytoplasm are not well understood but may be related to the enormous size of these cells. The presence of multiple nuclei, therefore, may be seen as a mechanism to avoid diffusion and translocation limitations of nuclear products [7, 169]. In this sense, each myonucleus may be responsible for the maintenance of a limited volume of cytoplasm, a concept referred as ''DNA unit'' [49] or "myonuclear domain" [91, 159], which is defined as the quantity of cytoplasm regulated by a single myonucleus [7, 79, 94]. Myonuclear domain size appears to differ between different fiber types, with fibers from predominantly slow muscles evidencing a greater myonuclear density than fibers from predominantly fast muscles [41, 169, 211]. The reason for the difference in myonuclear domain size between slow and fast fibers is not exactly known but may be related to the greater quantity of enzyme synthesis necessary in highly oxidative fibers and/or the greater protein turnover of slow vs. fast muscles [7, 41, 169]. It is also unclear whether these relationships are maintained during muscle adaptation within distinct adult fiber types. Modulations in the amount of available DNA may be a factor in regulating the cytoplasmic volume of muscle fibers in response to changes in neuromuscular activity [6, 7, 94]. The number of myonuclei and thus the size of the myonuclear domains appears to play a critical role in the adaptative potential of skeletal muscle to alter its phenotype [79].

Skeletal muscle fibers are characterized by the precise organization of contractile proteins into myofibrils, resulting in repeating units arranged in series, the sarcomeres, which can be viewed at as a system of interdigitating thick and thin filaments and associated titin and nebulin filaments extending between successive Z disks (figure 1) [26, 176]. Thick and thin filaments and Z disks contain a variety of protein components, most of which exist as multiple isoforms [37, 176]. Myofibrillar protein isoforms generally show a specific tissue distribution and can be used as markers for fiber types in skeletal muscle [37, 167, 175].

Figure 1. Electron micrograph taken from a longitudinal section of a control mouse *soleus* muscle, evidencing several myofibrils within a myofibre with a regular striated pattern; the sarcomeres have a normal structure with good definition of bands (I, A, and H) and lines (M and Z). Several intermyofibrillar mitochondria, components of sarcoplasmic reticulum, and T-tubes can also be observed.

Individual adult skeletal muscles are composed of a mixture of myofibers with different contractile and metabolic properties, ranging from a slow-contracting/fatigue-resistant type to a fast-contracting/non-fatigue-resistant type [48]. Based on histochemical ATPase staining, antibody staining, and electrophoresis, two major fiber types were recognized in mammalian species: type I or slow twitch fibers and type II or fast twitch fibers, which may be further divided into three subtypes (IIA, IIB and IIX or IID), according to their constitution of myosin heavy chain (MHC) isoforms [152, 155]. The predominance of distinct muscle fiber phenotypes is one of the main determinants of the muscle performance *in vivo* [37, 44].

Despite having different physiological properties, the basic mechanism of muscle contraction is similar in all fiber types and is the result of a "sliding mechanism" of the myosin-rich thick filament over the actin-rich thin filament after neuronal activation [for review, see 48]. The MHC is both an important structural and regulatory protein comprising the contractile apparatus, and once it can be expressed in different isoforms, it has an impact on the functional diversity of the muscle [20, 26, 37, 176]. The regulation of the myosin gene family is under the control of a complex set of processes including (but not necessarily limited to) activity, hormonal, and metabolic factors [19, 43, 44, 55, 83, 89, 167]. So, this protein has been used as a cellular "marker" for studies of muscle plasticity in response to various mechanical perturbations in which the quantity and type of myosin isoforms, along with other important cellular proteins, are altered in expression [20, 167].

Endothelial Cells

Contractile activity depends on an active consumption of energy substrates and replenishment of their pools, and requires continuous removal of end products of the energetic metabolism [35, 101, 102, 115, 202], which may explain the intimate anatomical interrelations between muscle fibers and the microvascular bed [35]. Endothelial cells, not only in capillaries, represent a very stable population of cells with very low mitotic activity [69, 100]. Nevertheless, changing patterns of contractile work, environmental stresses, neurohormonal stimuli, or pathological conditions promotes the remodelling of the capillary network [69, 101, 202], thereby helping to match capillary density with the capacity of myofibers for oxidative metabolism. This involves the delivery of oxygen and substrates as well as the removal of metabolic end-products, which will determine the myofibers capacity for generating ATP by oxidative phosphorylation [199, 205]. Capillary growth in adult skeletal muscle has been described during endurance training, during long term electrical stimulation, and in skeletal muscle of animals treated with thyroid hormones [69, 80, 102].

The density of capillaries within the microvasculature of skeletal muscles varies markedly between muscles according to their main fiber composition, being the red muscles (characterized by a highly aerobic metabolism), comparatively to the white muscles (with a main glycolytic metabolism), constituted by a much denser capillary network [80, 101]. The capillary-to-fiber (C/F) ratio, frequently used to evaluate the capacity for O_2 supply to skeletal muscle [80, 101], is higher in rat *soleus* muscle, which mainly contains slow oxidative fibers, than in *extensor digitorium longus*, predominantly composed of fast glycolytic fibers [80]. Moreover, capillary tortuosity is substantially larger in shortened than in extended muscles and in different muscles it seems to be a function of sarcomere length [124, 147]. However, it is still unclear what exactly induces development and adaptation of the capillary bed. Previous studies have suggested that stimuli promoting myofiber remodelling in skeletal muscles also lead to the elaboration of angiogenic growth factors, including the vascular endothelial growth factor (VEGF) [56, 69, 81, 199, 205], thereby providing a plausible explanation as to how fiber type remodelling can be linked to changes in capillary density. This mitogenic factor, acting through two tyrosine kinase receptors, KDR/Flk-1 and Flt-1, induces proliferation, migration, and survival of endothelial cells, stimulates vasodilation and enhances vascular permeability [81, 199, 202]. Capillary growth may also be controlled by fibroblasts either by paracrine secretions modifying the extracellular matrix or by acting as a source of newly differentiated pericytes [69]. The context of any stimulus seems to be crucial to the remodelling of the capillary network in skeletal muscle.

Connective Tissue

The elastic behaviour of skeletal muscle depends not only on its contractile components, but also on the intramuscular connective tissue [32, 107, 113, 148], which accounts to 1-10% of the skeletal muscle mass and varies substantially between muscles [113]. Intramuscular connective tissue is responsible for the support and conservation of the structural integrity of the tissue during muscle contraction and embraces the capillaries and nerves within the skeletal muscle to keep them in close proximity to the fibers [107, 113]. The dominant cell type in connective tissue is the fibroblast or fibrocyte, but other cell types also exist (e.g. macrophages, mastocytes). Under normal conditions, fibroblasts, a discrete fusiform or stellate cell population, are generally considered “static” units chiefly responsible for manufacturing the extracellular matrix [117, 118], of which composition greatly influences the passive and active mechanical properties of muscle [4, 133].

Collagen is the main fibrous protein in intramuscular connective tissue [4, 36, 113, 148, 192]. Several collagen types were identified, being the types I, III, IV and V the main forms in skeletal muscles [4, 36]. Fibrillar type I and III are the most stress tolerant collagens of the muscle. Type I collagen is the major collagen fibril, having high tensile strength and limited elasticity, typically associated with tissue stiffness (i.e., resistance to stretch). Type III collagen is the other main muscular collagen form, with a similar structure and arrangement like type I collagen, but is composed by thinner and more elastic fibers [4]. Its abundance correlates with tissue compliance (i.e., readily elongated) [133]. Type IV collagen, together with laminin and heparin sulphate, is mainly present in the muscle basal lamina [174].

Besides collagen type composition, muscle elasticity also depends on the intra- and intermolecular cross-links of collagen and on the arrangement of its fibrils [107, 148]. Changes in the abundance, phenotype, and degree of collagen cross-linking may have profound effects on the passive and mechanical properties of skeletal muscle [107, 133]. It is likely that the difference in relative content of connective tissue in specific muscles is coupled to muscular function and to the role of connective tissue [113]. For instance, in rats the higher collagen concentrations have been found in tonic contracting muscles (mainly composed by slow-twitch fibers) than in phasic muscles (essentially constituted by fast-twitch myofibers) [212]. Moreover, the expression of collagen isoforms is thought to be phenotype-dependent. For example, type I collagen is more expressed in slow-twitch muscles than type III collagen [133].

In skeletal muscle, collagen is found predominantly in three structures: the epimysium, which surrounds the entire muscle; the perimysium, which covers muscle fiber bundles and constitutes the bulk of muscle collagen; and the endomysium, which surrounds individual muscle fibers and contains the basal lamina [107, 148, 212]. The endomysial arrangement of collagen consisted of three components: (1) myocyte-myocyte connections; (2) myocyte-capillary connections; and (3) a weave network of collagen intimately associated with the basal lamina of the myofibers. Collagen fibrils in the endomysium therefore run in various directions during muscle relaxation, acquiring a predominantly longitudinal orientation to the major axis of the muscle fiber during muscle stretching. This reorientation of the fibrils may reflect the role of the endomysium in providing mechanical support to the fiber surface and its action as an elastic device for contraction–relaxation cycles; moreover, fibril arrangement in the endomysium is considered to be related to muscle elasticity [32, 113, 148]. The perimysium consists of large tendon-like bundles of interwoven collagen, which joins with the dense weave collagen that surrounded groups of muscles. The arrangement of the collagen in the perimysium and endomysium would explain differences in the mechanical properties of different muscles [32]. These structures of connective tissue can undergo a chronic rearrangement in response to pathophysiological stimuli, with repercussions to the muscle performance [113, 148, 204].

Satellite Cells

Satellite cells (SC), first identified using electron microscopy by Mauro in frog's muscle [125], are resident muscle precursor cells that lie between the basal lamina and the plasma membrane of skeletal muscle fibers (figure 2). Satellite cells number in skeletal muscle is dependent on species, age [38, 50, 59], and muscle fiber type [50, 63, 93, 108]. These mononucleated cells are primarily metabolically and mitotically quiescent and undergo division very infrequently under normal conditions in the adult muscle. This quiescence is not merely an inactive basal state, but rather a state under active transcriptional control [11, 48, 52, 63, 93, 95, 137, 164, 206]. These precursor cells can be easily identified in electronic microscopy due either to their distinct location or to its own morphological features (figure 2). Remaining in G_0 state of the cell cycle [48, 93, 123], SC are characterized by a reduced organelle content in a small amount of cytoplasm and a smaller nucleus displaying increased amounts of heterochromatin compared with nearby myonuclei [9, 38, 48, 170, 180]. Once activated in response to different insults, they enter the cell cycle, proliferate and their progenies progress down the myogenic lineage pathway to become fusion-competent myoblasts [52, 63, 132]. In mice, the satellite cell number constitutes ~30% of the muscle nuclei in the neonate and decreases with age to ~4% in the adult and ~2% in the senile (29-30 months) [29, 48, 93, 181, 182]. The huge decrease in the pool of satellite cells observed during the earliest phases of extra uterine life results from the continuous fusion of SC to adjacent fibers, which allows the post-natal growth with a progressive increase of nuclear domains in muscle fibers [48]. In addition, the amount of these cells type depends on the fiber type to which they are linked, e.g., the percentage of SC in adult *soleus* muscle is around two to threefold higher than in adult fast *tibialis anterior* or *extensor digitorum longus* muscles [48, 178], suggesting a relation between the quantity of SC and the metabolic or contractile characteristics of muscle fibers. The pool of SC also shows a preferential proximity to capillaries and neuromuscular junctions [156, 198, 200]. So, in oxidative fibers, characterized by a higher density of neuromuscular junctions and capillaries, a higher number of SC can be observed then in glycolytic fibers [93].

Quiescent SC express a variety of proteins that have been used for their identification and purification [63, 206]. Among the most widely studied of these are CD34, M-cadherin, Pax7, syndecan-3, syndecan-4, and c-met [48, 63, 206]. None of these markers is unique for quiescent satellite cells, although it is likely that a particular combination of some markers exclusively identifies the muscle stem cell population [63]. Upon exposure to signals resultant to environmental stresses, quiescent SC are activated (figure 2) and start to proliferate resulting in myogenic precursor cells and adult myoblasts.

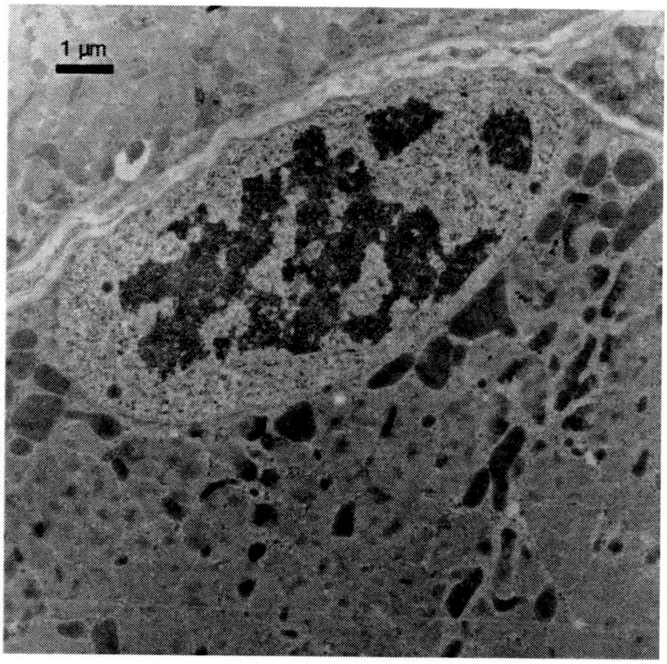

Figure 2. Electron micrograph taken from a cross section of a control mouse *soleus* muscle, illustrating a satellite cell located between the sarcolemma and the basal lamina of the adjacent mature myofiber; the nuclear envelop disappearance and the chromatin condensation (with developed chromosomes) suggest that this satellite cell is entering in the M phase of its cell cycle.

At the molecular level, activation of myogenic precursor cells is characterized by the rapid upregulation of two myogenic regulatory factors (MRF), *Myf5* and *MyoD*, normally undetectable in quiescent SC [10, 11, 48, 54, 186]. After the proliferation phase of myogenic precursor cells, expression of *Myogenin* and *MRF4* (members of MRF family) is upregulated in cells beginning their terminal differentiation program, which is then completed with the activation of muscle specific proteins, such as MHC, and the fusion of these myogenic precursor cells [10, 11, 48, 50, 191].

Changes in Cellular Proliferation in Response to Altered States of Muscle Activity and Injury

Skeletal muscle structure and function can be continuously changed as a consequence of stimuli that modify their contractile activity (e.g. endurance exercise, electrical stimulation, denervation), their load (e.g. resistance training, microgravity), their substrate supply (e.g. nutritional interventions, fasting) or by other environmental factors such as hypoxia or thermal stress. Moreover, skeletal muscle tissue undergoes gradual changes with age. Such adaptability is an important feature of skeletal muscle and allows for an economic design of muscle tissue sufficiently adaptable to the changes imposed by muscular use or by the environment [79, 96, 194]. This malleability of muscle, usually referred as "muscle plasticity", indicates that the terminally differentiated muscle fiber is not a fixed unit but represents a highly versatile entity [152]. The adaptations of skeletal muscle involves an integrated response of fibers, i.e. cytoarchitecture and composition [19, 24, 31, 51, 97, 176] and fiber-associated structures, i.e. nerves and capillaries [61, 196]. So, altered states of loading or other environmental stimuli may disturb the current muscle homeostasis with consequent modifications of its cellular turnover.

Cellular Events in Skeletal Muscle Submitted to Increased Activity or Load and Trauma

Adult mammalian skeletal muscle is a stable tissue with little turnover of nuclei [58] and the minor lesions inflicted by daily wear and tear require only a slow turnover of its constituents [48, 95]. It is estimated that in normal adult rat muscle, no more than 1-2% of myonuclei are replaced every week [48, 177]. Nevertheless, mammalian skeletal muscle is often subjected to mechanical stress that may result in considerable damage. Although considered a well differentiated tissue, skeletal muscle is capable of efficient regeneration [45, 50, 85]. Adult muscle possesses a remarkable regenerative capacity, and large numbers of new myotubes can emerge in only a few days after acute muscle damage [203]. These capacity for self-repair is due to the exceptional pool of undifferentiated myogenic precursor cells (mpc), the satellite cells [85, 86, 150, 183]. This muscle regeneration requires a coordinated repair response involving several cellular and molecular players such as the inflammatory system, myogenic precursors cells and vascular components [85]. The repair process furthermore depends on an orchestrated response between the inductive signals of cytokines, growth factors, and the extracellular matrix [53, 85, 203].

Regeneration of skeletal muscle can be observed *in vivo* after tissue damage induced by mechanical, chemical or thermic agents [45, 46, 93, 156, 203, 208]. The molecular and cellular sequences of events during regeneration have been well characterized for the entire muscle [70]. The cellular events that characterize the regenerative process of injured muscle are generally grouped in three phases: (i) the initial intrinsic degeneration phase; (ii) the extrinsic degeneration stage, and finally (iii) the muscular regeneration phase (figure 3) [46, 48, 74, 131].

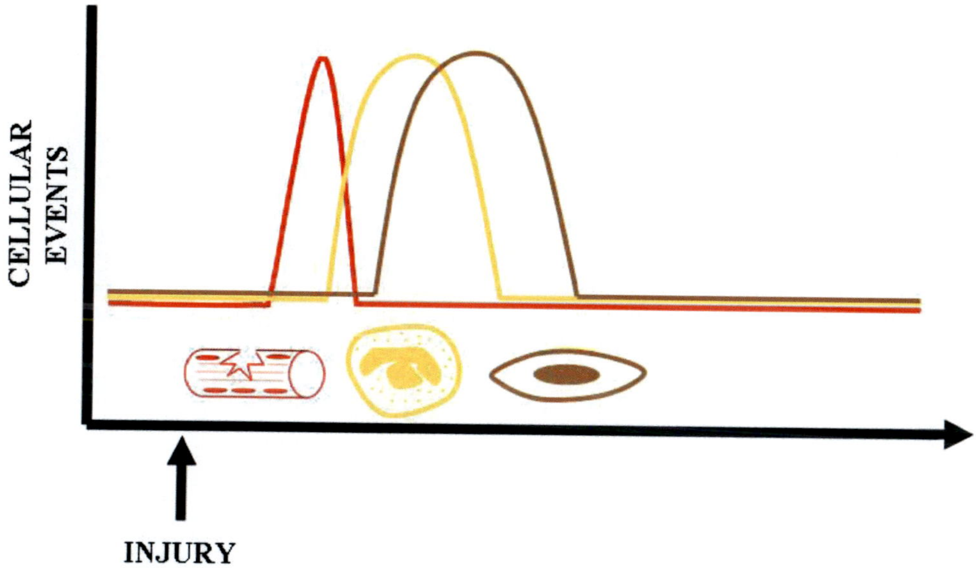

Figure 3. Hypothetical time-course of cellular events during the regenerative process. First, it is observed an intrinsic degeneration phase followed by an extrinsic one characterized by leukocyte tissue infiltration. Stem cell activation dominates the third phase of regeneration.

The first phase of muscle degeneration is mainly focused on the muscle fiber and is characterized by the loss of sarcolemma functionality and, consequently, by the increase of fiber permeability [46]. After damage of the sarcolemma or sarcoplasmic reticulum, the sarcoplasmic calcium influx is enhanced, resulting in a loss of calcium homeostasis (as illustrated in figure 4, after two days of mechanical injury) and in an increase of calcium-dependent proteolysis that will drive tissue degeneration [16, 23, 48]. Calpains are calcium-activated proteases that can cleave myofibrillar and cytoskeletal proteins and hence are implicated in this process [111, 190, 213]. Studies *in vitro* showed that calpains may cleave a variety of substrates, such as troponins T and I, tropomyosin, and C proteins, and therefore impair the stability of the thin and thick filaments [87]. In a similar way, many of the cytoskeletal proteins involved in linking the cytoskeleton to the plasma membrane, such as talin, spectrin, desmin, and vinculin, seem to be rapidly cleaved by calpains [23, 98, 172]. Besides the activation of neutral proteases, like calpains, other enzymes (e.g. lysosomal proteases, phopholipase A2) were also implicated in this degenerative phenomenon [16, 66]. However, the major degenerative alterations observed immediately after damage seem to be attributable to non-lysosomal proteases activation with calpains playing a key role [23, 71, 190].

Following this intrinsic degeneration stage, a second phase of cell-mediated degeneration takes place, generally designated as extrinsic degeneration. This phase is characterized by a prominent inflammatory response with invasion of several immunologic cell types (figures 3 and 5) [45, 65]. These inflammatory cells are activated by trophic factors released by the injured muscle, which provide the chemotactic signals to the circulating leukocytes [28, 48, 195].

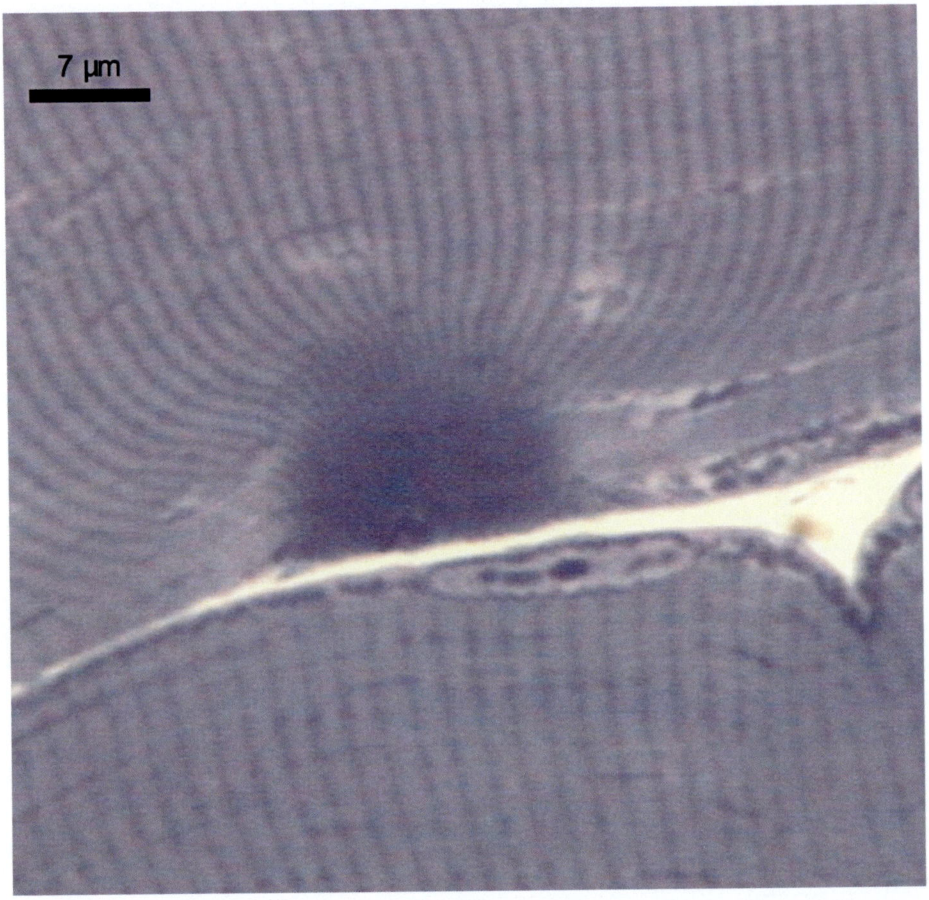

Figure 4. Light micrograph taken from a longitudinal section of *gastrocnemius* muscle from a mouse sacrificed 2 days after mechanical damage (crush), evidencing in the upper side fibre a punctual alteration of the striated pattern with disappearance of I bands nearby sarcolemma, which is suggestive of a local homeostatic calcium alteration induced by a focal dysfunction of cellular membrane.

Among these cells, neutrophils are the first to invade the injured muscle, with a significant increase in number observed in the first 1–6 hours after experimentally induced muscle damage [48, 84, 99, 127]. After the infiltration of neutrophils, a large number of monocytes/macrophages also invade the damaged muscles [45, 127] to phagocyte cellular debris and to promote muscle regeneration by activating myogenic cells [132, 181, 203]. The infiltration of leukocytes and macrophages in the damaged area of a *soleus* muscle after exhaustive eccentric exercise or of a crushed mouse *gastrocnemius* muscle is depicted in Figure 5A and 5B, respectively. During this process, they may also release hydrolytic enzymes stored in their granules, which can promote the recruitment and activation of further monocytes/macrophages [17, 195]. This vital role of macrophages in the muscle regeneration process is supported by the observation that myogenesis was markedly impaired when macrophage infiltration is inhibited [121].

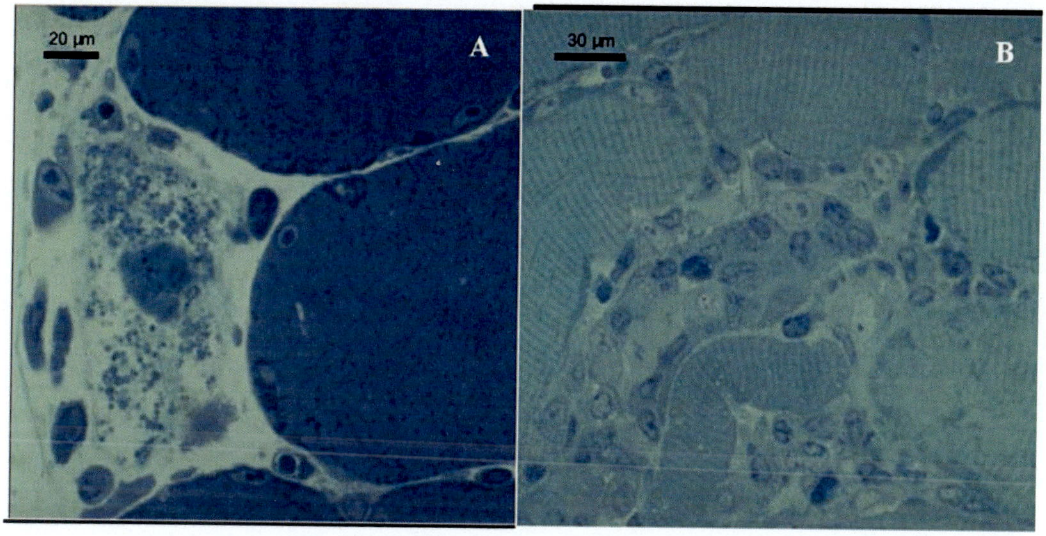

Figure 5. Light micrographs from a cross section of mouse *soleus* muscle after 2 days of exhaustive eccentric exercise (A) and from a longitudinal section of mouse *gastrocnemius* muscle after 2 days of crush injury (B). In A it can be observed a necrotic fibre (left side), surrounded by apparently normal myofibres, with cellular debris and several infiltrated cells within its basal lamina. In B, an extended necrotic area invaded by leukocytes and macrophages can be noted.

The influx of invading leucocytes appears to be stimulated by the release of growth factors from the damaged extracellular matrix, such as fibroblast growth factor (FGF) and platelet-derivated growth factor (PDGF) [45]. Moreover, macrophages also secrete a variety of growth factors, such as FGF, transforming growth factor β (TGF-β), interleukins, and leukaemia inhibitory factor (LIF), which activates satellite cells [120, 181, 200]. These growth factors can be enriched in autologous serum which in return can be injected into the injured site of a muscle, thereby activating the cascade of regeneration and shortening the time-course of the repair process [207, 208].

The regeneration stage arises almost simultaneously with the peak of macrophage invasion and is characterized by the activation and proliferation of satellite cells (figure 3) [45, 156]. To evaluate satellite cell proliferation several methodologies have been used to label satellite cells in regenerating skeletal muscle. Among them, 5-bromo-2-deoxyuridine (BrdU) labelling in combination with antibody-based detection methods has constitute the most popular technical approach used to measure cells that enter in the S phase of the cell cycle [39, 103, 114, 140, 158]. Attending to the position of the labelled cells, we quantitatively have evaluated the contribution of this cell population to the regenerative process during 20 days post trauma induced by muscle crush (table 1). Satellite cell proliferation was notorious after 2 days and reached maximal values at day 6, followed in time by an increased number of labelled central myonuclei. However, the time-course of satellite cell activation depends on the muscle type and damaged stimulus analyzed [14, 57].

Recent studies [48, 184] have clearly established that, besides satellite cells, other cell populations participate in these regenerative events. These stem cell populations either reside in skeletal muscle or may be recruited from circulating pools of nonmuscle origin in response to signals associated with injury and regeneration.

Table 1. BrdU-labelled satellite cells and central nuclei (percentage of total muscle labelled nuclei) in *gastrocnemius* muscle during 20 days post trauma induced by crush. Values are mean ± standard deviation

Experimental Group	% of labelled nuclei	
	Satellite cells	Central nuclei
Control	9,6 ± 5,70	7,7 ± 3,13
0 days	9,9 ± 4,35	7,2 ± 5,93
2 days	19,2 ± 3,56*	8,3 ± 2,93
4 days	24,5 ± 3,50*	10,9 ± 2,88
6 days	18,5 ± 3,33*	13,5 ± 3,5
8 days	19,2 ± 1,98*	16,1 ± 3,41*
10 days	13,1 ± 5,31	6,0 ± 4,17
12 days	15,7 ± 9,94	8,6 ± 7,79
14 days	16,8 ± 12,17	11,4 ± 7,63
16 days	12,3 ± 9,41	8,6 ± 8,14
18 days	16,9 ± 6,28	5,9 ± 8,40
20 days	11,5 ± 7,60	7,7 ± 6,55

*p<0,05 vs control

For example, bone marrow is a source of stem cell populations that has a myogenic potential. Using genetically labelled mouse models, marrow-derived cells have been shown to be recruited from circulation for injured muscle to generate fully differentiated myofibers [72, 184]. A population of pluripotent stem cells from adult skeletal muscle termed side population cells has also been recently identified [18, 93, 184]. This stem cell population increases in number following muscle injury, presents a distinct molecular signature and also participates in muscle regeneration [130, 184].

Although the molecular mechanisms that regulate the activation of stem cells and their entry into the cell cycle remain to be elucidated [9], it has been shown that several growth factors might be involved in this process, some of which are released by the adjacent damaged muscle fibers, and others are secreted by the infiltrating macrophages [201]. Several studies [27, 132, 201, 210] have implicated an extensive number of growth factors in the maintenance of the balance between growth and differentiation of satellite cells, in an attempt to restore the normal muscular architecture. However, the majority of these studies have examined the effects of growth factors on satellite cell biology using satellite cell cultures. In general, they analyzed the effects of growth factors alone or in combination, and with this approach, valuable insight into the regulation of the satellite cell was achieved [9, 22, 78, 120, 173, 188]. For example, IGF-I and IGF-II are able to alter MRFs expression and to promote both the proliferation and the differentiation/fusion of myoblasts [93]. *In vivo* and *in vitro* studies have showed an increase of muscle mass due to an augmentation of muscle protein and DNA content, when the levels of IGF-I raised within muscle cells [1, 3, 47]. Another growth factor, the Hepatocyte growth factor (HGF), has been proposed for the regulation of satellite cell activity. Among its multiple roles, the potent chemotactic action can be pointed out, as well as the activation of satellite cell, and the inhibition of myoblast differentiation [48, 50]. Similarly, FGF induces satellite cell proliferation and attenuates satellite cell differentiation to myofibers [93, 134]. On the other hand, TGFβ family members inhibit satellite cell proliferation and differentiation by silencing the transcriptional activation of MyoD family members. TGFβ negatively regulates the G_1 to S progression and thus maintains the quiescent status of satellite cells [126]. Experimental studies using TGFβ null mice generated by gene targeting showed a dramatic and widespread increase in skeletal muscle mass [129]. In general, growth factors have an important role in the regulation of satellite cell proliferation, differentiation, and motility [93].

In response to locally released growth factors from injured muscle fibers and macrophages, the satellite cells proliferate extensively within 2-3 days after injury [93]. Many of these myoblasts are able to fuse with the adjacent affected myofibres, preventing these fibers from complete degeneration. At the same time, fibroblasts invade the degeneration area and begin to produce excess extracellular matrix to restore the framework of the connective tissue (figure 6) [131]. In cases of severe muscular trauma, this proliferation of fibroblasts can quickly lead to the excessive formation of dense scar tissue that may hamper muscle regeneration and result in an incomplete muscle recovery [151].

After this proliferation phase, the progeny of satellite cells leave the cell cycle and fuse to adjacent myofibers or to other myoblasts, forming multinucleated myotubes which eventually replace the damaged muscle fibers [15, 46]. Such activity is promoted by IGF and myogenic regulatory factors, like myogenin [45]. This myogenic cell differentiation and the new myotube formation are usually observed between 5 to 6 days after damage [93, 208]. By 10 days post injury, the overall architecture of the muscle is almost restored, although most of the regenerated myofibers are smaller and present central myonuclei [48]. The nuclei migration to a peripheral position in the myofiber characterizes the final differentiation step, which results in mature muscle fibers [45]. This newly formed myofibers are often basophilic (reflecting high protein synthesis) and express embryonic forms of MHC [48, 112]. The phenotype presented by the new myofibers depends on the duration and type of damaging stimuli. In this sense, specialized myofiber phenotypes remain plastic and vary as a function of contractile load, hormonal milieu, and systemic diseases [166, 209]. For example, functional overload or endurance exercise training results in transformation of pre-existing fast fibers to a slow twitch oxidative phenotype [153, 154].

In summary, skeletal muscle has a remarkable regenerative capacity due to resident and non-resident stem cell populations. Despite the recognized importance of satellite cells and other stems cells in the regenerative process, it should not be forgotten that other muscular and nonmuscular cell types are involved (figures 3 and 5). Indeed, regeneration is compromised when the activation of one of this cell type is attenuated.

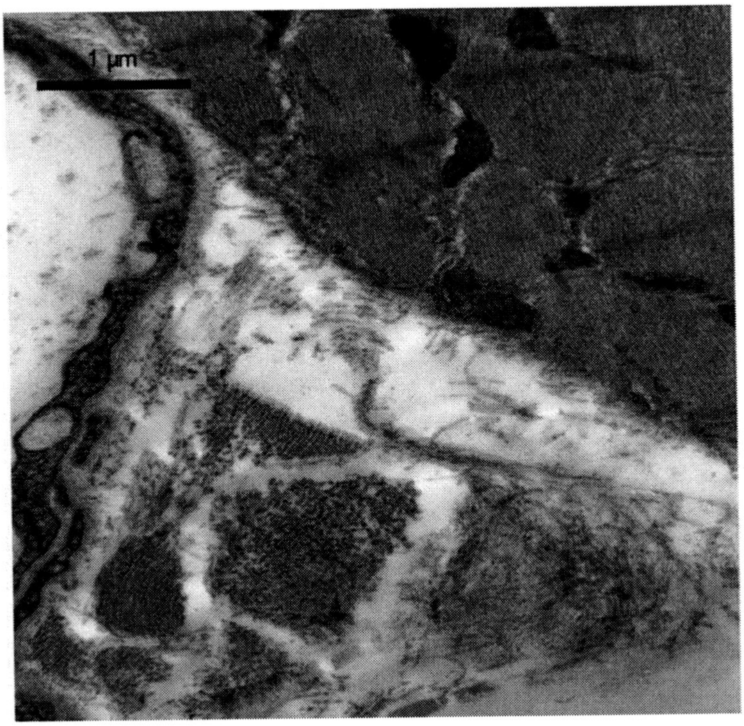

Figure 6. Electron micrograph taken from an oblique section of *gastrocnemius* muscle from a mouse sacrificed 16 days after mechanical damage (crush injury), showing an apparently normal myofibre (at right superior corner), a capillary wall (at the left side) and a great amount of collagen fibres placed between the muscle fibre and the capillary (at the bottom).

Cellular Events in Conditions of Reduced Activity and/or Load

In contrast to the effects of the resistance training, any reduction of neuromuscular activity and/or loading results in a rapid and substantial atrophy of skeletal muscle fibers [5, 105, 157, 169]. An example is the effects of unloading on skeletal muscle function in astronauts during spaceflight [77] and in individuals subjected to bed rest [25]. The adaptation of the musculoskeletal system to such conditions is well documented by using a number of animal models, including spinal cord transection, hindlimb suspension, and microgravity exposure [42, 44, 76, 92]. These paradigms demonstrate a reduction in the whole muscle size attributed to muscle fiber atrophy, loss of functional capacity, impaired locomotor coordination, and decreased resistance to fatigue in the antigravity muscles of the lower limbs [13, 30, 77, 105, 164, 187]. Microscopically, there are some noticeable changes including sarcomere dissolution, streamed Z lines, damaged mitochondria, endothelial degradation, and increase in connective tissue around muscle fibers [12, 13, 30, 35, 122, 147], which are more or less accentuated depending on the muscle fiber type and to the type of atrophic stimulus [13, 20, 75, 77, 90].

Regarding the myofiber, the decrease of its size, particularly in predominantly slow extensor muscles, is often accompanied by an increase in the percentage of fibers expressing a fast phenotype based on their MHC profile [4, 8, 21, 44, 64, 68, 79, 92, 142, 143]. The presence of hybrid fibers expressing a combination of MHC isoforms increases with the atrophic response [8, 64, 92]. These adaptations chiefly involve a downregulation of the slow type I MHC with the concomitant *de novo* expression of the fast type IIx MHC [2, 19]. Such alterations are obviously associated with significant changes in the contractile properties of whole muscles or isolated single fibers (that is, faster contractile properties) but with reduced force- and power generating profiles [19]. This so-called speeding of the contractile apparatus is accompanied by changes in the expression of proteins involved in the release and sequestering of calcium, whereby greater amounts of these calcium cycling proteins are expressed in the atrophied muscle fibers [20].

The marked reduction in the number of mitochondria [88, 141] and in the number of myonuclei may be at least partially responsible for the decreased fiber size (figure 7) [7, 8, 62, 143, 177]. A fiber type–specific response in myonuclear loss has been noted [7, 177] and seems to depend on the atrophic conditions imposed [5, 33, 79, 144]. For example, nuclear domains, whether expressed as nucleus *per* unit cross-sectional area (proportional to fiber volume) or perimeter (proportional to fiber surface), are much smaller in denervated than in innervated muscles [177]. After space flight, a loss of myonuclei and fiber atrophy has been described to be closely coupled and nuclear domains defined as myonuclei *per* unit fiber perimeter unchanged [94]. Others, however, have found that disuse atrophy may be more pronounced than the accompanying nuclear loss [177].

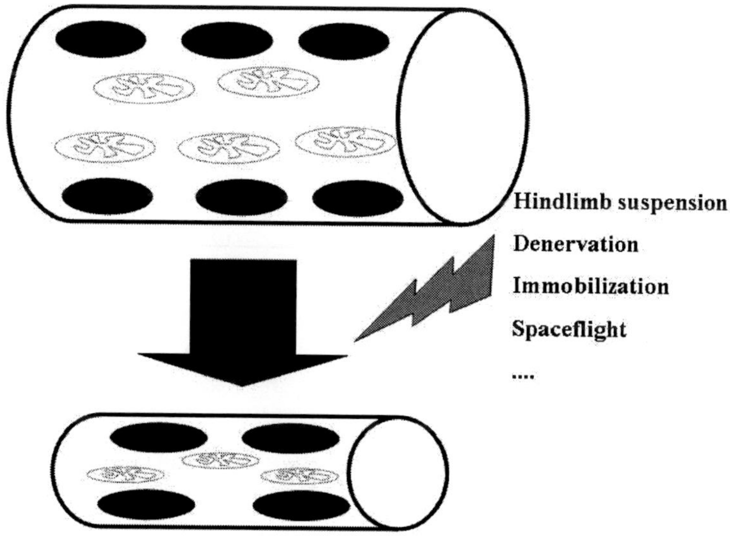

Figure 7. Schematic illustration of the muscle fiber changes induced by atrophic stimuli.

Nuclei elimination from the atrophying myofibers, apparently through an apoptotic mechanism [5, 7, 67, 119, 136, 185, 187, 193], can be seen as a cellular strategy to compensate the decreased volume and transcriptional/translational requirements of the myofiber (figure 7). This compensatory myonuclear loss may be related to the concept that each myonucleus has a tightly regulated nuclear domain [6, 8, 136] and individual myonuclei can express different proteins from neighbouring myonuclei, suggesting that these myonuclei receive distinct stimuli and respond accordingly [7]. Such dynamic modulation of myonuclear number suggests that a prolonged exposure to altered demands on muscle stimulates a rather long-term adaptive strategy than does the immediate up- or downregulation of mRNAs and translational events [145].

Because of the postmitotic nature of myonuclei, growth of atrophied myofibers depends on resident muscle precursor cells that exist outside the myofibers [135, 136]. A decreased number of mpc were observed regardless of the atrophy-inducing stimulus [57, 108, 109, 136]. Prolonged hindlimb suspension induced atrophy appears to result in long-lived deficits in the mpc, manifested by an inability of the muscle tissue to grow or to regenerate [47, 57, 136, 138]. Increased levels of catabolic factors such as myostatin [116, 160, 168] and TNFα [161, 189] as well as decreased contents of trophic factors are generally induced by several atrophic stimuli. The presence or absence of such factors could negatively affect mpc content or activity [126, 189]. However, studies of satellite cells in motor nerve transected rat muscles [108, 122, 136, 165, 177] have shown an early increase in satellite cell number followed by a steady decline to considerably lower than control values by six to seven months of denervation. In conditions of neuromuscular inactivity, the loss of electric or trophic regulation of the motor nerve appears to stimulate satellite cells to leave the quiescent stage and became activated [34], morphologically manifested by an elongated cytoplasmatic body of this cell population [122]. Concerning unloading-induced skeletal muscle atrophy, the majority of studies have described a decrease in the absolute number of satellite cells [57, 93, 108, 109, 136, 138, 179]. However, it was observed that also in hindlimb suspended muscle, satellite cells undergo an initial phase of rapid proliferation [73] in a similar way as reported for denervation. This apparent contradiction may be explained by the fact that the above-mentioned studies have analyzed different periods of unloading. The experimental time-point at which the mitotic activity is evaluated seems to be of major importance in the recognition of activated satellite cells. In atrophying conditions related to Duchenne muscular dystrophy and aging [10, 38, 82, 162], the potential for satellite cell replication has been demonstrated to be markedly diminished, which indirectly deteriorates the contractile function of skeletal muscle. It is conceivable that satellite cell depletion will impair the muscle ability to form new myofibers or to repair the existing ones [108]. Apoptosis may play a causative role in the depletion of satellite cells, thereby impairing the regenerative response to injury. Many variables, including satellite cell senescence, diminishing satellite cell numbers, increased satellite cell susceptibility to apoptosis, and changing factors within the local environment, all contribute to the impaired regenerative capacity of aged or dystrophic skeletal muscle [108]. In aged muscle, satellite cells also show a long latent period between any activation stimulus and their proliferation [11, 57].

Upon cessation of the atrophic stimulus, the myofiber eventually recovers its size and myonuclear number, which in part depends, on mpc proliferation, differentiation, and subsequent fusion to the adjacent myofibers [136]. However, under situations of simulated or real microgravity, muscles from laboratory animals and humans seem to become more vulnerable to damage upon reloading [64, 139]. Although some damage may occur in real or simulated weightlessness, the majority occurs upon reloading of the weakened muscle. The slow type I fibers seem to be preferentially affected, which has been attributed to the selective recruitment of this fiber type [77, 163]. The lesions are similar in appearance to those observed following eccentric contractions. The etiology of the increased sensitivity to cell damage following unloading is unknown, but may be caused by the cell atrophy and reduced actin content and/or by the selective loss of other proteins such as titin, desmin, and dystrophin [77, 104].

How other cells or tissue components of muscle (besides myofibers) participate in the process of muscle remodelling during disuse or inactivity, has received comparatively little considerations [67, 73, 80, 107]. For instance, atrophic conditions also result in qualitative and quantitative changes in the intramuscular connective tissue [34, 60, 107, 110, 122, 148, 177, 196], and specifically in collagen expression [133]. Its volume increases substantially and the normal three-dimensional orientation of the collagen fibers is disrupted. Although the epimysium (fascia) does not undergo these changes, all of the components of the connective tissue network within the skeletal muscle (endomysium, perimysium and myotendinous junction) are involved [107, 196]. The increase in the endomysial collagen fibers is especially prominent in areas connecting two adjacent muscle fibers (i.e., the perpendicularly oriented fibers), which most likely would restrict the mobility of individual muscle fibers [107].

A decrease of muscle elasticity and a reduction in the range of joint motion has been described after immobilization by plaster cast or after use of an orthosis, for example. This condition is considered to be caused by an augmentation of intramuscular connective tissue (figure 8) [13, 148], which then affects the collagen fibril arrangement. This change in arrangement may have important repercussions on muscle functionality during the late stage of immobilization [148].

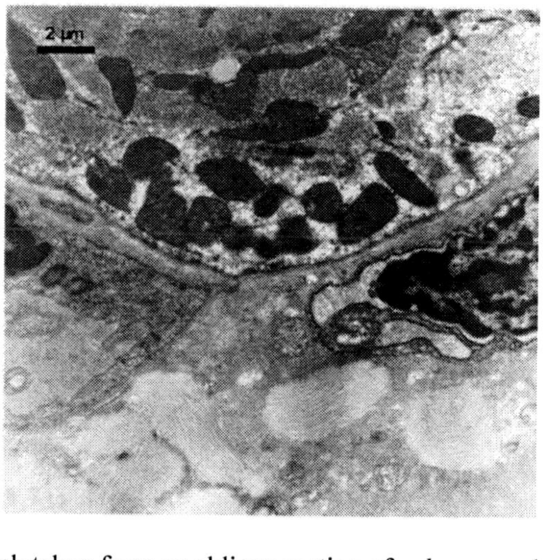

Figure 8. Electron micrograph taken from an oblique section of *soleus* muscle from a mouse sacrificed at the 8th day of hindlimb immobilization, illustrating a myofibre (above), a capillary wall (left side), an apparently active fibroblast (right side), and abundant collagen fibres in the bottom area.

A decrease in the proportion of type I collagen and an increase of type III collagen has been observed in unloaded *soleus* muscle and is consistent with the decrease in the stiffness of slow-twitch muscle [133]. Since fibroblasts are responsible for the synthesis of collagen in skeletal muscle [36], and collagen type I is diminished with hindlimb suspension [133], it is conceivable that muscle fibroblasts may undergo apoptosis [67]. Others, however, have reported an increase in the number of proliferating interstitial non-endothelial cells, mainly fibroblasts, with the time-course of hindlimb suspension [73, 107, 177].

With the increase in interstitial collagen, capillaries and other small blood vessels become surrounded by substantial layers of dense tissue, thereby isolating them from the muscle fibers [13, 107, 122, 196]. Such a connective tissue barrier is likely to impair the blood supply of the muscle fibers, eventually resulting in a vicious cycle of events with the result of a further increase in muscle fiber atrophy [13, 35, 60, 80, 107]. The outcome of atrophic-induced changes in capillarity in a given muscle seems to depend on blood flow [13, 80, 128, 197], which decreases, for example, after chronic and acute hindlimb unloading [128]. A reduced capillary-to-fiber ratio [35, 60, 80] and loss of capillaries *per* fascicle have also been observed after nerve transection [35]. In immobilized muscle the occurrence of capillaries with fenestrations has been described [146], possibly as a yet obsolete attempt to attenuate the diffusion barrier. The characteristic tortuosity of muscular capillaries seems to be replaced by a straight configuration in hindlimb unloading conditions, which may decrease the ability of oxygen delivery [80]. Based on electronic microscopic analysis, an alteration of the wave-like structure of capillaries towards a straight configuration has been also described in immobilized muscle [147]. This change parallels the decrease in the number and diameter of capillaries, possibly through an apoptotic process of endothelial cells [80]. Capillary death is a key cause of the reduction of the microcirculatory bed in chronically denervated muscles [35, 122]. In hindlimb suspended muscle, endothelial cells undergoing apoptosis were also identified and appeared to enter this process when muscle atrophy was measurable [67].

The accumulation of interstitial connective tissue affects the exchange of substrates between the vascular bed and muscle fibers [107], as well as the extent of axonal ingrowth during reinnervation [122]. Borisov and Carlson [35] observed that the increase in diffusion distances between muscle fibers and capillaries frequently exceeds 70% after long-term denervation. This discrete interstitial fibrosis can lead to hypoxic conditions in myofibers. Specifically, both decrease of capillary-to-fiber ratio and increase of the distance between capillaries and muscle mitochondria are considered the leading factors contributing to hypoxia in skeletal muscle [35, 62]. The lower susceptibility of oxidative muscle fibers to atrophy during the first weeks after denervation may be explained, among other factors, by their better vascularization and/or better capacity to retain capillaries after denervation as compared to fast fibers [35].

In brief and despite the outcome of skeletal muscle atrophy, there are some cellular adaptative occurrences that go outside the muscle fiber and involve not only the activation of degenerative events (e.g. death of capillaries) but also proliferating ones (e.g. increase of connective tissue).

Conclusion

The plasticity of skeletal muscle to adjust to stimuli such as contractile activity (e.g. exercise, denervation) or loading conditions (e.g. resistance training, real or stimulated microgravity) involves not only the myofiber but other cell types present in muscle tissue (e.g. capillaries, connective tissue, and satellite cells). The underlying structural-functional changes to such circumstances are due, at least partially, to alterations in the balance between cellular mechanisms of proliferation and death, in favour of one or other mechanism depending on the stimulus, muscle phenotype and cell type considered. For instance, states of increased mechanical load that results in muscle injury entails a coordinated regenerative response, which involves several cellular and molecular players such as the inflammatory system, growth factors, myogenic precursors cells and vascular components. On the other hand, the adaptations of skeletal muscle tissue to conditions of reduced activity and/or load implies not only degenerative events that culminate in smaller myofibers and in the decrease of capillarity but also in the activation of proliferation mechanisms that results in interstitial fibrosis. Future research will therefore focus on the signalling pathways governing the cellular machinery of proliferation and death that control the muscle plasticity in states of altered load and/or contractile activity.

References

[1] Adams, G. R. and Haddad, F. (1996). The relationships among IGF-1, DNA content, and protein accumulation during skeletal muscle hypertrophy. *J. Appl. Physiol., 6,* 2509-2516.

[2] Adams, G. R., Haddad, F., McCue, S. A., Bodell, P. W., Zeng, M., Qin, L., Qin, A. X. and Baldwin, K. M. (2000). Effects of spaceflight and thyroid deficiency on rat hindlimb development. II. Expression of MHC isoforms. *J. Appl. Physiol., 3,* 904-916.

[3] Adams, G. R. and McCue, S. A. (1998). Localized infusion of IGF-I results in skeletal muscle hypertrophy in rats. *J. Appl. Physiol., 5,* 1716-1722.

[4] Ahtikoski, A. M., Koskinen, S. O., Virtanen, P., Kovanen, V. and Takala, T. E. (2001). Regulation of synthesis of fibrillar collagens in rat skeletal muscle during immobilization in shortened and lengthened positions. *Acta Physiol. Scand., 2,* 131-140.

[5] Allen, D. L., Linderman, J. K., Roy, R. R., Bigbee, A. J., Grindeland, R. E., Mukku, V. and Edgerton, V. R. (1997). Apoptosis: a mechanism contributing to remodeling of skeletal muscle in response to hindlimb unweighting. *Am. J. Physiol., 2 Pt 1,* C579-587.

[6] Allen, D. L., Monke, S. R., Talmadge, R. J., Roy, R. R. and Edgerton, V. R. (1995). Plasticity of myonuclear number in hypertrophied and atrophied mammalian skeletal muscle fibers. *J. Appl. Physiol., 5,* 1969-1976.

[7] Allen, D. L., Roy, R. R. and Edgerton, V. R. (1999). Myonuclear domains in muscle adaptation and disease. *Muscle Nerve, 10,* 1350-1360.

[8] Allen, D. L., Yasui, W., Tanaka, T., Ohira, Y., Nagaoka, S., Sekiguchi, C., Hinds, W. E., Roy, R. R. and Edgerton, V. R. (1996). Myonuclear number and myosin heavy chain expression in rat soleus single muscle fibers after spaceflight. *J. Appl. Physiol., 1,* 145-151.

[9] Allen, R. E. and Rankin, L. L. (1990). Regulation of satellite cells during skeletal muscle growth and development. *Proc. Soc. Exp. Biol. Med., 2,* 81-86.

[10] Anderson, J. E. (2006). The satellite cell as a companion in skeletal muscle plasticity: currency, conveyance, clue, connector and colander. *J. Exp. Biol., Pt 12,* 2276-2292.

[11] Anderson, J. E. and Wozniak, A. C. (2004). Satellite cell activation on fibers: modeling events in vivo--an invited review. *Can J Physiol Pharmacol, 5,* 300-310.

[12] Appell, H. J. (1986). Skeletal muscle atrophy during immobilization. *Int. J. Sports Med., 1,* 1-5.

[13] Appell, H. J. (1990). Muscular atrophy following immobilisation. A review. *Sports Med., 1,* 42-58.

[14] Armand, A. S., Launay, T., Gaspera, B. D., Charbonnier, F., Gallien, C. L. and Chanoine, C. (2003). Effects of eccentric treadmill running on mouse soleus: degeneration/regeneration studied with Myf-5 and MyoD probes. *Acta. Physiol. Scand., 1,* 75-84.

[15] Armand, A. S., Launay, T., Pariset, C., Della Gaspera, B., Charbonnier, F. and Chanoine, C. (2003). Injection of FGF6 accelerates regeneration of the soleus muscle in adult mice. *Biochim. Biophys Acta, 1-2,* 97-105.

[16] Armstrong, R. B., Warren, G. L. and Warren, J. A. (1991). Mechanisms of exercise-induced muscle fibre injury. *Sports Med., 3,* 184-207.

[17] Arnhold, J. (2004). Properties, functions, and secretion of human myeloperoxidase. *Biochemistry (Mosc), 1,* 4-9.

[18] Asakura, A., Seale, P., Girgis-Gabardo, A. and Rudnicki, M. A. (2002). Myogenic specification of side population cells in skeletal muscle. *J. Cell Biol., 1,* 123-134.

[19] Baldwin, K. M. and Haddad, F. (2001). Effects of different activity and inactivity paradigms on myosin heavy chain gene expression in striated muscle. *J. Appl. Physiol., 1,* 345-357.

[20] Baldwin, K. M. and Haddad, F. (2002). Skeletal muscle plasticity: cellular and molecular responses to altered physical activity paradigms. *Am. J. Phys. Med. Rehabil., 11 Suppl,* S40-51.

[21] Baldwin, K. M., Herrick, R. E., Ilyina-Kakueva, E. and Oganov, V. S. (1990). Effects of zero gravity on myofibril content and isomyosin distribution in rodent skeletal muscle. *Faseb J, 1,* 79-83.

[22] Barbero, A., Benelli, R., Minghelli, S., Tosetti, F., Dorcaratto, A., Ponzetto, C., Wernig, A., Cullen, M. J., Albini, A. and Noonan, D. M. (2001). Growth factor supplemented matrigel improves ectopic skeletal muscle formation--a cell therapy approach. *J. Cell. Physiol., 2,* 183-192.

[23] Belcastro, A. N., Shewchuk, L. D. and Raj, D. A. (1998). Exercise-induced muscle injury: a calpain hypothesis. *Mol. Cell Biochem., 1-2,* 135-145.

[24] Berchtold, M. W., Brinkmeier, H. and Muntener, M. (2000). Calcium ion in skeletal muscle: its crucial role for muscle function, plasticity, and disease. *Physiol. Rev., 3,* 1215-1265.

[25] Berg, H. E., Larsson, L. and Tesch, P. A. (1997). Lower limb skeletal muscle function after 6 wk of bed rest. *J. Appl. Physiol., 1,* 182-188.

[26] Billeter, R., Oetliker, H., Hoppeler, H. (1994). *Advances in Veterinary Sciences and Comparative Medicine: Structural Basis of Muscle performance* ST: Academic Press

[27] Bischoff, R. (1990). Control of satellite cell proliferation. *Adv. Exp. Med. Biol.,* 147-157; discussion 157-148.

[28] Bischoff, R. (1997). Chemotaxis of skeletal muscle satellite cells. *Dev. Dyn., 4,* 505-515.

[29] Bischoff, R. and Heintz, C. (1994). Enhancement of skeletal muscle regeneration. *Dev. Dyn., 1,* 41-54.

[30] Boonyarom, O. and Inui, K. (2006). Atrophy and hypertrophy of skeletal muscles: structural and functional aspects. *Acta Physiol. (Oxf), 2,* 77-89.

[31] Booth, F. W. and Thomason, D. B. (1991). Molecular and cellular adaptation of muscle in response to exercise: perspectives of various models. *Physiol. Rev., 2,* 541-585.

[32] Borg, T. K. and Caulfield, J. B. (1980). Morphology of connective tissue in skeletal muscle. *Tissue Cell, 1,* 197-207.

[33] Borisov, A. B. and Carlson, B. M. (2000). Cell death in denervated skeletal muscle is distinct from classical apoptosis. *Anat. Rec., 3,* 305-318.

[34] Borisov, A. B., Dedkov, E. I. and Carlson, B. M. (2001). Interrelations of myogenic response, progressive atrophy of muscle fibers, and cell death in denervated skeletal muscle. *Anat. Rec., 2,* 203-218.

[35] Borisov, A. B., Huang, S. K. and Carlson, B. M. (2000). Remodeling of the vascular bed and progressive loss of capillaries in denervated skeletal muscle. *Anat. Rec., 3,* 292-304.

[36] Bosman, F. T. and Stamenkovic, I. (2003). Functional structure and composition of the extracellular matrix. *J. Pathol., 4,* 423-428.

[37] Bottinelli, R. and Reggiani, C. (2000). Human skeletal muscle fibres: molecular and functional diversity. *Prog Biophys Mol. Biol., 2-4,* 195-262.

[38] Brack, A. S., Bildsoe, H. and Hughes, S. M. (2005). Evidence that satellite cell decrement contributes to preferential decline in nuclear number from large fibres during murine age-related muscle atrophy. *J. Cell. Sci., Pt 20,* 4813-4821.

[39] Brotchie, D., Davies, I., Ireland, G. and Mahon, M. (1995). Dual-channel laser scanning microscopy for the identification and quantification of proliferating skeletal muscle satellite cells following synergist ablation. *J. Anat.,* 97-102.

[40] Bruusgaard, J. C., Liestol, K., Ekmark, M., Kollstad, K. and Gundersen, K. (2003). Number and spatial distribution of nuclei in the muscle fibres of normal mice studied in vivo. *J Physiol, Pt 2,* 467-478.

[41] Burleigh, I. G. (1977). Observations on the number of nuclei within the fibres of some red and white muscles. *J. Cell Sci.,* 269-284.

[42] Caiozzo, V. J., Baker, M. J. and Baldwin, K. M. (1998). Novel transitions in MHC isoforms: separate and combined effects of thyroid hormone and mechanical unloading. *J. Appl. Physiol., 6,* 2237-2248.

[43] Caiozzo, V. J., Baker, M. J., Huang, K., Chou, H., Wu, Y. Z. and Baldwin, K. M. (2003). Single-fiber myosin heavy chain polymorphism: how many patterns and what proportions? *Am. J. Physiol. Regul. Integr. Comp. Physiol., 3,* R570-580.

[44] Caiozzo, V. J., Haddad, F., Baker, M. J., Herrick, R. E., Prietto, N. and Baldwin, K. M. (1996). Microgravity-induced transformations of myosin isoforms and contractile properties of skeletal muscle. *J. Appl. Physiol., 1,* 123-132.

[45] Carlson, B. M. (2001). Regeneration of muscle. *Encyclopedia of life sciences,* 1-6.

[46] Carlson, B. M. (2003). Muscle regeneration in amphibians and mammals: passing the torch. *Dev Dyn, 2,* 167-181.

[47] Chakravarthy, M. V., Davis, B. S. and Booth, F. W. (2000). IGF-I restores satellite cell proliferative potential in immobilized old skeletal muscle. *J. Appl. Physiol., 4,* 1365-1379.

[48] Charge, S. B. and Rudnicki, M. A. (2004). Cellular and molecular regulation of muscle regeneration. *Physiol. Rev., 1,* 209-238.

[49] Cheek, D. B. (1985). The control of cell mass and replication. The DNA unit--a personal 20-year study. *Early Hum. Dev., 3,* 211-239.

[50] Chen, J. C. and Goldhamer, D. J. (2003). Skeletal muscle stem cells. *Reprod. Biol. Endocrinol.,* 101.

[51] Chopard, A., Pons, F. and Marini, J. F. (2001). Cytoskeletal protein contents before and after hindlimb suspension in a fast and slow rat skeletal muscle. *Am. J. Physiol. Regul. Integr. Comp. Physiol., 2,* R323-330.

[52] Collins, C. A., Olsen, I., Zammit, P. S., Heslop, L., Petrie, A., Partridge, T. A. and Morgan, J. E. (2005). Stem cell function, self-renewal, and behavioral heterogeneity of cells from the adult muscle satellite cell niche. *Cell, 2,* 289-301.

[53] Conboy, I. M. and Rando, T. A. (2002). The regulation of Notch signaling controls satellite cell activation and cell fate determination in postnatal myogenesis. *Dev. Cell, 3,* 397-409.

[54] Cornelison, D. D. and Wold, B. J. (1997). Single-cell analysis of regulatory gene expression in quiescent and activated mouse skeletal muscle satellite cells. *Dev Biol, 2,* 270-283.

[55] Cros, N., Muller, J., Bouju, S., Pietu, G., Jacquet, C., Leger, J. J., Marini, J. F. and Dechesne, C. A. (1999). Upregulation of M-creatine kinase and glyceraldehyde3-phosphate dehydrogenase: two markers of muscle disuse. *Am. J. Physiol., 2 Pt 2,* R308-316.

[56] Dai, Q., Thompson, M. A., Pippen, A. M., Cherwek, H., Taylor, D. A. and Annex, B. H. (2002). Alterations in endothelial cell proliferation and apoptosis contribute to vascular remodeling following hind-limb ischemia in rabbits. *Vasc. Med., 2,* 87-91.

[57] Darr, K. C. and Schultz, E. (1989). Hindlimb suspension suppresses muscle growth and satellite cell proliferation. *J. Appl. Physiol., 5,* 1827-1834.

[58] Decary, S., Mouly, V., Hamida, C. B., Sautet, A., Barbet, J. P. and Butler-Browne, G. S. (1997). Replicative potential and telomere length in human skeletal muscle: implications for satellite cell-mediated gene therapy. *Hum. Gene Ther., 12,* 1429-1438.

[59] Dedkov, E. I., Borisov, A. B., Wernig, A. and Carlson, B. M. (2003). Aging of skeletal muscle does not affect the response of satellite cells to denervation. *J. Histochem. Cytochem., 7,* 853-863.

[60] Dedkov, E. I., Kostrominova, T. Y., Borisov, A. B. and Carlson, B. M. (2002). Resistance vessel remodeling and reparative angiogenesis in the microcirculatory bed of long-term denervated skeletal muscles. *Microvasc. Res., 1,* 96-114.

[61] Deschenes, M. R., Giles, J. A., McCoy, R. W., Volek, J. S., Gomez, A. L. and Kraemer, W. J. (2002). Neural factors account for strength decrements observed after short-term muscle unloading. *Am. J. Physiol. Regul. Integr. Comp. Physiol., 2,* R578-583.

[62] Desplanches, D., Kayar, S. R., Sempore, B., Flandrois, R. and Hoppeler, H. (1990). Rat soleus muscle ultrastructure after hindlimb suspension. *J. Appl. Physiol., 2,* 504-508.

[63] Dhawan, J. and Rando, T. A. (2005). Stem cells in postnatal myogenesis: molecular mechanisms of satellite cell quiescence, activation and replenishment. *Trends Cell Biol., 12,* 666-673.

[64] di Prampero, P. E. and Narici, M. V. (2003). Muscles in microgravity: from fibres to human motion. *J. Biomech., 3,* 403-412.

[65] Duarte, J. A., Carvalho, F., Bastos, M. L., Soares, J. M. and Appell, H. J. (1994). Do invading leucocytes contribute to the decrease in glutathione concentrations indicating oxidative stress in exercised muscle, or are they important for its recovery? *Eur J Appl Physiol Occup. Physiol., 1,* 48-53.

[66] Duarte, J. A., Soares, J. M. and Appell, H. J. (1992). Nifedipine diminishes exercise-induced muscle damage in mouse. *Int. J. Sports Med., 3,* 274-277.

[67] Dupont-Versteegden, E. E., Strotman, B. A., Gurley, C. M., Gaddy, D., Knox, M., Fluckey, J. D. and Peterson, C. A. (2006). Nuclear translocation of EndoG at the initiation of disuse muscle atrophy and apoptosis is specific to myonuclei. *Am. J. Physiol. Regul. Integr. Comp. Physiol., 6,* R1730-1740.

[68] Edgerton, V. R., Roy, R. R., Allen, D. L. and Monti, R. J. (2002). Adaptations in skeletal muscle disuse or decreased-use atrophy. *Am. J. Phys. Med. Rehabil., 11 Suppl,* S127-147.

[69] Egginton, S., Zhou, A. L., Brown, M. D. and Hudlicka, O. (2001). Unorthodox angiogenesis in skeletal muscle. *Cardiovasc. Res., 3,* 634-646.

[70] Esser, K. A. and White, T. P. (1995). Mechanical load affects growth and maturation of skeletal muscle grafts. *J. Appl. Physiol., 1,* 30-37.

[71] Farges, M. C., Balcerzak, D., Fisher, B. D., Attaix, D., Bechet, D., Ferrara, M. and Baracos, V. E. (2002). Increased muscle proteolysis after local trauma mainly reflects macrophage-associated lysosomal proteolysis. *Am. J. Physiol. Endocrinol. Metab., 2,* E326-335.

[72] Ferrari, G., Cusella-De Angelis, G., Coletta, M., Paolucci, E., Stornaiuolo, A., Cossu, G. and Mavilio, F. (1998). Muscle regeneration by bone marrow-derived myogenic progenitors. *Science, 5356,* 1528-1530.

[73] Ferreira, R., Neuparth, M. J., Ascensao, A., Magalhaes, J., Vitorino, R., Duarte, J. A. and Amado, F. (2006). Skeletal muscle atrophy increases cell proliferation in mice gastrocnemius during the first week of hindlimb suspension. *Eur. J. Appl. Physiol., 3,* 340-346.

[74] Fisher, B. D., Renne, S., Warren, S., Magee, D. and Koh, J. (2000). The effects of low power laser therapy on muscle healing following acute blunt trauma. *J. Phys. Ther. Sci,* 49-55.

[75] Fitts, R. H., Metzger, J. M., Riley, D. A. and Unsworth, B. R. (1986). Models of disuse: a comparison of hindlimb suspension and immobilization. *J Appl Physiol, 6,* 1946-1953.

[76] Fitts, R. H., Riley, D. R. and Widrick, J. J. (2000). Physiology of a microgravity environment invited review: microgravity and skeletal muscle. *J. Appl. Physiol., 2,* 823-839.

[77] Fitts, R. H., Riley, D. R. and Widrick, J. J. (2001). Functional and structural adaptations of skeletal muscle to microgravity. *J. Exp. Biol., Pt 18,* 3201-3208.

[78] Floss, T., Arnold, H. H. and Braun, T. (1997). A role for FGF-6 in skeletal muscle regeneration. *Genes Dev., 16,* 2040-2051.

[79] Fluck, M. and Hoppeler, H. (2003). Molecular basis of skeletal muscle plasticity--from gene to form and function. *Rev. Physiol. Biochem. Pharmacol.,* 159-216.

[80] Fujino, H., Kohzuki, H., Takeda, I., Kiyooka, T., Miyasaka, T., Mohri, S., Shimizu, J. and Kajiya, F. (2005). Regression of capillary network in atrophied soleus muscle induced by hindlimb unweighting. *J. Appl. Physiol., 4,* 1407-1413.

[81] Germani, A., Di Carlo, A., Mangoni, A., Straino, S., Giacinti, C., Turrini, P., Biglioli, P. and Capogrossi, M. C. (2003). Vascular endothelial growth factor modulates skeletal myoblast function. *Am. J. Pathol., 4,* 1417-1428.

[82] Gibson, M. C. and Schultz, E. (1982). The distribution of satellite cells and their relationship to specific fiber types in soleus and extensor digitorum longus muscles. *Anat Rec, 3,* 329-337.

[83] Giger, J. M., Haddad, F., Qin, A. X., Zeng, M. and Baldwin, K. M. (2005). Effect of unloading on type I myosin heavy chain gene regulation in rat soleus muscle. *J. Appl. Physiol., 4,* 1185-1194.

[84] Gleeson, M., Almey, J., Brooks, S., Cave, R., Lewis, A. and Griffiths, H. (1995). Haematological and acute-phase responses associated with delayed-onset muscle soreness in humans. *Eur. J. Appl. Physiol. Occup. Physiol., 2-3,* 137-142.

[85] Goetsch, S. C., Hawke, T. J., Gallardo, T. D., Richardson, J. A. and Garry, D. J. (2003). Transcriptional profiling and regulation of the extracellular matrix during muscle regeneration. *Physiol. Genomics, 3,* 261-271.

[86] Goldring, K., Partridge, T. and Watt, D. (2002). Muscle stem cells. *J Pathol, 4,* 457-467.

[87] Goli, D. E., Thompson, V. F., Li, H., Wei, W. and Cong, J. (2002). The Calpain System. *Physiol. Rev.,* 731-801.

[88] Goto, Y. I., Komaki, H., Igarashi, F. and Nonaka, I. (2000). Muscle mitochondrial changes by experimental immobility and hindlimb suspension. *J. Gravit. Physiol., 2,* P109-110.

[89] Haddad, F., Qin, A. X., Zeng, M., McCue, S. A. and Baldwin, K. M. (1998). Interaction of hyperthyroidism and hindlimb suspension on skeletal myosin heavy chain expression. *J. Appl. Physiol., 6,* 2227-2236.

[90] Haddad, F., Roy, R. R., Zhong, H., Edgerton, V. R. and Baldwin, K. M. (2003). Atrophy responses to muscle inactivity. I. Cellular markers of protein deficits. *J. Appl. Physiol., 2,* 781-790.

[91] Hall, Z. W. and Ralston, E. (1989). Nuclear domains in muscle cells. *Cell, 5,* 771-772.

[92] Harrison, B. C., Allen, D. L., Girten, B., Stodieck, L. S., Kostenuik, P. J., Bateman, T. A., Morony, S., Lacey, D. and Leinwand, L. A. (2003). Skeletal muscle adaptations to microgravity exposure in the mouse. *J. Appl. Physiol., 6,* 2462-2470.

[93] Hawke, T. J. and Garry, D. J. (2001). Myogenic satellite cells: physiology to molecular biology. *J. Appl. Physiol., 2,* 534-551.

[94] Hikida, R. S., Van Nostran, S., Murray, J. D., Staron, R. S., Gordon, S. E. and Kraemer, W. J. (1997). Myonuclear loss in atrophied soleus muscle fibers. *Anat. Rec., 3,* 350-354.

[95] Hill, M., Wernig, A. and Goldspink, G. (2003). Muscle satellite (stem) cell activation during local tissue injury and repair. *J. Anat., 1,* 89-99.

[96] Hoppeler, H. and Fluck, M. (2002). Normal mammalian skeletal muscle and its phenotypic plasticity. *J. Exp. Biol., Pt 15,* 2143-2152.

[97] Hoppeler, H. and Weibel, E. R. (1998). Limits for oxygen and substrate transport in mammals. *J. Exp. Biol., Pt 8,* 1051-1064.

[98] Huang, J. and Forsberg, N. E. (1998). Role of calpain in skeletal-muscle protein degradation. *Proc. Natl. Acad. Sci. U S A, 21,* 12100-12105.

[99] Huard, J., Li, Y. and Fu, F. H. (2002). Muscle injuries and repair: current trends in research. *J. Bone Joint Surg. Am., 5,* 822-832.

[100] Hudlicka, O. (1982). Growth of capillaries in skeletal and cardiac muscle. *Circ Res, 4,* 451-461.

[101] Hudlicka, O. (1985). Development and adaptability of microvasculature in skeletal muscle. *J Exp Biol,* 215-228.

[102] Hudlicka, O., Brown, M. and Egginton, S. (1992). Angiogenesis in skeletal and cardiac muscle. *Physiol. Rev., 2,* 369-417.

[103] Hurme, T. and Kalimo, H. (1992). Activation of myogenic precursor cells after muscle injury. *Med. Sci. Sports Exerc., 2,* 197-205.

[104] Itai, Y., Kariya, Y. and Hoshino, Y. (2004). Morphological changes in rat hindlimb muscle fibres during recovery from disuse atrophy. *Acta Physiol. Scand., 2,* 217-224.

[105] Jackman, R. W. and Kandarian, S. C. (2004). The molecular basis of skeletal muscle atrophy. *Am. J. Physiol. Cell Physiol., 4,* C834-843.

[106] Jarvinen, T. A., Jarvinen, T. L., Kaariainen, M., Kalimo, H. and Jarvinen, M. (2005). Muscle injuries: biology and treatment. *Am. J. Sports Med., 5,* 745-764.

[107] Jarvinen, T. A., Jozsa, L., Kannus, P., Jarvinen, T. L. and Jarvinen, M. (2002). Organization and distribution of intramuscular connective tissue in normal and immobilized skeletal muscles. An immunohistochemical, polarization and scanning electron microscopic study. *J. Muscle Res. Cell Motil., 3,* 245-254.

[108] Jejurikar, S. S. and Kuzon, W. M., Jr. (2003). Satellite cell depletion in degenerative skeletal muscle. *Apoptosis, 6,* 573-578.

[109] Jejurikar, S. S., Marcelo, C. L. and Kuzon, W. M., Jr. (2002). Skeletal muscle denervation increases satellite cell susceptibility to apoptosis. *Plast. Reconstr. Surg., 1,* 160-168.

[110] Kauhanen, S., Leivo, I. and Michelsson, J. E. (1993). Early muscle changes after immobilization. An experimental study on muscle damage. *Clin. Orthop., 297,* 44-50.

[111] Kawiak, J., Brzoska, E., Grabowska, I., Hoser, G., Streminska, W., Wasilewska, D., Machaj, E. K., Pojda, Z. and Moraczewski, J. (2006). Contribution of stem cells to skeletal muscle regeneration. *Folia Histochem. Cytobiol., 2,* 75-79.

[112] Keller, H. L., St Pierre Schneider, B., Eppihimer, L. A. and Cannon, J. G. (1999). Association of IGF-I and IGF-II with myofiber regeneration in vivo. *Muscle Nerve, 3,* 347-354.

[113] Kjaer, M. (2004). Role of extracellular matrix in adaptation of tendon and skeletal muscle to mechanical loading. *Physiol. Rev., 2,* 649-698.

[114] Knobloch, J., Kunz, W. and Grevelding, C. G. (2002). Quantification of DNA synthesis in multicellular organisms by a combined DAPI and BrdU technique. *Dev. Growth. Differ., 6,* 559-563.

[115] Kubinova, L., Janacek, J., Ribaric, S., Cebasek, V. and Erzen, I. (2001). Three-dimensional study of the capillary supply of skeletal muscle fibres using confocal microscopy. *J. Muscle Res. Cell Motil., 3,* 217-227.

[116] Lalani, R., Bhasin, S., Byhower, F., Tarnuzzer, R., Grant, M., Shen, R., Asa, S., Ezzat, S. and Gonzalez-Cadavid, N. F. (2000). Myostatin and insulin-like growth factor-I and -II expression in the muscle of rats exposed to the microgravity environment of the NeuroLab space shuttle flight. *J. Endocrinol., 3,* 417-428.

[117] Langevin, H. M., Cornbrooks, C. J. and Taatjes, D. J. (2004). Fibroblasts form a body-wide cellular network. *Histochem. Cell Biol., 1,* 7-15.

[118] Langevin, H. M., Storch, K. N., Cipolla, M. J., White, S. L., Buttolph, T. R. and Taatjes, D. J. (2006). Fibroblast spreading induced by connective tissue stretch involves intracellular redistribution of alpha- and beta-actin. *Histochem. Cell Biol., 5,* 487-495.

[119] Leeuwenburgh, C., Gurley, C. M., Strotman, B. A. and Dupont-Versteegden, E. E. (2005). Age-related differences in apoptosis with disuse atrophy in soleus muscle. *Am. J. Physiol. Regul. Integr. Comp. Physiol., 5,* R1288-1296.

[120] Lefaucheur, J. P. and Sebille, A. (1995). Muscle regeneration following injury can be modified in vivo by immune neutralization of basic fibroblast growth factor, transforming growth factor beta 1 or insulin-like growth factor I. *J. Neuroimmunol., 1-2,* 85-91.

[121] Lescaudron, L., Peltekian, E., Fontaine-Perus, J., Paulin, D., Zampieri, M., Garcia, L. and Parrish, E. (1999). Blood borne macrophages are essential for the triggering of muscle regeneration following muscle transplant. *Neuromuscul. Disord., 2,* 72-80.

[122] Lu, D. X., Huang, S. K. and Carlson, B. M. (1997). Electron microscopic study of long-term denervated rat skeletal muscle. *Anat .Rec., 3,* 355-365.

[123] Maier, F. and Bornemann, A. (1999). Comparison of the muscle fiber diameter and satellite cell frequency in human muscle biopsies. *Muscle Nerve, 5,* 578-583.

[124] Mathieu-Costello, O., Ellis, C. G., Potter, R. F., MacDonald, I. C. and Groom, A. C. (1991). Muscle capillary-to-fiber perimeter ratio: morphometry. *Am. J. Physiol., 5 Pt 2,* H1617-1625.

[125] Mauro, A. (1961). Satellite cell of skeletal muscle fibers. *J. Biophys Biochem. Cytol.,* 493-495.

[126] McCroskery, S., Thomas, M., Maxwell, L., Sharma, M. and Kambadur, R. (2003). Myostatin negatively regulates satellite cell activation and self-renewal. *J. Cell Biol., 6,* 1135-1147.

[127] McCroskery, S., Thomas, M., Platt, L., Hennebry, A., Nishimura, T., McLeay, L., Sharma, M. and Kambadur, R. (2005). Improved muscle healing through enhanced regeneration and reduced fibrosis in myostatin-null mice. *J. Cell Sci., Pt 15,* 3531-3541.

[128] McDonald, K. S., Delp, M. D. and Fitts, R. H. (1992). Effect of hindlimb unweighting on tissue blood flow in the rat. *J. Appl. Physiol., 6,* 2210-2218.

[129] McPherron, A. C. and Lee, S. J. (1997). Double muscling in cattle due to mutations in the myostatin gene. *Proc. Natl. Acad. Sci. U S A, 23,* 12457-12461.

[130] Meeson, A. P., Hawke, T. J., Graham, S., Jiang, N., Elterman, J., Hutcheson, K., Dimaio, J. M., Gallardo, T. D. and Garry, D. J. (2004). Cellular and molecular regulation of skeletal muscle side population cells. *Stem Cells, 7,* 1305-1320.

[131] Menetrey, J., Kasemkijwattana, C., Day, C. S., Bosch, P., Vogt, M., Fu, F. H., Moreland, M. S. and Huard, J. (2000). Growth factors improve muscle healing in vivo. *J. Bone Joint Surg. Br., 1,* 131-137.

[132] Merly, F., Lescaudron, L., Rouaud, T., Crossin, F. and Gardahaut, M. F. (1999). Macrophages enhance muscle satellite cell proliferation and delay their differentiation. *Muscle Nerve, 6,* 724-732.

[133] Miller, T. A., Lesniewski, L. A., Muller-Delp, J. M., Majors, A. K., Scalise, D. and Delp, M. D. (2001). Hindlimb unloading induces a collagen isoform shift in the soleus muscle of the rat. *Am. J .Physiol. Regul. Integr. Comp. Physiol., 5,* R1710-1717.

[134] Mitchell, P., Steenstrup, T. and Hannon, K. (1999). Expression of fibroblast growth factor family during postnatal skeletal muscle hypertrophy. *J. Appl. Physiol., 1,* 313-319.

[135] Mitchell, P. O. and Pavlath, G. K. (2001). A muscle precursor cell-dependent pathway contributes to muscle growth after atrophy. *Am. J. Physiol. Cell Physiol., 5,* C1706-1715.

[136] Mitchell, P. O. and Pavlath, G. K. (2004). Skeletal muscle atrophy leads to loss and dysfunction of muscle precursor cells. *Am. J. Physiol. Cell Physiol., 6,* C1753-1762.

[137] Morgan, J. E. and Partridge, T. A. (2003). Muscle satellite cells. *Int. J. Biochem. Cell Biol., 8,* 1151-1156.

[138] Mozdziak, P. E., Pulvermacher, P. M. and Schultz, E. (2000). Unloading of juvenile muscle results in a reduced muscle size 9 wk after reloading. *J. Appl. Physiol., 1,* 158-164.

[139] Mozdziak, P. E., Pulvermacher, P. M. and Schultz, E. (2001). Muscle regeneration during hindlimb unloading results in a reduction in muscle size after reloading. *J. Appl. Physiol., 1,* 183-190.

[140] Mozdziak, P. E., Pulvermacher, P. M., Schultz, E. and Schell, K. (2000). Hoechst fluorescence intensity can be used to separate viable bromodeoxyuridine-labeled cells from viable non-bromodeoxyuridine-labeled cells. *Cytometry, 2,* 89-95.

[141] Mujika, I. and Padilla, S. (2001). Muscular characteristics of detraining in humans. *Med. Sci. Sports Exerc., 8,* 1297-1303.

[142] Musacchia, X. J., Steffen, J. M., Fell, R. D., Dombrowski, M. J., Oganov, V. W. and Ilyina-Kakueva, E. I. (1992). Skeletal muscle atrophy in response to 14 days of weightlessness: vastus medialis. *J. Appl. Physiol., 2 Suppl,* 44S-50S.

[143] Ohira, M., Hanada, H., Kawano, F., Ishihara, A., Nonaka, I. and Ohira, Y. (2002). Regulation of the properties of rat hind limb muscles following gravitational unloading. *Jpn. J. Physiol., 3,* 235-245.

[144] Ohira, Y., Yoshinaga, T., Nomura, T., Kawano, F., Ishihara, A., Nonaka, I., Roy, R. R. and Edgerton, V. R. (2002). Gravitational unloading effects on muscle fiber size, phenotype and myonuclear number. *Adv. Space Res., 4,* 777-781.

[145] Ohira, Y., Yoshinaga, T., Ohara, M., Nonaka, I., Yoshioka, T., Yamashita-Goto, K., Shenkman, B. S., Kozlovskaya, I. B., Roy, R. R. and Edgerton, V. R. (1999). Myonuclear domain and myosin phenotype in human soleus after bed rest with or without loading. *J. Appl. Physiol., 5,* 1776-1785.

[146] Oki, S., Desaki, J., Matsuda, Y., Okumura, H. and Shibata, T. (1995). Capillaries with fenestrae in the rat soleus muscle after experimental limb immobilization. *J. Electron. Microsc. (Tokyo), 5,* 307-310.

[147] Oki, S., Itoh, T., Desaki, J., Matsuda, Y., Okumura, H. and Shibata, T. (1998). Three-dimensional structure of the vascular network in normal and immobilized muscles of the rat. *Arch. Phys. Med. Rehabil., 1,* 31-32.

[148] Okita, M., Yoshimura, T., Nakano, J., Motomura, M. and Eguchi, K. (2004). Effects of reduced joint mobility on sarcomere length, collagen fibril arrangement in the endomysium, and hyaluronan in rat soleus muscle. *J. Muscle Res. Cell Motil., 2,* 159-166.

[149] Ordahl, C. P. (1999). Myogenic shape-shifters. *J. Cell Biol., 4,* 695-698.

[150] Partridge, T. (2003). The role of satellite cells and circulating stem cells in regeneration of skeletal muscle. *Basic. Appl. Myol., 6,* 277-280.

[151] Peng, H. and Huard, J. (2004). Muscle-derived stem cells for musculoskeletal tissue regeneration and repair. *Transplant immunology,* 311-319.

[152] Pette, D. (2001). Historical Perspectives: plasticity of mammalian skeletal muscle. *J Appl Physiol, 3,* 1119-1124.

[153] Pette, D. (2002). The adaptive potential of skeletal muscle fibers. *Can. J. Appl. Physiol., 4,* 423-448.

[154] Pette, D. and Staron, R. S. (1997). Mammalian skeletal muscle fiber type transitions. *Int. Rev. Cytol.,* 143-223.

[155] Pette, D. and Staron, R. S. (2000). Myosin isoforms, muscle fiber types, and transitions. *Microsc. Res. Tech., 6,* 500-509.

[156] Phillips, G. D. and Knighton, D. R. (1990). Skeletal muscle regenerates. *Wounds: A compendium of clinical Research,* 82-94.

[157] Powers, S. K., Kavazis, A. N. and DeRuisseau, K. C. (2005). Mechanisms of disuse muscle atrophy: role of oxidative stress. *Am. J. Physiol. Regul. Integr. Comp. Physiol., 2,* R337-344.

[158] Putman, C. T., Dusterhoft, S. and Pette, D. (2000). Satellite cell proliferation in low frequency-stimulated fast muscle of hypothyroid rat. *Am. J. Physiol. Cell Physiol., 3,* C682-690.

[159] Ralston, E. and Hall, Z. W. (1989). Transfer of a protein encoded by a single nucleus to nearby nuclei in multinucleated myotubes. *Science, 4908,* 1066-1069.

[160] Reardon, K. A., Davis, J., Kapsa, R. M., Choong, P. and Byrne, E. (2001). Myostatin, insulin-like growth factor-1, and leukemia inhibitory factor mRNAs are upregulated in chronic human disuse muscle atrophy. *Muscle Nerve, 7,* 893-899.

[161] Reid, M. B. and Li, Y. P. (2001). Tumor necrosis factor-alpha and muscle wasting: a cellular perspective. *Respir. Res., 5,* 269-272.

[162] Reimann, J., Irintchev, A. and Wernig, A. (2000). Regenerative capacity and the number of satellite cells in soleus muscles of normal and mdx mice. *Neuromuscul. Disord., 4-5,* 276-282.

[163] Riley, D. A., Bain, J. L., Thompson, J. L., Fitts, R. H., Widrick, J. J., Trappe, S. W., Trappe, T. A. and Costill, D. L. (2000). Decreased thin filament density and length in human atrophic soleus muscle fibers after spaceflight. *J. Appl. Physiol., 2,* 567-572.

[164] Robinson, C. E., Keshavarzian, A., Pasco, D. S., Frommel, T. O., Winship, D. H. and Holmes, E. W. (1999). Determination of protein carbonyl groups by immunoblotting. *Anal. Biochem., 1,* 48-57.

[165] Rodrigues Ade, C. and Schmalbruch, H. (1995). Satellite cells and myonuclei in long-term denervated rat muscles. *Anat. Rec., 4,* 430-437.

[166] Rosser, B. W. and Bandman, E. (2003). Heterogeneity of protein expression within muscle fibras. *J. Anim.Sci, 2,* E94-E101.

[167] Rossini K, S. M., Corrado, R., Destro C, Dell'Aica, I., Don M, Carraro U. (1999). Molecular markers of muscle plasticity, damage, regeneration and repair. *Basic Applied Myology, 6,* 293-300.

[168] Roth, S. M. and Walsh, S. (2004). Myostatin: a therapeutic target for skeletal muscle wasting. *Curr. Opin. Clin. Nutr. Metab. Care, 3,* 259-263.

[169] Roy, R. R., Monke, S. R., Allen, D. L. and Edgerton, V. R. (1999). Modulation of myonuclear number in functionally overloaded and exercised rat plantaris fibers. *J. Appl. Physiol., 2,* 634-642.

[170] Russell, B., Dix, D. J., Haller, D. L. and Jacobs-El, J. (1992). Repair of injured skeletal muscle: a molecular approach. *Med.Sci. Sports Exerc., 2,* 189-196.

[171] Russell, B., Motlagh, D. and Ashley, W. W. (2000). Form follows function: how muscle shape is regulated by work. *J. Appl. Physiol., 3,* 1127-1132.

[172] Saido, T. C., Sorimachi, H. and Suzuki, K. (1994). Calpain: new perspectives in molecular diversity and physiological-pathological involvement. *Faseb J, 11,* 814-822.

[173] Sakuma, K., Watanabe, K., Sano, M., Uramoto, I. and Totsuka, T. (2000). Differential adaptation of growth and differentiation factor 8/myostatin, fibroblast growth factor 6

and leukemia inhibitory factor in overloaded, regenerating and denervated rat muscles. *Biochim. Biophys. Acta, 1,* 77-88.

[174] Sanderson, R. D., Fitch, J. M., Linsenmayer, T. R. and Mayne, R. (1986). Fibroblasts promote the formation of a continuous basal lamina during myogenesis in vitro. *J. Cell Biol., 3,* 740-747.

[175] Schiaffino, S. and Reggiani, C. (1994). Myosin isoforms in mammalian skeletal muscle. *J. Appl. Physiol., 2,* 493-501.

[176] Schiaffino, S. and Reggiani, C. (1996). Molecular diversity of myofibrillar proteins: gene regulation and functional significance. *Physiol. Rev., 2,* 371-423.

[177] Schmalbruch, H. and Lewis, D. M. (2000). Dynamics of nuclei of muscle fibers and connective tissue cells in normal and denervated rat muscles. *Muscle Nerve, 4,* 617-626.

[178] Schultz, E. (1989). Satellite cell behavior during skeletal muscle growth and regeneration. *Med. Sci. Sports Exerc., 5 Suppl,* S181-186.

[179] Schultz, E., Darr, K. C. and Macius, A. (1994). Acute effects of hindlimb unweighting on satellite cells of growing skeletal muscle. *J. Appl. Physiol., 1,* 266-270.

[180] Schultz, E. and McCormick, K. M. (1994). Skeletal muscle satellite cells. *Rev. Physiol. Biochem. Pharmacol.,* 213-257.

[181] Seale, P. and Rudnicki, M. A. (2000). A new look at the origin, function, and "stem-cell" status of muscle satellite cells. *Dev. Biol., 2,* 115-124.

[182] Seale, P., Sabourin, L. A., Girgis-Gabardo, A., Mansouri, A., Gruss, P. and Rudnicki, M. A. (2000). Pax7 is required for the specification of myogenic satellite cells. *Cell, 6,* 777-786.

[183] Shefer, G., Partridge, T. A., Heslop, L., Gross, J. G., Oron, U. and Halevy, O. (2002). Low-energy laser irradiation promotes the survival and cell cycle entry of skeletal muscle satellite cells. *J.Cell Sci., Pt 7,* 1461-1469.

[184] Shi, X. and Garry, D. J. (2006). Muscle stem cells in development, regeneration, and disease. *Genes Dev., 13,* 1692-1708.

[185] Siu, P. M. and Alway, S. E. (2005). Mitochondria-associated apoptotic signalling in denervated rat skeletal muscle. *J. Physiol., Pt 1,* 309-323.

[186] Smith, C. K., 2nd, Janney, M. J. and Allen, R. E. (1994). Temporal expression of myogenic regulatory genes during activation, proliferation, and differentiation of rat skeletal muscle satellite cells. *J. Cell Physiol., 2,* 379-385.

[187] Smith, H. K., Maxwell, L., Martyn, J. A. and Bass, J. J. (2000). Nuclear DNA fragmentation and morphological alterations in adult rabbit skeletal muscle after short-term immobilization. *Cell Tissue Res., 2,* 235-241.

[188] Spangenburg, E. E. and Booth, F. W. (2002). Multiple signaling pathways mediate LIF-induced skeletal muscle satellite cell proliferation. *Am. J. Physiol. Cell Physiol., 1,* C204-211.

[189] Stewart, C. E. and Rittweger, J. (2006). Adaptive processes in skeletal muscle: molecular regulators and genetic influences. *J. Musculoskelet Neuronal Interact., 1,* 73-86.

[190] Sultan, K. R., Dittrich, B. T. and Pette, D. (2000). Calpain activity in fast, slow, transforming, and regenerating skeletal muscles of rat. *Am. J. Physiol. Cell Physiol., 3,* C639-647.

[191] Sumariwalla, V. M. and Klein, W. H. (2001). Similar myogenic functions for myogenin and MRF4 but not MyoD in differentiated murine embryonic stem cells. *Genesis, 4,* 239-249.

[192] Tanzer, M. L. (2006). Current concepts of extracellular matrix. *J. Orthop. Sci., 3,* 326-331.

[193] Tews, D. S., Behrhof, W. and Schindler, S. (2005). Expression patterns of initiator and effector caspases in denervated human skeletal muscle. *Muscle Nerve, 2,* 175-181.

[194] Thompson, L. V. (2002). Skeletal muscle adaptations with age, inactivity, and therapeutic exercise. *J. Orthop. Sports Phys. Ther., 2,* 44-57.

[195] Tidball, J. G. (2005). Inflammatory processes in muscle injury and repair. *Am. J. Physiol.Regul Integr. Comp. Physiol., 2,* R345-353.

[196] Tyml, K. and Mathieu-Costello, O. (2001). Structural and functional changes in the microvasculature of disused skeletal muscle. *Front Biosci.,* D45-52.

[197] Tyml, K., Mathieu-Costello, O., Cheng, L. and Noble, E. G. (1999). Differential microvascular response to disuse in rat hindlimb skeletal muscles. *J. Appl. Physiol., 4,* 1496-1505.

[198] Umnova, M. M. and Seene, T. P. (1991). The effect of increased functional load on the activation of satellite cells in the skeletal muscle of adult rats. *Int. J. Sports Med., 5,* 501-504.

[199] van Weel, V., Deckers, M. M., Grimbergen, J. M., van Leuven, K. J., Lardenoye, J. H., Schlingemann, R. O., van Nieuw Amerongen, G. P., van Bockel, J. H., van Hinsbergh, V. W. and Quax, P. H. (2004). Vascular endothelial growth factor overexpression in ischemic skeletal muscle enhances myoglobin expression in vivo. *Circ. Res., 1,* 58-66.

[200] Vierck, J., O'Reilly, B., Hossner, K., Antonio, J., Byrne, K., Bucci, L. and Dodson, M. (2000). Satellite cell regulation following myotrauma caused by resistance exercise. *Cell Biol. Int., 5,* 263-272.

[201] Villena, J. and Brandan, E. (2004). Dermatan sulfate exerts an enhanced growth factor response on skeletal muscle satellite cell proliferation and migration. *J. Cell Physiol., 2,* 169-178.

[202] Wagatsuma, A., Tamaki, H. and Ogita, F. (2005). Capillary supply and gene expression of angiogenesis-related factors in murine skeletal muscle following denervation. *Exp. Physiol., 3,* 403-409.

[203] Wagers, A. J. and Conboy, I. M. (2005). Cellular and molecular signatures of muscle regeneration: current concepts and controversies in adult myogenesis. *Cell, 5,* 659-667.

[204] Williams, P. E. and Goldspink, G. (1984). Connective tissue changes in immobilised muscle. *J. Anat.,* 343-350.

[205] Williams, R. S. and Annex, B. H. (2004). Plasticity of myocytes and capillaries: a possible coordinating role for VEGF. *Circ. Res., 1,* 7-8.

[206] Wozniak, A. C., Kong, J., Bock, E., Pilipowicz, O. and Anderson, J. E. (2005). Signaling satellite-cell activation in skeletal muscle: markers, models, stretch, and potential alternate pathways. *Muscle Nerve, 3,* 283-300.

[207] Wright-Carpenter, T., Klein, P., Schaferhoff, P., Appell, H. J., Mir, L. M. and Wehling, P. (2004). Treatment of muscle injuries by local administration of autologous conditioned serum: a pilot study on sportsmen with muscle strains. *Int. J. Sports Med., 8,* 588-593.

[208] Wright-Carpenter, T., Opolon, P., Appell, H. J., Meijer, H., Wehling, P. and Mir, L. M. (2004). Treatment of muscle injuries by local administration of autologous conditioned serum: animal experiments using a muscle contusion model. *Int. J. Sports Med., 8,* 582-587.

[209] Wu, H., Gallardo, T., Olson, E. N., Williams, R. S. and Shohet, R. V. (2003). Transcriptional analysis of mouse skeletal myofiber diversity and adaptation to endurance exercise. *J. Muscle Res. Cell Motil. 8,* 587-592.

[210] Zhao, P. and Hoffman, E. P. (2004). Embryonic myogenesis pathways in muscle regeneration. *Dev. Dyn., 2,* 380-392.

[211] Zhong, H., Roy, R. R., Siengthai, B. and Edgerton, V. R. (2005). Effects of inactivity on fiber size and myonuclear number in rat soleus muscle. *J. Appl. Physiol., 4,* 1494-1499.

[212] Zimmerman, S. D., McCormick, R. J., Vadlamudi, R. K. and Thomas, D. P. (1993). Age and training alter collagen characteristics in fast- and slow-twitch rat limb muscle. *J. Appl. Physiol., 4,* 1670-1674.

[213] Zimowska, M., Szczepankowska, D., Streminska, W., Papy, D., Tournaire, M. C., Gautron, J., Barritault, D., Moraczewski, J. and Martelly, I. (2001). Heparan sulfate mimetics modulate calpain activity during rat Soleus muscle regeneration. *J. Cell Physiol., 2,* 178-187.

In: Cell Growth Processes: New Research
Editor: Daiki Kimura, pp. 89-121

ISBN: 978-1-60456-132-6

Chapter III

On the Biological Significance of Tumor Growth

Raúl A. Ruggiero [1,a]***, Oscar D. Bustuoabad*** [1,b]***, Paula Chiarella***[1]***, Juan Bruzzo***[1] ***and Roberto P. Meiss***[2]

[1] División Medicina Experimental. Instituto de Investigaciones Hematológicas
[2] Centro de Estudios Oncológicos
Academia Nacional de Medicina de Buenos Aires
Pacheco de Melo 3081, 1425 Buenos Aires, Argentina

Abstract

Despite their wide differences, most theories about cancer proposed during the past century agree that cancer is a *biological nonsense* for the organism in which it originates since cancer cells are believed to be *autonomous* (or relatively autonomous), meaning that they are not subject to the rules and regulations that control normal cell proliferation and differentiation in the organism.

In this chapter we have challenged this interpretation on the basis that, throughout the animal kingdom, cancer seems to arise only in organs and tissues that display lost or diminished regenerative ability. In these organs, any injury causing loss of cells or cellular function cannot be compensated by cellular division and in consequence, the original size and function cannot be restored.

We suggest that this situation induces a *crisis*, which, through putative danger signals resulting from retardation of tissue repair, acceleration of cell loss and functional compromise, might promote some degree of variability in the remaining live but arrested cells of the injured organ. The outcome of this situation would be the emergence, by chance, of a genetically or epigenetically modified cell variant bearing mitotic ability to respond to the reparative signal. This new variant would begin to divide and if it were poorly functional or non-functional, the organ would be numerically but not functionally restored. In consequence, it would not score the regeneration as effective and it would

[a] Corresponding author. E-mail: ruloruggiero@yahoo.com.ar .
[b] E-mail: busdao@yahoo.com.ar.

continue to send mitotic signals to restore the lost or diminished organ function. As a result, the new variant would grow over and over and the outcome would be a tumor.

According to this hypothesis, cancer would *not be an autonomous* entity disobeying the mechanisms controlling cell proliferation; in fact, in the injured organ which has its regenerative ability lost or diminished, a tumor cell would be the only one able to respond to the demand to proliferate surrounded by "normal" but arrested cells that cannot respond to that signal. According to this interpretation, cancer would have a profound *biological sense*: it would be the ultimate attempt to restore organ functions and structures that have been lost or altered by aging and noxious environmental agents. However, unlike normal structures, cancer would have *no physiological value*, because the usually poor or non-functional nature of its cells would make their reparative task unattainable.

The hypothesis advanced in this chapter might have significant practical implications. All conventional therapies against cancer attempt to kill all tumor cells. However, according to our hypothesis, the problem might not be solved even if all the tumor cells were eradicated. In effect, if the organ failure remained, new tumor cells would emerge and the tumor would reinitiate its progressive growth in response to the permanent regenerative signal of the non-restored organ. Therefore, efficient anti-cancer therapy should combine an attack against the tumor cells themselves with the correction of the organ failure, which, according to this hypothesis, would be in the core of the cancer problem.

Background

The Antiquity of Cancer

Cancer as well as benign neoplasias are very old diseases which have afflicted animals since long before man appeared on earth [1, 2] and human beings since prehistoric times [1, 3]. In effect, the fossil record shows that benign bone tumors (osteomas) occurred occasionally in aquatic vertebrates living during the Paleozoic Era more than 300 million years ago. There are also many well-documented cases of benign neoplasms such as osteomas, osteochondromas, hemangiomas and odontomas in mosasaurus and dinosaurs from the Jurassic and Cretaceous - about 200-70 million years ago - and in fossil elephants, mammoths, ungulates, fishes and walrus from the Tertiary and Quaternary Eras. In the same way, malignant tumors were noted in the Mesozoic, Tertiary and Quaternary fossil record, including a parostotic osteosarcoma in a Triassic amphibian, a chondrosarcoma in the humerus of a terapod dinosaur from the Jurassic, a metastatic cancer consisting of multiple lytic lesions with cortical bone invasion in a fossil bone fragment from a large-sized dinosaur from the Jurassic, multiple myeloma affecting the bones of some Cretaceous dinosaurs, osteosarcomas affecting some mammals such as a Pleistocene buffalo, a Pleistocene cave bear and a Holocene Capra and chondrosarcomas reported in some species of fossil Canidae [1, 2, 4-6]. Neoplastic diseases have affected humans since their first appearance on the Earth. The earliest known example is a pathological growth reported to be a possible Burkitt lymphoma or an ossifying sarcoma observed in a fragment of mandibular ramous attributable to a Homo erectus which lived in Kenya 1.5 million years ago. Later, almost all types of modern neoplastic diseases have been documented in ancient human remains. For example, meningiomas have been reported in the fossil bones pertaining to a Homo Steinheimensis and to a Neanderthal man which lived in Germany 365,000 and 35,000 years ago, respectively, and in skeletons from Ancient Egypt and pre-historic America. Other benign tumors such as hemangiomas, osteoclastomas, histiocytomas and osteomas have been documented since prehistory in Europe and North Africa. Malignant primary bone tumors such as osteosarcomas, chondrosarcomas, hemangiosarcomas and Ewing's sarcoma have been reported in ancient populations of Europe and Egypt and prehistoric populations of Peru, Chile and Hawaii. Paleopathological studies have also revealed the existence of bone metastasis of nasopharingeal and breast carcinoma and lytic lesions due to multiple myeloma and melanoma in the skeleton of individuals of prehistoric and ancient populations of Europe, Iran, Egypt and Pre-Columbian America. Furthermore, paleopathological investigations of naturally or artificially mummified human bodies, excavated in Egypt, Nubia, Peru, Chile, Alaska, China and Europa have revealed the existence of some benign and malignant primary tumors of bones and soft tissues including carcinomas of the prostate and rectum, naso-orbital cancer, rhabdomyosarcomas, nasopharingeal carcinoma and multiple myeloma among others [1, 3, 4, 7].

The fact that tumors have been present, at least among vertebrates, for hundred millions of years, suggests that natural selection does not seem to have evolved mechanisms to limit their emergence even though many apparently less harmful traits have been eliminated during species evolution. This argument has often been challenged on the ground that most tumors develop after reproduction has ceased and that, in consequence, negative selection would not be operative against tumor-bearing individuals. However, although many species show a decline in reproductive function with age, the female menopause occurs only in humans and a few other primates [8].

Brief History of Cancer Research

Written records concerning cancer can be traced to the Ancient Egypt. In effect, the Edwin Smith and the Ebers papyruses were written approximately 1,600 years BC and they are among the oldest written records concerned with cancer that have survived our inspection. Both contain references to possible cases of breast, uterus and stomach cancer and Kaposi's sarcoma as well as descriptions of practical, empirical and palliative treatments. Many years later, theories of tumor formation together with therapeutic methods against cancer were codified by leading physicians of antiquity such as Hippocrates (460-370 BD), Galen (129-199 AD) and others. During the 18th century, some authors focused the attention on the fact that malignant neoplasms may arise as a result of particular ways of life (for example the higher incidence of breast cancer in nuns than in other women) or as a result of extensive exposure to occupational or chemical factors (for example, the high incidence of scrotal cancer in former chimney sweeps attributed to their extensive exposure to the carcinogenic effects of soot, or the development of nasal cancers due to the use of tobacco in the form of snuff) [4, 9].

However, there is consensus that only in the last 100-150 years, a truly scientific approach to malignant diseases emerged as a result of the mounting and concerted efforts of clinical physicians, experimentalists and theoretical scientists. Since 1863, when Virchow extended his aphorism *omnis cellula e cellula* to include cancer cells, many researchers have proposed different theories aimed to understand the nature of cancer and to answer the question of how tumors appear to begin with [4,10-12]. Since the late 1970, different alterations in cellular genes as well as in several intracellular transducing signaling pathways have been identified in cancer cells and on this basis a unified genetic theory of carcinogenesis has been advanced [13-16].

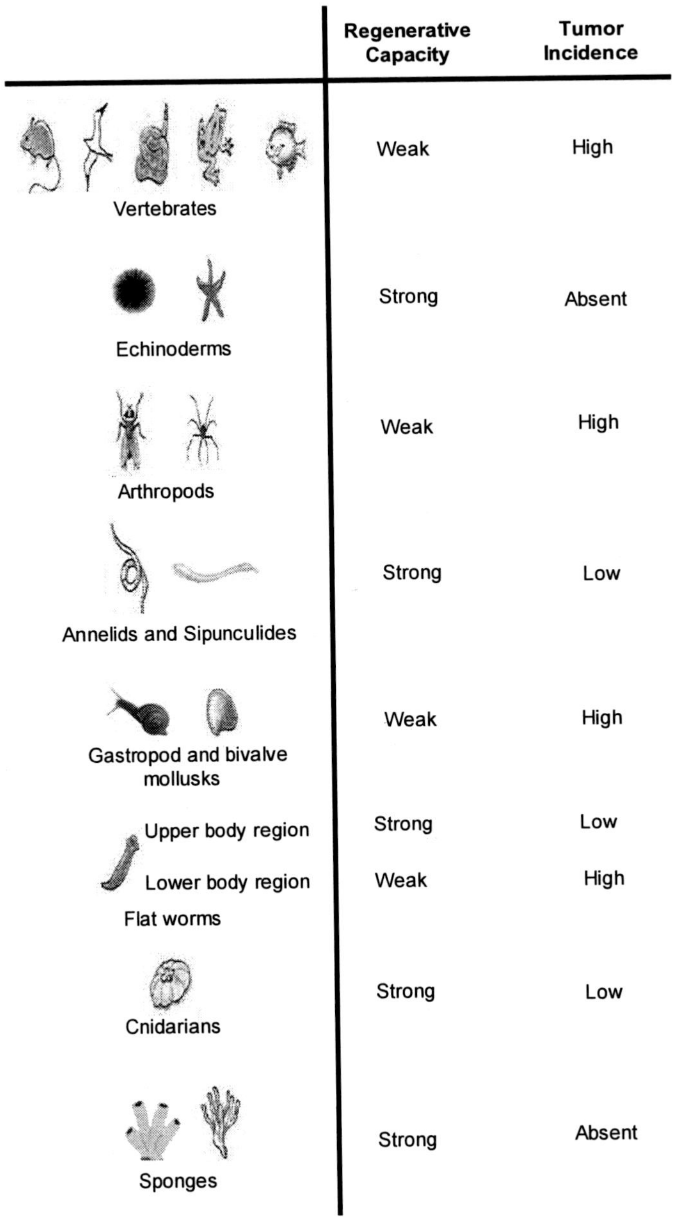

Figure 1. General relationship between regenerative capacity and tumor incidence among the major metazoa groups. Exceptions are considered in the text.

This theory states that cancer starts and ends with the malignant cell, in which genetic changes lead to constitutive activation of some genes (oncogenes) and/or inactivation of others (anti-oncogenes or tumor suppressor genes) allowing that cell to evade – in all or in some microenvironments – the mechanisms controlling cell proliferation. These genetic changes would define the molecular and cellular attributes of the cancer cell, which, in turn, should be the target of specific therapies against cancer. This theory has the enormous merit of unifying, through an immediate common cause, the numerous different mediate causes of cancer such as chemicals, radiation, viruses, etc. However it has some theoretical difficulties, which have been addressed by some authors [17-19] which have also emphasized that cancer remains a major cause of morbidity and mortality, despite the explosive development of our knowledge about the molecular mechanisms associated with the control of cell cycle and survival [20]. Of course, these theoretical difficulties and the persistent failure in treating cancer do not necessarily imply that the unified genetic theory of carcinogenesis is incorrect. However, they encourage us to explore other possible theoretical approaches.

In a recent paper [21], on the basis of ideas advanced by Prehn, Zajicek, Bissell, Duesberg, Sonnenschein and Soto among others [17, 22-25], we have proposed a hypothesis of cancer that does not consider it an autonomous entity disobeying the mechanisms controlling cell proliferation, but one dependent on a reparative signal originating in the particular environment of an injured tissue with diminished or exhausted reparative ability. In this chapter, we have reviewed and extended the arguments of that hypothesis. Hopefully, this hypothesis might help to reconcile some apparently contradictory approaches entailed in the unified genetic and some alternative theories of carcinogenesis, improving our understanding of the relationship among ageing, regeneration and cancer. In turn, this understanding might help to find a plausible explanation for the intriguing fact that the feature "sensitivity to tumor" has survived for, at least, three geological eras.

Postulates

This Hypothesis is Based on Three Postulates:

1) Throughout the animal kingdom, cancer is, in general, rarely produced in body regions displaying *strong* regenerative ability, "strong" meaning the ability to regenerate complex structures such as a whole limb. These regions can encompass the whole body as it occurs in sponges, cnidarians, echinoderms, annelids, sipunculides, [26-29], etc. or parts of the body as in the upper body regions of Planaria, phylum Platyhelminthes [30]; limbs of urodele amphibians [22, 31], etc. Conversely, cancer is, in general, relatively frequent in animals that display *weak* regenerative ability throughout their bodies such as vertebrates others than urodele amphibians, arachnids, insects, gastropod and bivalve mollusks [22, 28, 32-39], etc., "weak" meaning the ability to repair or regenerate relatively simple structures only, as in compensatory hyperplasia of the liver, skin regeneration, etc. A similar relatively high frequency of tumors has been observed in the body regions of urodele amphibians that cannot regenerate [40, 41] and in the lowest body region of Planaria

where the regenerative ability gradient is minimal or absent [30]. It is worth noting that the association between strong or weak regenerative ability with low or high, respectively, tumor incidence, even though observed in most cases, is not an absolute rule. In effect some corals, that exhibit strong regenerative ability, display a relatively high – instead of low – tumor incidence; conversely, the cephalopod mollusks, that exhibit weak regenerative ability, display low – instead of high - tumor incidence [42, 43] (Figure 1).

2) In animals in which cancer is relatively frequent, cancer incidence rises exponentially with age [44, 45]. In addition, when cancer develops in young animals, it is usually associated with injured organs and tissues such as cirrhotic liver, gastric tissues exhibiting chronic atrophic gastritis, radiation- damaged skin, colon displaying ulcerative colitis, breast of nulliparous women, non-secreting prostate alveoli, testicles of cryptorchidism-afflicted persons, etc., which may have exhausted or diminished their regenerative ability [22, 46-48].
3) In animals displaying a strong regenerative ability, reparative or/and regenerative mechanisms tend to remain fairly efficient throughout life [49]. On the other hand, in most of the animals displaying a weak regenerative ability, reparative or/and regenerative mechanisms are efficient mainly during youth; as these animals age, cellular loss increases and those mechanisms wane progressively [50]. In the rare cases of animals exhibiting both, a strong regenerative ability and a high tumor incidence, tumors seem to be associated with the presence of potent environmental stressors producing degenerative forces that overcome the regenerative ability of these animals [51]. In the rare cases of animals exhibiting both a weak regenerative ability and a low tumor incidence, these animals seem to have evolved some particular mechanisms that prevent or ameliorate the injuring action of noxious products [43] thus allowing them to preserve their regenerative abilities (even weak) fairly efficient throughout their life.

Corollaries

1) Throughout the animal kingdom, cancer is rarely – if ever - induced in organs (or tissues) displaying an efficient reparative or regenerative mechanism, "efficient" meaning the ability of organs and tissues to regenerate themselves numerically and functionally. In effect, when these mechanisms remain fairly efficient throughout life - even under the action of putative noxious agents – as they do in most of the animals displaying strong regenerative ability, cancer never (or almost never) occurs. When they remain efficient only during youth – and even during youth, some noxious agents can deplete them – as they do in most of the animals displaying weak regenerative ability, cancer occurs mainly in aging individuals and also in injured organs from young individuals that may have exhausted their regenerative ability because of the action of those noxious agents.
2) Homeostasis in organs or tissues with mitotic potential would be maintained by regulatory fields, "regulatory field" meaning the existence of inhibitory and

stimulatory signals for cell proliferation and differentiation within the space of an organ or tissue. Both types of signal, regardless of their molecular nature, would not be symmetric. In effect, when a reparative or regenerative mechanism is efficient, all cellular loss is compensated by cellular division until the organ attains its original size and function, after which all new mitoses are inhibited. This inhibitory signal, associated with the "right" number of normal functional cells located in the "right" place, must be obeyed not only by the normal cells of the organ but also by all putative anomalous cells which could have emerged within the organ by chance, injury or other cause. In effect, if these anomalous cells could disobey the inhibitory signal and grow autonomously, cancer could develop rather easily in an organ exhibiting an efficient reparative or regenerative mechanism, contradicting *Corollary 1*. In contrast, the mere existence of an organ displaying an inefficient reparative mechanism, means that some or all of their cells could occasionally be non responsive to the stimulatory signal associated with (or produced by) the "less than right" number of functional cells of that organ. The concept of the "right" number of cells in the "right" place can be elucidated by the following example: when a liver is intact, no proliferation of hepatocytes occurs; when it is partially excised and regenerative ability is normal, proliferation occurs until the liver attains its original size and function. The number of hepatocytes in the intact liver would be the "right" number of functional cells which would induce or produce an inhibitory signal(s) for the hepatocytes. Proliferation of hepatocytes after partial hepatectomy would not be prevented by ectopic implantation of liver cells, meaning that these ectopic cells would not be in the "right" place for sending inhibitory signals for preventing hepatocyte proliferation in the remnant liver.

Origin of Tumor Cells

What, According to this Hypothesis, is the Putative Origin of Cancer?

We have said that cancer would not be induced in organs (or tissues) exhibiting an efficient regenerative mechanism. However, when an organism becomes aged and its regenerative ability is progressively lost, any injury causing loss of cells or cellular function can not be compensated by cellular division and in consequence, the original size and function of the organ can not be restored.

We suggest that this situation induce a "crisis" which, through putative danger signals resulting from a retardation of tissue repair, acceleration of cell loss and functional compromise, might create an environment capable of promoting some degree of variability in the remaining live but arrested cells of the injured organ. The outcome of this situation would be the emergence of some genetically and/or epigenetically modified cell variants. Most of these would still lack the ability to divide in response to the organ demand, but sooner or later, a variant bearing that mitotic ability would emerge by chance. This new variant would begin to divide; and if it were poorly functional or non-functional, the organ would be numerically but not functionally restored. In consequence, it would not score the regeneration as effective and it would continue to send mitotic signals to restore the lost or diminished organ function. As a result, the new variant would grow over and over and the outcome would be a tumor. On the other hand, if that emergent new variant were functionally active, normal function might be restored and this "restored" organ might, in most cases, mimic the negative regulatory field associated with the intact organ, after which further mitosis would be halted. In a few cases, however, the new variant – even if functional – might be unable to mimic that negative regulatory field (for example, because of aberrant cellular features not directly related to cellular function) and in such cases, a tumor would also be produced. In the case of poorly functional or non-functional variants, the tumor would be poorly functional or non-functional, as most tumors are. On the other hand, in the special cases of functional variants producing tumors, they would be functioning ones, such as some adenomas or some papillary and follicular carcinomas of the thyroid.

Many authors have highlighted the critical importance of injury in the development of cancer [47, 52-55] and the idea that cancer actually behaves as a wound healing process has been suggested by Dvorak [56]. Others have challenged this interpretation [57, 58], but a critical examination of their data reveals that they scored only massive necrosis and overt degenerative changes as "injury", dismissing less evident injuries such as lost or diminished function of the whole organ or a part of the organ, apoptosis, cellular senescence, etc., which are as relevant as the former for this hypothesis, because both demand a regenerative response.

Cellular heterogeneity, and a genomic instability phase during stages of high-grade dysplasia prior to the acquisition of a frankly malignant phenotype, are two well- documented (though so far unexplained) phenomena [50, 59]. Similarly well-documented are the picture of a tumor arising in a tissue surrounded by "normal" arrested cells, and the existence of factors involved in organ and tissue regeneration that enhance or are necessary for tumor growth [24,54, 60]. Moreover, under certain conditions, the immune response might play a role in tissue regeneration, and in that case it would stimulate rather than inhibit tumor growth [61, 62].

In summary, according to this hypothesis, cancer would originate on the basis of three conditions:

1) An injury of the affected organ (or tissue), "injury" meaning not only partial removal of the organ, massive necrosis or extensive degenerative change but also less evident deleterious effects such as lost or diminished function of the whole organ or a part of the organ, apoptosis, cellular senescence, etc.
2) The impossibility to restore the injury of that organ, and the consequent existence of a permanent reparative signal to the remaining live cells.
3) The existence or the emergence of atypical cells able to respond to the mitotic reparative signal of the injured organ but unable to mimic the negative regulatory field associated with the intact organ.

Our hypothesis about the origin of cancer seems to work regardless of which hypothesis we adopt for the control of the cell proliferation. In effect, if we adopt the stimulatory or positive hypothesis [63], the regenerative signals will be represented by different kinds of growth factors depending on the tissue or organ involved. In the same way, the diminished or lost expression of at least one of the numerous molecular steps in the growth factor signaling pathway in normal aged cells - and conversely, the existence of a responsive pathway in cancer cells - might explain why the latter can proliferate in an organ where normal aged cells cannot. On the other hand, if we adopt the inhibitory or negative hypothesis [63], the regenerative signals will be represented by the absence of some kinds of inhibitory factors (chalones, TGF-β among others). In the same way, the constitutive expression of at least one step in the inhibitory signaling pathway in normal aged cells – and, conversely, the absence of such constitutive expression in tumor cells - might explain why tumor cells can proliferate while normal aged cells cannot.

A plausible objection may be raised about the origin of cancer postulated by this hypothesis. If cancers originate in injured organs or tissues that have exhausted or diminished regenerative capacities, they should be much more frequent in organs or tissues that display poor or null regenerative ability from birth. An obvious example is neuronal tissue in the human brain; however, this tissue actually exhibits fewer tumors than other organs and tissues such as colon, breast, lung or skin [20, 64].

The answer to this objection might be as follows: as stated in *Corollary 2*, "regulatory fields" seem to be necessary to control the proliferation of cells with mitotic potential, which are found in almost all body organs and tissues. However, the theory does not require that "regulatory fields" control the proliferation of postmitotic cells such as brain neurons, because they would not proliferate on their own, as shown by their inability to re-enter the cell cycle even upon stimulation [65]. Therefore, while the neuronal tissue of the brain remains intact, no extracellular inhibitory signals seem to be necessary to keep its cells arrested. On the other hand, when that tissue is injured, probably no stimulatory signals will be generated. In consequence, according to the hypothesis, no primary condition exists for tumor initiation.

In all the above considerations, we have suggested a putative origin of tumor cells by assuming that any injury causing loss of cells or cellular function in an aged organ cannot be compensated by cellular division because the *lost or diminished capacity of individual cells to respond to the regenerative signal* and that in consequence, emerging tumor cells as a result of the "crisis" would be the only ones *bearing the ability to respond to that signal.* At first sight, we might also imagine the alternative view that the absence of regeneration in an aged organ can also be due to the *lost capacity of the organ to produce a restorative signal* and that in consequence, emerging tumor cells as a result of the "crisis" would be those *bearing the ability to grow when restorative cues are absent.* This alternative view, however, does not seem to be compatible with the hypothesis presented herein. In effect, if tumor cells could grow when restorative cues are absent, it means that they could grow in the face of inhibitory signals (because the concept of "regulatory field" implies the existence of mutually excluding stimulatory and inhibitory signals for cell proliferation, regardless of which hypothesis – positive or negative – we adopt) contradicting *Corollary 2.* This does not mean that the absence of regeneration can never be explained by a lost capacity of an organ to produce stimulatory signals of cell proliferation, but if it were the case, cancer would not be, according to our hypothesis, the outcome of that resulting "crisis".

Properties of Tumor Growth

Since tumor growth does not restore the negative regulatory field associated with the intact organ, the "crisis" would persist and, as a consequence, new variants would be forced to emerge continuously by chance in the "normal" resting tissue as well as within the growing tumor. In fact, new cellular variants have been found in the "normal" tissue surrounding a tumor [66, 67].

In the same way, new variants continuously emerging in the tumor itself, could account for cellular heterogeneity typically observed in both experimental and clinical tumors [68].

In addition, since the speed of regeneration of a partially removed organ or tissue is greatest at the outset of the process, when the lack of function is maximal [69], our hypothesis would predict that the more undifferentiated and non-functional the tumor cell, the faster its growth, because for all practical purposes, "regeneration" by non-functional tumor cells would always simulate the outset of the normal regeneration process. The faster growth of more undifferentiated tumors compared with more differentiated ones, is a common but not yet satisfactorily explained phenomenon in tumor biology [64, 70].

The Nature of the Tumor Cell

The most intriguing consequence of this hypothesis concerns the nature of the tumor cell itself.

During the past century, many quite different theories and hypothesis about cancer have been proposed [reviewed in 63, 64, 69]. Despite their wide differences, most of these accounts agree that a frank or true tumor cell is autonomous meaning that it is not subject to the rules and regulations that control normal cell proliferation and survival. The concept of autonomy was originally enunciated in a biological sense (classical definition of Ewing [71]) but the main goal of experimental oncology has been "to understand it in the molecular sense", that is "to elucidate the molecular definition of the cancer cell regardless of its environment" [64].

With the help of the new molecular technologies, several intracellular transducing pathways have been elucidated in the last 25 years and progress in dissecting these pathways "has begun to lay out a circuitry that will likely mimic electronic integrated circuits in complexity and finesse, where transistors are replaced by proteins (e.g. kinases and phosphatases) and the electrons by phosphates and lipids, among others" [14]. Some of these pathways transmit stimulatory growth signals from the extracellular medium to the nucleus, such as the mitogen-activated protein kinase (MAP-kinase) cascade. Others transmit inhibitory signals (most of them funneled through the retinoblastoma protein, pRB, and its two relatives, p107 and p130), death signals (such as that initiated by Fas L), survival signals (such as that initiated by IGF-1), etc [14, 72].

In this context, the constitutive expression of any step(s) in the stimulatory and/or survival signaling pathways (most of them related to the expression of some known "protooncogenes") or the constitutive blockade of any step(s) in the inhibitory and/or death signaling pathways (most of them related to the expression of some known "antioncogenes") or a combination of both, would confer the capacity for autonomous growth on the cell.

Some authors have claimed that this autonomy is not absolute but relative, meaning that the expression of some oncogenes or the silencing of some antioncogenes may generate cancer in some but not all environments. This contention was originally suggested by the classical experiments of Brinster and Mintz and Illmense, demonstrating that the malignant potential of teratocarcinoma cells could be constrained if they were injected in the blastocyst and that the resulting mice contained tumor-free tissues derived from the teratocarcinoma cells [24]. Further evidence is available to support this claim. For example, infection of adult chickens with Rous Sarcoma Virus (RSV) leads to malignant transformation associated with the expression of the oncogene v-src; however, infection of chick embryos in *ovo* with RSV does not lead to malignant transformation, even though v-src is both expressed and active [24]. In the same way, expression of v-myc and c-myc is typical of some tumors, but myc is also expressed in echinoderms which never develop tumors [26]. In any case, irrespective of whether a tumor cell is considered absolutely or relatively autonomous, there is a consensus that it has molecular anomalies that allow it to escape – in all or in some environments – from the regulatory mechanisms that inhibit normal cell proliferation in those environments.

However, if the hypothesis advanced in this paper were true, a tumor cell would not be one ignoring the mechanisms that control normal cell proliferation. In fact, in the injured organ where tumor originates, the tumor cell would be the only one able to respond to the organ demand to proliferate, surrounded by "normal" aged cells that cannot respond to that signal. *In this way, any attempt to find the molecular definition of the cancer cell, meaning the molecular anomalies that allow the tumor cell to escape from the inhibitory signals of normal cell proliferation, might be an attempt to find something that does not exist.* Of course, there are many reported genetic and even hereditable epigenetic changes in different tumors [73, 74], but these changes might not be, the origin of cancer. Instead, they could be reinterpreted as adaptations of cancer cells that enable them to respond more accurately to the demand of the aged organ to proliferate in response to injury. Claims that several putatively oncogenic mutations could be the result rather than the cause of cancer are available in the literature [17-19, 75].

According to our hypothesis any non-functional (and a few aberrant functional) but mitotically active variant present in an injured "aged" organ – with exhausted or diminished regenerative capacity – could behave as a tumor cell. But the same cell put into a "young" organ with an intact regenerative capacity, would behave as a normal cell.

Moreover, in very special situations, even absolutely normal functional cells could behave as tumor cells. For example, when an inert foreign body (such as a glass cylinder) is subcutaneously implanted in a mouse, tissue homeostasis is disrupted and, in consequence, a regenerative signal must be produced. If the tissue is "young", absolutely normal cells will proliferate to repair it, but the presence of the foreign body would not allow the repair to be effected. Therefore, the regenerative signal would continue (presumably because although there are sufficient normal functional cells to heal the injury, they are in the "wrong" place), and a tumor-like proliferation of exclusively normal cells would result. The "crisis" generated by the unresolved disruption of homeostasis would persist, and eventually new non-functional variants would emerge, better adapted to respond to the regenerative stimulus;

these would be the origin of the late sarcomas observed in such cases [76, 77]. The existence of a tumor-like proliferation of normal mesenchymal cells, relatively early after foreign body implantation, is a well documented observation [76].

Our suggestion that a tumor cell is not be autonomous but dependent on a reparative or regenerative signal originating in an "aged" organ or tissue, seems heretical because it contradicts the classical definition of Ewing ("A neoplasm is an autonomous, or relatively autonomous, growth of tissue") which has guided cancer research for the last 60 or more years [71].

However, closer examination of Ewing's proposition reveals that it is a postulate rather than a true definition. First, pathologists do not use it as an operational tool to diagnose the presence of a tumor; in fact, "the means to diagnose cancer have not changed that much since"... the 19th century, "when pathologists began describing the histological pattern of tumors using the light microscope" [63]. Second, if nobody knows exactly what the mechanisms control normal cell proliferation [63], how can anyone be absolutely sure that cancer cells are disobeying those mechanisms? Some years ago, Dr Joseph Aub suggested that the "ugly word of autonomy" be dropped with the understanding that while one can prove dependency, one is never certain of autonomy [78].

The Riddle of the Blue Whale and the Mouse

The unified genetic as well as some (but not all) alternative theories of carcinogenesis share the idea that the malignant cell is the physiological and anatomical unit of cancer disease. Implicit in this contention is the assumption that the probability of origin of an aberrant, neoplastic cell lineage is the same per unit of cell population, regardless of species or cell type concerned.

However, this assumption evokes one of the most intriguing riddles in cancer research, which remain unsolved. This riddle, stated by Dawe [29] some years ago, asks: "Why don't extremely large animals develop neoplasms with a much higher incidence than very small ones since the cell population at risk is greater by several orders of magnitude?". As an extreme example, let us consider the blue whale and the mouse. "If one takes the weight of the mouse as 30 g and that of the blue whale as 100 tons, the whale is equivalent to 3,030,303 mice. Then, if one accounts for differences of lifespan (65 years for the blue whale, 3 years for the mouse), the ratio of weight-year units per whale to weight-year units per mouse is about 66,670,000" [29]. We should therefore expect the blue whale to develop neoplasia about $3x10^6$ and $6.6x10^7$ times more often than the mouse per unit time and per lifespan, respectively. Considering that, about 40% of wild mice kept under laboratory observation develop spontaneous neoplasias during their lives [79], we should expect each blue whale to develop about $2.6x10^7$ neoplasms per lifespan (Figure 2). It is clear that these expectations do not match reality; the incidence of neoplasia in whales as in most mammals, is roughly similar to that in mice. Therefore, the incidence of neoplasia is not a simple function of protoplasm mass at risk per unit time. In fact, the greater the body size of the animal, the greater seems to be its resistance to oncogenesis on a unit weight per unit time basis.

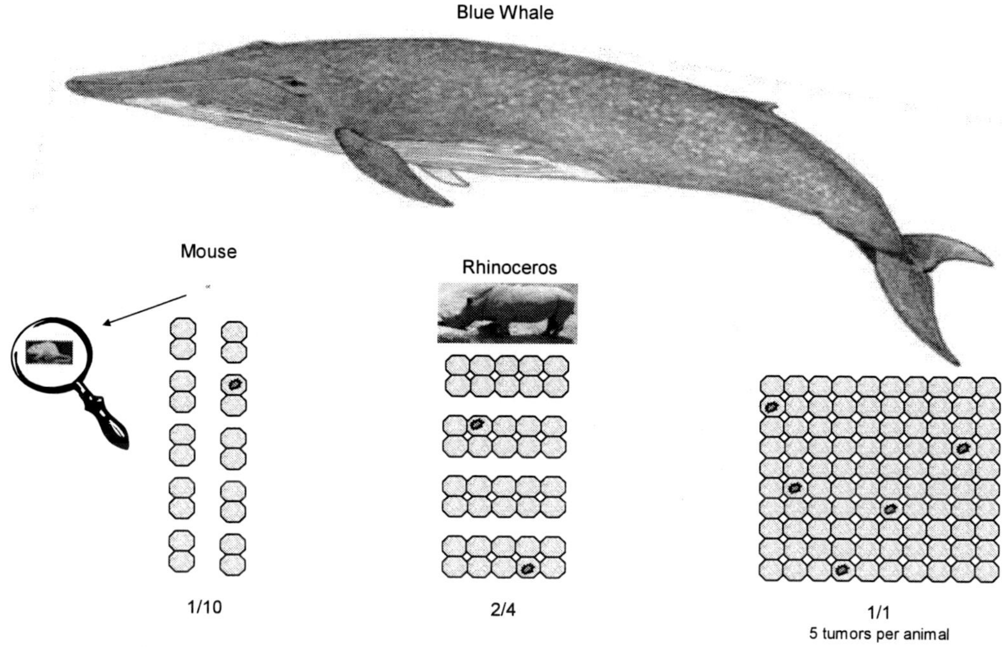

Figure 2. Relative size of mouse (~ 30 g), rhinoceros (~ 2 tons) and blue whale (~ 100 tons) and theoretical influence that size of whole body, organ or tissue would have on tumor incidence per unit time on the assumption that the individual cell in an organ or tissue is the unit at risk of carcinogenesis. We have arbitrarily assumed that a carcinogenic mutation occurs at a rate of 1 per 20 cell units per unit time (tumor cells are identified by) and in consequence, animals with 2, 10 and 100 cells should develop, respectively, 1 neoplasm in every 10 animals, 2 neoplasms in every 4 animals and 5 neoplasms per animal, per unit time. The correspondence between animals (mouse, rhinoceros and blue whale) and number of cells (2, 10 and 100, respectively) is only illustrative. The figure was inspired in Dawe (quotation 29).

Some *ad hoc* hypothesis have been invoked to account for this fact on the assumption that *the individual cell in an organ or tissue is the unit at risk of carcinogenesis.* For example, the animal fat depots might sequester fat-soluble carcinogens with an efficiency proportional to animal's size and thereby proportionally diminish the exposure of other tissues. Another possibility is that the efficiency of defenses against neoplasia such as mechanisms of DNA repair, cellular resistance to metabolism and mutagenic activation of putative carcinogens, immunological surveillance, etc., could be proportional to animal size. While these invoked mechanisms remain largely undemonstrated as general rules [80-82], the hypothesis of cancer that we present in this paper could offer a relatively easy solution of the riddle (although not necessarily excluding other interpretations [80, 83], which in fact might complement ours), by assuming that *the true basic unit at risk of carcinogenesis is the tissue or organ as a whole rather than the individual cell itself.* In effect, according to the hypothesis, cancer originates in organs or tissues that have exhausted or diminished their regenerative capacities, and this would occur when all or a critical proportion of their cells have partially or wholly lost that capacity. In such a case, if an organ were x times larger than another one, the probability that its regenerative capacity is critically diminished would be x times lower, because an x times higher number of cells would have to be affected to depress that capacity. This lower probability would balance the proportionally higher number of their cells that could be transformed. As a result, if the unit at risk is, for example, one liver rather than 10^9 (mouse) as opposed to $3x10^{15}$ (blue whale) liver cells, then the whale will be at no greater risk of developing liver cancer than the mouse, or any other animal with an equally efficient defense mechanism against neoplasia.

The idea that cancer would be an organ or tissue disease rather than a cellular one has been advocated by Waddington, Smithers and others many years ago [11, 84], and, more recently, by the group of Sonnenschein and Soto [63].

Tumor Progression. Invasion and Metastases

Sooner or later, tumor growth will be restrained by the rather rigid architecture of the organ or tissue in which the tumor originated (first tissue). However, the persistent "crisis" will force the emergence of new variants with the ability to disrupt that architecture, so growth can be re-initiated. When these new variants reach the basal membrane, they would eventually be able to disrupt that, allowing the tumor cells to invade another tissue (second tissue). The claim that cancer cells can produce enzymes that destroy the matrix barriers surrounding the tumor permitting invasion into surrounding tissues, has significant experimental support [85, 86].

Assuming that the second tissue is not injured and that its regenerative capacity is intact, the invading tumor cells would face an inhibitory signal from the second tissue which – according to *Corollary 2* – they could not disobey. At that point, the tumor cells might remain arrested indefinitely.

Alternatively, the arrested tumor cells might produce – directly, by releasing inhibitory factors, or indirectly, by attracting inflammatory cells that in turn release inhibitory factors – a lowering of the regenerative capacity of the second tissue. If an injury were incurred in the second tissue, simultaneously or subsequently – most probably associated with the pre-acquired ability of the tumor cells to disrupt the architecture of the first tissue – a stimulatory signal would appear, aimed at repairing the injured tissue. Since the regenerative capacity of the tissue would thereby become exhausted or diminished, the tumor cells would have a selective advantage over normal cells to proliferate. Examples of this selective advantage have been documented [87, 88].

However, the tumor cells did not originate in the second tissue, and since repair or regeneration processes in different tissues are, in general, independent of each other [63], stimulatory signals from one tissue would not usually induce the proliferation of cells from another. Why, then, could the growth of tumor cells from the first tissue actually be stimulated by the stimulatory signal of the second tissue?

We suggest that the less the tumor resembles the primary tissue (presumably the more undifferentiated it is), the more likely it would be to respond to the stimulatory signal of the second tissue and thus, to grow in it. The same procedure could also explain why tumor cells can grow in distant organs (metastases), assuming that they can reach those organs.

On the other hand, more differentiated tumor cells from the first tissue could hardly grow in the second tissue unless the stimulatory signal from the first had reached the orbit of the second. In that case, the tumor cells would grow in the injured second tissue under the guidance of the stimulatory signal from the first. This particular case can be illustrated by the behavior of so-called hormonally-conditioned tumors growing in secondary tissues or organs [78, 89].

Tumor Dormancy

Tumors can occasionally remain dormant for several years, even decades, but suddenly, often in association with surgical stress or another injury, they can awake and resume progressive growth [64, 90].

Some hypothesis have been advocated to explain this phenomenon [91] but its nature remains obscure.

According to the hypothesis presented in this paper, cancer would originate in injured organs or tissues with exhausted or diminished regenerative capacities. However, if this exhausted or diminished capacity could sometimes be recovered, normal cells could reassume their mitotic potential and divide in response to the regenerative signal. Of course, tumor cells would also divide in response to that signal, but as the organ attained its "right size and function" – as a result of the growth of normal functionally active cells – all new mitosis would be stopped including that in tumor cells, according to *Corollary 2*. That could be the mechanism underlying the induction of a dormant tumor. The hypothesis could also offer a plausible explanation for the awakening of the dormant tumor. In effect, after years or decades of dormancy, the organ could become aged, and therefore, its regenerative ability could decrease irreversibly. In that situation, any injury would induce a reparative signal to which the hitherto "dormant tumor cells" could respond. They would thus resume their progressive growth.

Our hypothesis could operate not only for primary tumors but also for dormant metastases, the main clinical problem. In effect, as stated in the preceding section ("Tumor progression. Invasion and metastases"), when tumor cells invade a second intact tissue, they would face an inhibitory signal that they could not disobey. At that point, if these invading tumor cells were not able by themselves to injure and deplete the regenerative ability of that second tissue, they might remain arrested indefinitely, behaving as dormant metastases. Dormant metastases may awaken as a dormant primary does, even years or decades after the tumor cells were seeded in the second tissue, when this tissue becomes aged and loses its regenerative ability.

The induction of tumor dormancy in secondary tumor implants in the presence of a primary growing tumor (concomitant resistance phenomenon [92-94]) might also be interpreted according to this hypothesis, by assuming that the local regenerative signal(s), promoting tumor growth, generated at the site of secondary tumor implantation, could be counteracted by a diffusible inhibitory factor(s) produced or induced by the large primary tumor [95].

Transplantability of Tumors

The hypothesis advanced in this paper postulates that a tumor cell is never autonomous even in the case of invasive and metastatic tumors. In effect, the mere existence of hereditable changes (genetic and/or epigenetic) that endow a cell with the ability to evade the rules controlling normal cell proliferation would mean that these changes could appear by chance in a normal cell within an organ with intact regenerative capacity. But if it were possible, cancer could develop rather easily in that organ, contradicting the *Corollary 1*.

In this section, we consider an apparently fatal objection to this hypothesis, which is one of the milestones in the development of the conventional ideas about cancer: the transplantability of experimental tumors.

In 1877, Novinsky successfully transplanted tumors from adults to young dogs for the first time. These experiments were reproduced in 1888 by Moreau and later by Loeb and Jensen, using rat and murine tumors [69]. These pioneer experiments, which became universal laboratory practice for more than a century, demonstrated that only a small fragment of a tumor or a relatively small number of tumor cells dispersed in physiological saline will suffice to transplant that tumor from a donor to a recipient host. This would mean that the growth of a tumor does not need to be supported by any tissue, organ or organismic pathological condition but only by the nature of the tumor cells themselves. In other words, tumor cells would be autonomous and this claim means that the statements of our hypothesis would be false.

However, the whole of this apparently fatal objection pivots on the ambiguity of the word "autonomy".

We can accept that tumors cells are deemed "autonomous" if their inoculation into an appropriate recipient host is enough to induce a new tumor growth (the first meaning of autonomy). But this does not contradict our hypothesis because the new tumor growth need not be accomplished by evading the rules controlling normal cell proliferation in the recipient host (the second meaning of autonomy). That is, we can accept that tumor cells are autonomous in the first sense, but not in the second sense.

According to our hypothesis, the mechanisms involved in tumor transplantation would not differ markedly from those used by a tumor to invade adjacent or distant organs or tissues within its primary host. In neither case would the tumor cells be autonomous in the second sense of "autonomy" because they would have to injure the recipient organ or tissue and to eliminate or reduce its regenerative capacity as a prerequisite for regenerative signals produced by the injured organ or tissue to promote tumor growth.

Our contention concerning the mechanisms underlying tumor transplantation have significant experimental support:

a) Benign tumors, which are not invasive and commonly produce little damage to host tissues, seldom – if ever - grow when transplanted into another host [96].
b) In chickens, tumors induced by Rous sarcoma virus (RSV) typically form at the viral injection site but not at distant sites; the wound associated with the injection seems to be required for local tumor growth, because additional tumors can be induced at distant sites simply by wounding the infected birds [24].

c) The liver of a young rat, but not of an aged rat in which regenerative ability is diminished or lost, can normalize the morphology and growth capacity of transplanted hepatocarcinoma cells. The most successful normalization occurred when cells were transplanted into the spleen and filtered as solitary cells into the liver without disrupting normal liver architecture. On the other hand, when this architecture was disrupted by transplanting a greater number of malignant cells directly into the liver, normalization was less likely to occur [97].
d) Upon transplantation, tumors usually grow into anatomically correct (orthotopic) organs better than in heterotopic ones [98]. This observation can be interpreted by assuming that an invasive and transplantable tumor, even if quite different from the organ of origin, tends to be more similar to that organ than to others; in consequence, it would respond to a regenerative signal from the former better than to one from the latter, resulting in faster tumor growth.

Carcinogenesis in Vitro

Carcinogenesis *in vitro* can also be considered an objection to our hypothesis. In effect, when "transformed cells" are produced in culture – spontaneously or induced by a given carcinogen – they are assumed to be endowed with the ability to evade normal inhibitory signals when implanted in the organism. If this were true, it would be contradictory to *Corollary 2*, because, according to that corollary, no body cell can evade such signals.

However, this conclusion is not unavoidable. It could alternatively be proposed that so-called carcinogenesis *in vitro* produces cells with particular features that enable them to disrupt homeostasis in the organ or tissue into which they are eventually implanted. This situation would initiate regenerative signals, which could be detected and utilized by the *in vitro* "transformed" cells, promoting growth in a setting in which normal cells would have been prevented from growing.

That is, the putative objection of "carcinogenesis *in vitro*" could be reducible to the objection of "transplantability of tumors" which we addressed in the preceding section.

Carcinogens

In this section we will consider another apparently fatal objection to the hypothesis presented in this chapter: the existence of carcinogens.

As Miller and Miller proposed: "a carcinogen is an agent whose administration to previously untreated animals leads to a statistically significant increased incidence of malignant neoplasms as compared with that in appropriate control animals"[64].

The most prevalent interpretation of this definition, mainly based on the putative mode of action of chemicals, radiation and oncogenic viruses, suggests that most carcinogens exert their critical effects by inducing genetic changes that endow the affected cells with the ability to grow independently of the mechanisms controlling normal cell proliferation.

If this were the case, a cancer cell could emerge in the middle of an otherwise normal organ or tissue, directly contradicting Corollary 1.

However, closer examination of the available data suggests that this prevalent view is not as straightforward as is usually thought. In effect, cancer development with chemicals, radiation, DNA viruses and retroviruses that lack oncogenes, in humans and animals, is a very prolonged process, often lasting one third to two thirds of the life span of the organism. This long period of development is associated with many adaptive cellular proliferative responses that may show a slow evolution to cancer [53]. For example, after treatment of rats with many different types of chemical hepatocarcinogens, rapid inhibition of cell proliferation and cellular death was observed in the liver. This early effect was followed by the appearance of clones of resistant hepatocytes which proliferated vigorously in response to a proliferative stimulus in the hostile environment created by the carcinogen, in which the vast majority of hepatocytes, the non-resistant ones, were inhibited or dead. The resistant hepatocyte nodules have physiological value; but later, as the carcinogen-mediated injury persists, they can evolve into fully transformed cells [53, 54]. Similarly, in Africa and Asia, infection with the hepatitis B DNA virus early in life is associated with the appearance of hepatocellular carcinomas 25 or 30 years later. Prevention of this disease has been achieved by a vaccine against the virus, thus preventing hepatitis and the resulting damage to the liver. This damage, caused by the cytolysis of virally-infected hepatocytes and the aberrant compensatory proliferation of the surviving hepatocytes, seems to be essential for the development of liver tumors since it is the common denominator of both virally- and non-virally-associated hepatocellular carcinomas [63, 99].

Similar considerations can be drawn, for example, from men and women who have developed leukemia since the atom bomb explosions in Hiroshima and Nagasaki. Survivors quite generally showed signs of severe damage to the blood-forming organs soon after the exposure to atomic radiation, and in fact, many of them died at that time. Thus, development of leukemia may be similar to hepatic carcinogenesis to the extent that in both cases a tissue or organ essential to survival is damaged so severely that its total functional capacity is greatly reduced, with the result that some of the victims died immediately whereas others entered a period of active repair, characterized by proliferation of the surviving cells in the damaged tissue. In some cases, during the period of active proliferation aimed to heal the damaged tissue, some malignant cells are produced, the presence of which is revealed only after sufficient time has elapsed to allow their multiplication to generate recognizable tumors [100].

On the other hand, carcinogenesis by retroviruses that carry oncogenes or v-onc genes, such as Abelson murine leukemia virus (Ab-MLV), Rous sarcoma virus (RSV), Avian erythroblastosis virus (AEV) etc., offers, at first glance, a very different picture, because of their ability to induce tumors rapidly and to transform cells in vitro. Reliable experiments, including the use of mutants lacking v-onc genes and transfection assays using cDNA of v-onc genes, have unambiguously demonstrated that v-onc genes are both necessary and sufficient for the transforming ability of such viruses. In addition, use of temperature-sensitive mutants has shown that the expression of protein(s) encoded by the v-onc gene(s) is essential for the expression of the neoplastic phenotype. Furthermore, several systems of regulation of gene expression in transgenic mice have allowed controlled models of neoplasia initiated by numerous oncogenes to be developed in a variety of tissues [24, 101]. Retroviruses that carry oncogenes are not a significant cause of naturally occurring tumors. However, most researchers, stimulated mainly by the discovery in normal cells of protooncogenes homologous to viral oncogenes, have assumed that in all cancers, independently of their etiology and the duration of the preneoplastic process, the critical step driving a normal cell into a neoplastic one, must be similar to that carried out by these retroviruses on their target cells [101]. If it were absolutely true, the hypothesis advanced in this paper would again have to be rejected, because that critical step would be a single intracellular event independent of the environment in which the affected cell resides. However, the final word may not have been said yet.

In effect, although signals from v-onc genes have a dominant role in transformation, changes in cellular genes are also required for transformation to occur. This contribution is highlighted by the fact that some v-onc genes fail to transform certain kinds of primary cell cultures but can transform established cell lines derived from them. Similarly, some cellular lineages can be both infected and transformed while others can be infected but not transformed, by a particular retrovirus carrying a v-onc gene [101].

Furthermore, transgenic animals are usually susceptible to spontaneous tumors involving the tissue (or tissues) in which the transgenic oncogene is expressed. However, in most cases, only a fraction of the animals develop tumors from only a small subset of cells in the infected tissue, and a long latent period is required, indicating that expression of the transgenic oncogene is not sufficient for tumor development. Similar conclusions can be drawn from studies in which a tumor suppressor gene has been selectively disrupted alone or in association with the constitutive expression of a transgenic oncogene [101-103].

A clue to understand the transforming effect of retroviruses carrying oncogenes on their target cells might be the existence of a common denominator among the different lineages that are both infected and transformed by different retroviruses. In all these lineages, expression of the particular v-onc gene interferes primarily with the normal differentiation of the cells that will be transformed. Conversely, when expression of the v-onc gene fails to arrest the differentiation of the infected cell, no transformation occurs [101]. For example, Abelson murine leukemia virus (Ab-MLV), a virus that normally arrests differentiation of pre-B cells, induces pre-B lymphomas from a small subset of the infected pre-B cells. In contrast, Ab-MLV infects erythroid precursors but does not arrest their differentiation and never induces transformation in this lineage. In fact, expression of v-abl gene (the v-onc gene of Ab-MLV) can stimulate erythropoeitin-independent differentiation of erythroid cells. Presumably, this reflects the ability of v-Abl protein to mimic signals normally transmitted via the Epo receptor in a situation where the oncoprotein cannot stimulate continued growth [101].

On the basis of the above considerations, we will now advance an interpretation of retroviral carcinogenesis according to the postulates of our hypothesis.

Consider a schematic representation of a single hematopoietic normal cell lineage comprising a stem cell, some undifferentiated mitotically active cells and some differentiated and functional postmitotic cells. The regulation of cell maturation and turnover in a lineage is not completely understood at the molecular level. Nevertheless, the differentiated cells of the lineage somehow control the proliferation of the less differentiated ones. [69, 104]. For example, when a differentiated cell dies, a restorative signal is generated that induces an undifferentiated cell to divide; one of the resulting cells will differentiate into a functional postmitotic cell while the other will remain undifferentiated, restoring the original function and structure of the lineage.

However, when a retrovirus carrying a v-onc gene infects undifferentiated cells and arrests their differentiation, the normal program of tissue regeneration will be damaged. In effect, although all differentiated functional cells die, promoting a strong regenerative signal, no undifferentiated cell can now differentiate into functional cells, meaning that this lineage would lose its regenerative capacity. Presumably, at early stages of infection, cells that fail to differentiate could only divide three or four times before dying. At that moment, the stem cell would begin to divide to compensate the loss of undifferentiated cells, but these new undifferentiated cells would again be infected with the virus, rendering them unable to differentiate. As a result, a "crisis" would generate a state of variability, and undifferentiated variants not committed to die after a few mitosis, would sooner or later emerge. These variants would divide over and over in response to the regenerative signal, thus resulting in the generation of a neoplastic growth. This suggests that the expression of a v-onc gene could be interpreted otherwise than as a single intracellular event that directly drives a normal cell into an autonomous one, as it usually is. Instead, this expression could be a powerful force primarily arresting normal cell differentiation. Only on that basis, would a tumor emerge in a subset of those arrested cells. That an impediment to normal cellular differentiation is an essential element in the formation of malignant tumors has recently been suggested by Harris [105].

All the above considerations raise the possibility that carcinogenesis induced by chemicals, radiation and oncogenic viruses, even retroviruses carrying viral oncogenes, considered as the paradigm of the unified genetic theory of cancer, might be reinterpreted according to the postulates of the hypothesis advanced in this chapter.

Two Intriguing Cases

The Rabbit Ear

Previous studies by several authors have shown that the rabbit ear is capable of repairing holes up to 1 cm in diameter by a process that appears to be analogous to the regeneration that restores the amputated leg of a urodele amphibian; a blastema is formed and the regeneration restores all of the tissues including the cartilage [106]. However, contrary to the expectations based on the urodele amphibian studies where induction of cancer is very difficult in anatomical areas that have *strong* regenerative ability (see *postulate 1*), the rabbit ear is susceptible to oncogenesis; in fact, it was the first mammalian tissue in which the induction of cancer by chemical agents was demonstrated [107]. Furthermore, if holes are made through a previously hydrocarbon-*initiated* rabbit ear, neoplasias arise preferentially in the hole-filling regenerates more than in other parts of the same initiated rabbit ear.

These observations would be reconciled if the process of regeneration in the rabbit ear depleted the local regenerative capacity so as to make the regenerate *per se,* the tissue in the rabbit ear with the least regenerative capacity. However, it was demonstrated some years ago that new holes made through the regenerated tissue in former ear-holes were repaired at least as well as were primary holes made through normal ear, indicating that the process of regeneration *per se* does not deplete the regenerative capacity of the rabbit ear [108].

At first glance, the experiments presented in that paper seems to contradict the *Corollary 1* of our hypothesis that states that "cancer is hardly... induced in organs (or tissues) displaying an efficient reparative or regenerative mechanism".

However, this apparent objection is based on a misinterpretation of the *Corollary 1.* In effect, the *corollary 1* does not say that a tissue which displays an efficient regenerative ability produces some tumor-inhibitory factors which could prevent a carcinogen to achieve its carcinogenic effect. All that it says is that while the regenerative capacity of a tissue remains intact, cancer will not be produced, but this does not mean that a putative noxious agent can not deplete the regenerative ability of that tissue and on this basis (only after depleting this regenerative ability) to allow a tumor to have a chance to emerge. This means that even though both, the normal (not *initiated*) rabbit ear and the hole-filling regenerate in a normal rabbit ear have an efficient regenerative capacity (as demonstrated in [108]) the *initiated* rabbit ear and the hole filling regenerate in an *initiated* rabbit ear could have their normal regenerative ability diminished or exhausted. The regenerative ability-depleting effect of some *initiators* has been reported [22, 69]. At first sight, the latter contention seems to contradict *Corollary 1*, because according to it, reparative mechanisms in animals exhibiting *strong* regenerative ability would remain fairly efficient even under the action of putative noxious agents. However, the critical point is that, even analogous to the regenerative ability of urodele limbs, the regenerative ability of the rabbit ear can not be considered *strong*, at least as *strong* as that of urodele limbs, upper regions of Planaria body, echinoderms, etc. In consequence, we can postulate, without contradicting *Corollary 1*, that a rabbit ear can lose its regenerative ability upon the action of some *initiator* agents.

On the basis on the above considerations, we suggest a putative explanation for the occurrence of neoplasias in the hole-filling regenerates more than in other parts of the same initiated rabbit ear. When a hole is made in a normal rabbit ear, normal cells will proliferate to restore the injured tissue. However, if a hole were made in an initiated rabbit ear, normal cells would have their regenerative ability diminished and in consequence, the initiated cells would have a selective advantage to proliferate in response to the regenerative signal. Therefore the regenerate would have a higher concentration of initiated cells than other parts of the ear. If these initiated variants were functionally active, normal function would be restored, after which additional mitosis would be halted. However, if these variants were not functionally active, the tissue would be numerically but not functionally restored and in consequence, it would not score the regeneration as effective and it would continue sending mitotic signals aimed to restore the diminished or altered tissue function. As a result, these initiated variants would grow over and over and the outcome would be a tumor. On the other hand, in other parts of the rabbit ear, the number of the initiated cells could be below a critical threshold and the surrounding normal cells could be enough to produce the inhibitory signal associated with the "right" number of functional cells located in the "right" place which, according to Corollary 2, would have to be obeyed by all normal and anomalous cells present in the tissue. This situation would lower the probability of induction of tumors as compared with the regenerate.

Cancer in Corals

In the last 25-30 years, some neoplastic diseases termed "calicoblastic epitheliomas" have been described in corals, especially (although not exclusively) in the genus Acropora in the Caribbean.

This fact seems to represent a fatal objection to the hypothesis presented in this chapter, on the basis on the following argument.

First; because corals (and cnidarians in general) show strong regenerative ability, the existence of neoplasms in corals would be an exception to the rather general rule enunciated in the postulate 1 of our hypothesis ("cancer is, in general, rarely produced in body regions displaying strong regenerative ability").

Second; since, according to postulate 3, in animals exhibiting strong regenerative ability, this ability tends to remain fairly efficient throughout life, neoplasms in corals would have emerged in the face of an efficient regenerative mechanism. This would directly contradict the Corollary 1 of our hypothesis.

In the following paragraphs, however, we will attempt to show how the above mentioned observations of neoplasms in corals can be reconciled with the statement of Corollary 1.

Before 1980, practically no neoplasms –spontaneous or induced- had been reported in corals as well as among other cnidarians [109-110].

On the contrary, as we said above, the existence of colonies of Acropora displaying abnormal proliferation of gatrovascular canals and associated calicoblastic epidermis, with loss of normal polyp structures and zooxanthellae – a condition termed calicoblastic ephitelioma – has frequently been reported over the last two decades. A similar neoplastic condition has been reported in the genus Montipora [42, 111].

Is this increase in the reported cases of neoplasias in corals, the outcome of an increased awareness and regular monitoring or, on the contrary, it is the result of particular conditions only, or specially, operating in the last years?

Continuous surveys of coral reefs worldwide have revealed that tropical coral reefs, particularly those of the Caribbean, have experienced unprecedent levels of bleaching and disease during the last three decades, including the tissue bleaching (TBL), the white and black band diseases (WBD and BBD), the white plague disease, the coralline lethal orange disease (CLOD) among many others. Declining health has been attributed to several biotic and abiotic stressors such as viruses, bacteria, fungi, exposures to elevated water temperatures, increased solar radiation and degraded water quality. While single stressors acting in temporal separation can cause impairment on a limited level, when they act synergistically, the intensity of these degrading effects becomes more distressing, thereby seriously affecting the chances of survival of coral colonies [51, 111]. This seems to have occurred in some coral reefs, especially in the Caribbean. In effect, paleontological data suggest that the recent mortality of Acropora spp in some locations of Caribbean region is without precedent in the late Holocene [112]. For example, in Belize, A. cervicornis was the primary ecological and geological constituent of reefs until the mid 1980s. After constructing reef framework for thousands of years, A. cervicornis was virtually eliminated from the area over a ten-year period [113].

According to our hypothesis, the sudden emergence of neoplastic diseases in corals in the last 25-30 years might be explained in the following way. Although corals show strong regenerative ability and in consequence, tend to maintain their regenerative ability fairly efficient throughout life, even upon the action of some noxious agents (*postulate 3 and Corollary 1*), when the action of such stressors becomes higher than a critical level, degenerative forces tend to prevail and the consequence is tissue loss and later, the death of the individual polyp, and eventually, of the entire coral colony. This situation might, according to our hypothesis, generate a "crisis" and the outcome of this crisis would be the emergence of tumor cells as an attempt –futile- to compensate the action of those degenerative forces and to restore the injured tissues.

Plant Tumors

It has long been known that the induction of crown gall tumors by Agrobacterium tumefaciens in a wide variety of plants depends on the existence of a wound, because inoculating the bacterium into intact plants rarely, if ever, causes tumors [114-116]. However, the precise role of wounding in each step of the tumorigenic process remains unclear.

The conventional interpretation states that the wound would be necessary for transformation but not for tumor growth itself. In effect, previous experiments have suggested that phenolic compounds released from the wound would trigger both the attachment of A. tumefaciens to plant cells and the expression of the *vir* regulon which is necessary for transferring the oncogenic T-DNA from the bacterium to the plant cells [114, 117]. However, no role in the proliferation of transformed plant cells has been attributed to wounding, since crown gall tumor growth has usually been assumed to depend only on the plant growth hormones produced by the proper transformed cells.

This interpretation contradicts the concept of tumor cell advocated in this chapter.

However, more recent evidence seems to offer a different picture. A. tumefaciens was inoculated in unwounded tobacco seedlings and new molecular technologies were used to demonstrate that *vir* gene induction, T-DNA transfer and plant cell transformation were produced as they are in wounded plants; but, differently from that occurred in wound sites, the transformed plant cells *could not produce tumors* [116] suggesting that, as long as tissue architecture is not disrupted, negative regulatory signals prevent growth of the transformed cells.

On the other hand, such negative regulatory signals would tend to be reduced at wound sites, and proliferation of transformed cells could be initiated in consequence. Since growing galls retard or inhibit the development of normal host tissues [118], transformed cells would have a selective advantage to proliferate, and in consequence, the wound would tend to be filled only with transformed cells, which (as opposed to normal wound-healing meristematic cells) display a limited ability to differentiate [114, 116]. From that moment, tumor growth could proceed as described in the section "Origin of tumor cells", suggesting that the hypothesis presented in this chapter might work even beyond the animal kingdom.

Anti-Cancer Treatments

Despite many years of basic and clinical research and trials of promising new therapies, most cancers are resistant to therapy at presentation or become resistant after an initial response [20, 119, 120].

All current conventional therapies against cancer attempt to kill *all* cancer cells with minimal toxic side effects. A similar aim is pursued by some of the new anti-cancer trials.

However, according to our hypothesis, even if all tumor cells were eradicated, the problem might not be solved. In effect, if the organ failure remained, new tumor cells would emerge and tumor would reinitiate its progressive growth in response to the permanent regenerative signal of the non-restored organ.

A theoretically attractive approach would be to make tumor cells functional because in that case, the organ function would be restored and no regenerative signal would remain to promote new cellular growth. This therapeutic schedule is exemplified by the successful treatment of acute promyelocytic leukemia by retinoic acid-based therapies, resulting in the induction of differentiation of promyelocytes to mature cells [121]. However, in general, attempts to differentiate tumor cells *in vitro* or *in vivo* by using differentiating molecules have proved very difficult [23, 70]. On the other hand, full-blown tumor cells become functional or behave as normal cells in the context of strong regenerative fields, stressing that the field in which tumor cells originate, and not only the tumor cells themselves should be an important objective of our studies [23]. The possibility of manipulating the microenvironment and therefore the growth status of the cell has been demonstrated, for example, with cells that overexpress the hyaluronan receptor [122].

Does this mean that an efficient anti-cancer treatment could be achieved by correcting the organ failure only?

Perhaps it depends on the stage of tumor progression. In effect, to correct organ failure means that normal cells can resume their previously lost regenerative capacities and normal organ function and/or structure can be restored. Theoretically, this correction would stop tumor growth because in that condition, the regenerative signals – which, according to our hypothesis, always guide that growth – would have disappeared.

However, that hopeful result could only be achieved if we faced tumor cells at the beginning of tumor growth. Later, when the tumor cells have evolved to become invasive and metastatic, they will have acquired particular features endowing them with the capacity to produce new organ failures. In consequence, to correct the organ failure only would be a transient solution; it would work only until tumor cells damaged the organ again, demanding new tumor growth.

A better putative therapy against full-blown tumor cells might combine the correction of organ failure with an attack against the tumor cells themselves. Perhaps, it would not be necessary to eradicate *all* the tumor cells; perhaps it would be enough to lower their number below a critical threshold (tumor dose 50?) by cytostatic rather than cytotoxic therapies, thus avoiding or reducing the occurrence of new organ injuries that would vitiate attempts to correct the organ failure. Below this threshold, cancer cells would presumably be unable to exert any deleterious effect on the organ. In consequence, if the organ failure were corrected, no new failure could be expected and any remaining tumor cells would therefore behave as normal ones.

However, to correct organ failure will demand understanding of the so far elusive nature of the regulatory fields operating in normal organs. Hopefully, the study of wound and wound healing-like phenomena in three-dimensional organotypic cultures [123], which maintain or mimic the natural organ structure, may provide valuable insights into the basic mechanisms of those fields.

It is worth noting, however, that the understanding of the nature of these fields, might prove to be more difficult to attain than usually thought. In effect, when we think about "regulatory fields", that is, networks of signals for cell proliferation and differentiation within the space of an organ or tissue, we usually imagine transducing pathways involving multiple and successive molecular steps transmitting information from the extra-cellular environment up to the nucleus of the cell. However, it is difficult to fit this image with the statement of *Corollary 2* proposing that the inhibitory signal - associated with the "right" number of normal functional cells located in the "right" place - can not be disobeyed by any cell –if normal or anomalous. In effect, if stimulatory or inhibitory signals for normal cell proliferation are transmitted through intracellular chains of activated or inactivated molecules, respectively, we can easily imagine the emergence of mutated cells bearing the constitutive expression of at least one activated molecule, thus mimicking a proliferative signal even though an inhibitory response is demanded. However, if the inhibitory signal can not be disobeyed in any case, the image of a causal relationship associated with a transducing pathway might not be the best representation of a regulatory field. Therefore, it is likely that the search for the nature and mode of action of the regulatory fields might be more than a technical task merely aimed to identify the molecules involved in each step within each transducing pathway. Possibly, the elaboration of concepts of unity and wholeness not reduced to the conventional causality by contact of multiple parts, might be necessary to achieve a better approach to the problem.

Conclusions

Despite their obvious differences and with few exceptions [23], most theories about cancer proposed during the last century share a common denominator: cancer is believed to be a biological nonsense for the organism in which it originates since cancer cells are believed to be ones that evade the rules controlling normal cell proliferation and differentiation. This prevalent concept of cancer is, in fact, based on a very old tradition which can be traced to Galen who classified all true neoplasms, benign and malignant as "contrary to nature" [4, 124]. According to that understanding, cancer cells have usually been considered as cells returning to a more primitive, unicellular, condition of life. Different features of tumor cells such as asocial behavior, reduced need for putative growth factors, ability to grow in overcrowded settings, lack of contact inhibition, etc., have been interpreted by most researchers as a confirmation of that primitive way of life.

In contrast, according to the hypothesis advanced in this chapter, cancer would have a profound biological sense: it would be the ultimate attempt to restore the organ function and structure that have been lost or altered by aging or environmental noxious agents, that is, an attempt to evade the aging and the death of the organ or the organism (if the organ is essential to survival). Therefore, the above mentioned features of tumor cells could be reinterpreted as progressive acquired adaptations for responding to a permanent regenerative signal in the context of tissue injury, just as several embryonic developmental stages such as morphological differentiation and modeling events could dependent on cellular damage and death together with disruption of topography field [125]. However, unlike normal structures, cancer would have no physiological value, because the usually poorly functional or non-functional nature of its cells would make their reparative task unattainable.

Under special circumstances, however, the attempt of tumors to correct organ failure or to evade death could have been successful. For example, fossil fish of the genus Pachylebias that lived 8 million years ago, adopted pachyostosis to facilitate immersion in the hyper-saline water of the Mediterrean Sea, "through the development of diffuse hyperostosis that did not differ from a neoplastic form of benign tumor originating from bone tissue" [1]. Similarly, mammals of the Sirenidae group from the Oligocene acquired tumor-like forms "in the axial skeleton to consent browsing on the bottom in shallow waters" [1].

In other cases, tumors appeared as a product of an inter-specific association between pairs of organisms. The classic examples are the insect-induced plant gall tumors, which serve the propagation of some groups of insects. They represent a re-differentiation and neo-formation of host tissues, characterized by morphological and histological changes elicited by the developing insect that are unique for both the inducing insect and the affected plant organ. In the more highly developed galls, these self-limiting neoplastic growths are almost comparable, in the determinate growth of their structures, to a leaf or a fruit [115].

The above considerations raise the possibility that some neoplasms could have been adopted by some organisms as a biological strategy to increase their adaptability to difficult environmental conditions, allowing the "pathology" to survive for millions of years.

Numerous predictions of the hypothesis advanced in this manuscript might be experimentally testable. For example: I) the close relationship between strong regenerative ability and absence of tumor throughout the animal kingdom; II) the existence of an injury and a decreased regenerative capacity in a whole organ or tissue, or in a part of that organ or tissue, before the emergence therein of a primary or secondary metastatic tumor; III) the existence of danger signals resulting from a retardation of tissue repair, acceleration of cell loss and functional compromise, inducing hereditable changes (genetic or epigenetic) of some kind in aging or injured cells with diminished or exhausted regenerative ability; IV) the existence of cellular heterogeneity and a genomic instability phase in an organ or tissue before the acquisition of a frankly malignant phenotype, and in normal tissues surrounding a tumor; V) the emergence of tumor cells de *novo,* after total extirpation of a tumor nodule VI) the ability of tumor cells, and the inability of surrounding normally aging cells, to respond to local mitogenic or regenerative signals of the tissue in which the tumor has emerged; VII) the capacity of transplantable tumor cells to injure and to diminish the regenerative ability of the organ or tissue in which they can grow; VIII) the injurious action of carcinogens on cells of organ and tissues as a prerequisite for inducing neoplastic growth, injury meaning not only partial removal of the organ, massive necrosis or extensive degenerative change, but also less evident deleterious effects such as lost or diminished function of the whole organ or a part of the organ, apoptosis, cellular senescence, etc.

Acknowledgements

We are grateful to Dr. Richmond T. Prehn, Dr. Carlos M. Galmarini and Dr. Paul S. Agutter for their helpful and critical discussion of the manuscript and to Miss Victoria Ruival for excellent technical assistance.

This article is dedicated to the memory of an intelligent, honest and extraordinarily generous man, Mr. Juan J. Portaluppi, the technician's chief of our laboratory for almost 50 years.

References

[1] Capasso LL. Antiquity of cancer. *Int.J. Cancer* 2005, 113: 2-13.

[2] Rothschild BM, Witzke BJ, Hershkovitz I. Metastatic cancer in the Jurassic. *Lancet* 1999, 354: 398.

[3] Czarnetzki A, Schwaderer E, Pusch CM. Fossil record of meningioma. *Lancet* 2003, 362: 408.

[4] Shimkin MB. Contrary to nature. Washington: Public Health Service, National Institute of Health; 1977.

[5] Gubin IuM, Petrovichev NN, Solov'ev IuN, Kochergina NV, Luk'ianchenko AB, Makov SM. *Vopr. Onkol.* 2001, 47: 449-455.

[6] Rothschild BM, Tanke D, Hershkovitz I, Schultz M. Mesozoic neoplasia: origins of haemangioma in the Jurassic age. *Lancet* 1998, 351: 1862.

[7] Pahl WM. Tumors of bones and soft tissues in ancient Egypt and Nubia: a synopsis of the detected cases. *Int. J. Anthropol.* 1986, 1: 267-276

[8] Woodruff M. Interaction of cancer and host. *Br.J.Cancer* 1982, 46: 313-322.

[9] Karpozilos A, Pavlidis N. The treatment of cancer in Greek antiquity. *Eur. J. Cancer* 2004, 40: 2033-2040.

[10] Boveri T. The origin of malignant tumors. Williams and Williams, Baltimore MD, 1929.

[11] Waddington CH. Cancer and the theory of organisers. *Nature* 1935, 135: 606-608.

[12] Warburg O. On the origin of cancer cells. *Science* 1956, 123: 309-314.

[13] Bishop JM. The molecular biology of RNA tumor viruses: a physician's guide. Basic science for clinicians. *N. Engl. J. Med.* 1980, 303: 675-682.

[14] Hanahan D, Weinberg RA. The hallmarks of cancer. *Cell* 2000, 100: 57-70.

[15] Bignold LP. The mutator phenotype theory can explain the complex morphology and behavior of cancer. *Cell Mol. Life Sci.* 2002, 59: 950-958.

[16] Frank SA. Age-specific incidence of inherited versus sporadic cancers: a test for the multistage theory of carcinogenesis. *Proc. Natl. Acad. Sci.* USA 2005, 102:1071-1075.

[17] Duesberg P, Rasnick D. Aneuplody, the somatic mutation that makes cancer a species of its own. *Cell Motility and the Cytoskeleton* 2000, 47: 81-107.

[18] Sonnenschein C, Soto AM. Somatic mutation theory of carcinogenesis: why it should be dropped and replaced. *Molecular Carcinogenesis* 2000, 29: 205-211.

[19] Prehn RT. The role of mutation in the new cancer paradigm. *Cancer Cell International* 2005, 5: 9-14.

[20] Bailar III JC, Gornik HL. Cancer undefeated. *N. Engl. Med.* 1997, 336: 1569-1574.

[21] Ruggiero RA, Bustuoabad OD. The biological sense of cancer: a hypothesis. *Theor. Biol. Med Mod.* 2006, 3: 43.

[22] Prehn RT. Regeneration versus neoplastic growth. *Carcinogenesis* 1997, 18: 1439-1444.

[23] Zajicek G. A new cancer hypothesis. *Medical Hypothesis* 1996, 47: 111-115.

[24] Kenny PA, Bissell MJ. Tumor reversion: correction of malignant behavior by microenvironmental cues. *Int. J. Cancer* 2003, 107: 688-695.

[25] Maffini MV, Soto AM, Calabro JM, Ucci AA, Sonnenschein C. The stroma as a crucial target in rat mammary gland carcinogenesis. *J. Cell Science* 2004, 117: 1495-1502.

[26] Walker CW, Boom JDG, Marsh AG. First non-vertebrate member of the *myc* gene family is seasonally expressed in an invertebrate testis. Oncogene 1992, 7: 2007-2012.

[27] Wellings SR. Neoplasia and primitive vertebrate phylogeny: echinoderms, prevertebrates and fishes. A review. *Natl. Cancer Inst. Monograph.* 1969, 31: 59-128.

[28] Sparks AK. Review of tumors and tumor-like conditions in Protozoa, Coelenterata, Platyhelminthes, Annelida, Sipunculida and Arthropoda excluding Insects. *Natl. Cancer Inst. Monograph* 1969, 31: 671-682.

[29] Dawe CJ. Phylogeny and oncogeny. *Natl. Cancer Inst. Monograph* 1969, 31: 1-40.

[30] Hall F, Morita M, Best JB. Neoplastic transformation in the planarian: Cocarcinogenesis and histopathology. *J. Exp. Zoology* 1986, 240: 211-227.

[31] Del Rio-Tsonis K, Tsonis PA. Amphibian tissue regeneration: a model for cancer regulation (review). *Int. J. Oncology* 1992, 1: 161-164.

[32] Mechler BM, Strand D, Kalmes A, Merz R, Schmidt M, Török I. Drosophila as a model system for molecular analysis of tumorigenesis. *Environ. Health Perspect.*1991, 93: 63-71.

[33] Taylor RL. Formation of tumor-like lesions in the cockroach *Leucophaea maderae* after decapitation. *Natl. Cancer Inst. Monograph* 1969, 31: 419-432.

[34] Lopez A. Tumoral pathology in spiders. *J. Invertebr. Patho*l. 1979, 34: 224-230.

[35] Caussinus E, Gonzalez C. Induction of tumor growth by altered stem-cell asymmetric division in Drosophila melanogaster. *Nat. Genet.* 2005, 37: 1027-1028.

[36] Smolarz K, Renault T, Wolowicz M. Ultrastructural study of neoplastic cells in Macomabalthica (Bivalvia) from the Gulf of Gdansk (Polland). *J.Invertebr.Pathol.* 2006, 92:79-84

[37] Cooper EL. Immunity and neoplasia in Mollusks. *Isr.J.Med.Sci.*1976, 12: 479-494.

[38] Sanchez J, Kuba L, Berón-vera B, Dans SL, Crespo EA, Van Bressem MF, Coscarella MA, García NA, Koen Alonso M, Pedraza SN, Mariotti PA. Uterine adenocarcinoma

with metastasis in a bottlenose dolphin *Tursiops truncatus* from northern Patagonia, Argentina. *Dis. Aquat. Org.* 2002, 48: 155-159.

[39] Ostrander GK, Cheng KC, Wolff JC, Wolfe MJ. Shark cartilage, cancer and the growing threat of pseudoscience. *Cancer Research* 2004, 64: 8485-8491.

[40] Zilakos NP, Tsonis PA, Del Rio-Tsonis K, Parchment RE. Newt squamous carcinoma proves phylogenetic conservation of tumors as caricatures of tissue renewal. *Cancer Res.* 1992, 52: 4858-4865.

[41] Okanoto M. Simultaneous demonstration of lens regeneration from dorsal iris and tumor production from ventral iris in the same newt eye after carcinogen administration. *Differentiation* 1997, 61: 285-292.

[42] Peters EC, Halas JC, McCarty HB. Calicoblastic neoplasms in Acropora palmata, with a review of reports on anomalies of growth and form in corals. *J. Natl. Cancer Inst.* 1986, 76: 895-912.

[43] Lee PG, Lu LJ, Salazar JJ, Holoubek Y. Absence of formation of benzo[a]pyrene/DNA adducts in the cuttlefish (Seppia officinalis, Mollusca: Cephalopoda). *Environ.Mol.Mutagen.* 1994, 23: 70-73.

[44] Kitagawa M, Utsuyama M, Kurata M, Yamamoto K, Yuasa Y, Ishikawa Y, Arai T, Hirokawa K. Cancer and aging: symposium of the 27th annual meeting of the Japanese society for biomedical gerontology, Tokio. Cancer Immunol. *Immunother.* 2005, 54: 623-634.

[45] Kosuda K. Ageing and temperature effects on the tumor development in Drosophila melanogaster females. *Gerontology* 1990, 36:121-125.

[46] Karin M, Lawrence T, Nizet V. Innate immunity gone awry: linking microbial infections to chronic inflammation and cancer. *Cell* 2006, 124: 823-836.

[47] Harrison SA, Bacon BR. Relation of hemochromatosis with hepatocellular carcinoma: epidemiology, natural history, pathophysiology, screening, treatment and prevention. *Med. Clin. North Am.* 2005, 89: 391-409.

[48] Raman JD, Nobert CF, Goldstein M. Increased incidence of testicular cancer in men presenting with infertility and abnormal semen analysis. *J. Urol.* 2006, 174: 1819-1822.

[49] Zamaraev VN. Does regenerative capacity decrease with age? *Sov. J. Dev. Biol.* 1974, 4: 495-503.

[50] Sharpless NE, DePinho RA. Telomeres, stem cells, senescence and cancer. *J. Clin. Invest.* 2004, 113: 160-168.

[51] Fisher WS, Santavy DL, Davis WP, Courtney LA. Regional monitoring of coral condition in the Florida Keys. In: USDA Forest Service Proceedings RMRS-P-42 CD, 2006.

[52] Haddow A. Cellular inhibition and the origin of cancer. *Acta Unio Int. Contra Cancrum.* 1938, 3: 342-352.

[53] Farber E, Rubin H. Cellular adaptation in the origin and development of cancer. *Cancer Res.* 1991, 51: 2751-2761.

[54] Farber E. Cell proliferation as a major risk factor for cancer: a concept of doubtful validity. *Cancer Res.* 1995, 55: 3759-3762.

[55] Simpson CJ, Bissell MJ, Werb Z. Mammary gland tumor formation in transgenic mice overexpressing stromelysin-1. *Semin. Cancer Biol.* 1995, 6: 159-163.

[56] Dvorak HF. Tumors: wounds that do not heal. Similarities between tumor stroma generation and wound healing. *N. Engl. J. Med.* 1986, 315: 1650-1691.

[57] Hoel DG, Haseman JK, Hogan MD, Huff J, Mc Connell E. The impact of toxicity on carcinogeneity studies: implications for risk assessment. *Carcinogenesis* 1988, 9: 2043-2052.

[58] Asare GA, Paterson AC, Kero MC, Khan S, Mossanda KS. Iron-free neoplastic nodules and hepatocellular carcinoma without cirrhosis in Wistar rats fed a diet high in iron. *J. Pathol.* 2006, 208: 82-90.

[59] Nery R. Carcinogenic mechanisms: a critical review and a suggestion that oncogenesis may be adaptive ontogenesis. *Chem. Biol. Interact.* 1976, 12: 145-169.

[60] Schindel DT, Grosfeld JL. Hepatic resection enhances growth of residual intrahepatic and subcutaneous hepatoma, which is inhibited by octreotide. *J. Pediatr. Surg.* 1997, 32: 995-998.

[61] Pliskin ME, Prehn RT. Stimulation of liver regeneration and compensatory kidney hyperplasia by passive transfer of spleen cells. *J. Reticuloendothel. Soc.* 1975, 17: 290-299.

[62] Prehn RT. An adaptive immune reaction may be necessary for cancer development. *Theor.Biol.Med .Model.* 2006, 3(1): 6.

[63] Sonneschein C, Soto AM. The society of cells. New York: Springer-Verlag; 1999

[64] Pitot HC. Fundamentals of Oncology. New York: Marcel Dekker Inc; 1978.

[65] Geneser F. Histología. Buenos Aires: Editorial Médica Panamericana; 2000 (translated into spanish from Histologi. Copenhagen: Munksgaard; 1999).

[66] Deng G, Lu Y, Zlotnikov G, Thor AD, Smith HS. Loss of heterozygosity in normal tissue adjacent to breast carcinoma. *Science* 1996, 274: 2057-2059.

[67] Mirsalis JC, Shimon JA, Johnson A, Fairchild D, Kanazawa N, Nguyen T, de Boer J, Glickman B, Winegar RA. Evaluation of mutant frequencies of chemically induced tumors and normal tissues in I/c II transgenic mice. *Environmental and Molecular Mutagenesis* 2005, 45: 17-35.

[68] Lengauer C, Kinzler KW, Vogelstein B. Genetic instabilities in human cancers. *Nature* 1998, 396: 643-649.

[69] Suss R, Kinzel V, Scribner JD. Cancer. Experiments and Concepts. New York, Heidelberg, Berlin: Springer-Verlag; 1973.

[70] Cotran RS, Kumar V, Collins T (eds.). Tissue repair: cellular growth, fibrosis and wound healing. In: Robbins Pathologic basis of disease, 6th ed. New York: WB Saunders; 1999: 89-112.

[71] Ewing J. Neoplastic disease: a treatise on tumors. Philadelphia: Saunders; 1940.

[72] Hahn WC, Weinberg RA. Modelling the molecular circuitry of cancer. *Nat. Rev. Cancer* 2002, 2: 331-341.

[73] Ionov Y, Peinado MA, Malkhosyan S, Shibata D, Perucho M. Ubiquitous somatic mutations in simple repeated sequences reveal a new mechanism for colonic carcinogenesis. *Nature* 1993, 363: 558-561.

[74] Jaffe LF. Epigenetic theories of cancer initiation. Adv. Cancer Res. 2004, 90: 209-230.

[75] Sato M, Vaughan MB, Girard L, Peyton M, Lee W, Shames DS, Ramirez RD, Sunaga N, Gazdar AF, Shay JW, Minna JD. Multiple oncogenic changes (K-RAS (v 12), p53

knockdown, mutant EGFRs, p16 bypass, telomerase) are not sufficient to confer a full malignant phenotype on human bronchial epithelial cells. *Cancer Res.* 2006, 66: 2116-2128.

[76] Brand KG. Cancer associated with asbestosis, schistosomiasis, foreign bodies and scars. In: Cancer: A comprehensive treatise. Edited by Becker FF. New York: Plenum Press; 1982: 661-692.

[77] Pasqualini CD, Ruggiero RA, Bustuoabad OD, Nepomnaschy I, Piazzon I. Experimental onco-immunology revisited. *Current Cancer Therapy Reviews* 2005, 1: 289-298.

[78] Furth J. Conditioned and autonomous neoplasms: a review. *Cancer Res.* 1953, 13: 477-492.

[79] Andervont HB, Dunn TB. Occurrence of tumors in wild house mice. *J. Natl. Cancer Inst.* 1962, 28: 1153-1163.

[80] Hahn WC, Weinberg RA. Rules for making human tumor cells. *N. Engl. J. Med.* 2002, 347: 1593-1603.

[81] Rudo K, Meyers WC, Dauterman W, Langenbach R. Comparison of human and rat hepatocyte metabolism and mutagenic activation of 2-acetylaminofluorene. *Cancer Res.* 1987, 47: 5861-5867.

[82] Hsu IC, Harris CC, Lipsky MM, Snyder S, Trump BF. Cell and species differences in metabolic activation of chemical carcinogens. *Mutat. Res.* 1987, 177: 1-7.

[83] Holliday R. Aging is no longer an unsolved problem in biology. *Ann. N. Y. Acad. Sci.* 2006, 1067: 1-9.

[84] Smithers DW. Cancer. An attack on cytologism. *Lancet* 1962, 1: 493-499.

[85] Coussens LM, Fingleton B, Matrisian LM. Matrix metalloproteinase inhibitors and cancer: trials and tribulations. *Science* 2002, 295: 2387-2392.

[86] Tan X, Egami H, Abe M, Nozawa F, Hirota M, Ogawa M. Involvement of MMP-7 in invasion of pancreatic cancer cells through activation of the EGFR mediated MEK-ERK signal transduction pathway. *J. Clin. Pathol.* 2005, 58: 1242-1248.

[87] Elias D, De Baere T, Roche A, Ducreux M, Leclere J, Lasser P. During liver regeneration following right portal embolization the growth rate of liver metastases is more rapid than that of the liver parenchyma. *Br. J. Surgery* 1999, 86: 784-788.

[88] García Alonso I, Palomares T, Alonso A, Portugal V, Castro B, Carames J, Mendez J. Effect of hepatic resection on development of liver metastases. *Rev. Esp. Enferm. Dig.* 2003, 95: 765-770.

[89] Li R, Wheeler T, Dai H, Frolov A, Thompson T, Ayala G. High level of androgen receptor is associated with aggressive clinicopathologic features and decreased biochemical recurrence-free survival in prostate: cancer patients treated with radical prostatectomy. *Am. J. Surg. Pathol.* 2004, 28: 928-934.

[90] El Saghir NS, Elhajj II, Geara FB, Hourani MH. Trauma-associated growth of suspected micrometastasis. *BMC Cancer* 2005, 5: 94.

[91] Glick AB, Yuspa SH. Tissue homeostasis and the control of the neoplastic phenotype in epithelial cancers. *Semin. Cancer Biol.* 2005, 15: 75-83.

[92] Gorelik E. Concomitant tumor immunity and resistance to a second tumor challenge. *Adv.Cancer Res.* 1983, 39: 71-120.

[93] Ruggiero RA, Bustuoabad OD, Cramer P, Bonfil RD, Pasqualini CD. Correlation between seric antitumor activity and concomitant resistance in mice bearing non-immunogenic tumors. *Cancer Res.* 1990, 50: 7159-7165.
[94] O'Reilly MS, Holmgren L, Sage EH, Folkman J. Angiostatin: a novel angiogenesis inhibitor that mediates the suppression of metastases by a Lewis lung carcinoma. *Cell* 1994, 79: 315-328.
[95] Prehn RT. Two competing influences that may explain concomitant tumor resistance. *Cancer Res*. 1993, 53: 3266-3269.
[96] Shimosato Y, Kameya T, Nagai K, Hirohashi S, Koide T, Hayashi H, Nomura T. Transplantation of human tumors in nude mice. *J. Natl. Cancer Inst.* 1976, 56: 1251-1260.
[97] Rubin H. What keeps cells in tissues behaving normally in the face of myriad mutations? *BioEssays* 2006, 28: 515-524.
[98] Nathanson SD, Nelson LT, Lee M. A spontaneous subcutaneous tumor in C57BL/6 mice that metastatize to the liver. *Clin. Exp. Metastasis* 1993, 11: 45-54.
[99] Maeda S, Kamata H, Luo JL, Leffert H, Karin M. IKKβ couples hepatocyte death to cytokine-driven compensatory proliferation that promotes chemical hepatocarcinogenesis. *Cell* 2005, 121: 977-990.
[100] Laird AK, Barton AD. Cell growth and the development of tumours. *Nature* 1959, 183: 1655-1657.
[101] Rosenberg N, Jolicoeur P. Retoviral Pathogenesis. In: Retroviruses. Edited by Coffin JM, Hughes SH and Varmus H. New York: Cold Spring Harbor Laboratory Press; 1997: 475- 585.
[102] Adams JM, Cory S. Transgenic models of tumor development. *Science* 1991, 254: 1161- 1167.
[103] Suzuki A, Itami S, Ohishi M, Hamada K, Inoue T, Komazawa N, Senoo H, Sasaki T, Takeda J, Manabe M, Wah Mak T, Nakano T. Keratinocyte-specific Pten deficiency results in epidermal hyperplasia, accelerated hair follicle morphogenesis and tumor formation. *Cancer Res* 2003, 63: 674-681.
[104] Pierce GB, Speers WC. Tumors as caricatures of the process of tissue renewal: Prospects for therapy by directing differentiation. *Cancer Res.* 1988, 48: 1996-2004.
[105] Harris H. A long view of fashions in cancer research. *BioEssays* 2005, 27: 833-838.
[106] Joseph J, Dyson M. Tissue replacement in the rabbit ear. *Br.J. Surg.* 1996, 53: 372-380.
[107] Yamagiwa K, Ishikawa K. Experimentelle studie ber die pathogenese der epithelgeschwiste. *Mitt.Med.Fak* 1915, 15: 295-344.
[108] Prehn RT, Rosenfeld ME, Ross R. On the lack of an inverse correlation between epimorphioc regenerative capacity in the rabbit ear and susceptibility to carcinogenesis. *The Cancer Journal* 1998, 11: 220-222.
[109] Dawe CJ. Comparative neoplasia. In: Holland JF and Frei E, eds. Cancer Medicine. Philadelphia: Lea and Febiger, 1982: 209-256.
[110] Squires DF. Neoplasia in a coral? *Science* 1965, 148: 503-505.
[111] Madl P, Antonius A. Chapter IV. In: The silent sentinels-the demise of tropical coral reefs, 2005. Available from URL: http://www.sbg.ac.at/ ipk/avstudio/ pierofun /reefs/ch4.htm.

[112] Garzon-Ferreira J, Moreno-Bonilla M, Valderrama-Vazquez JM. Current condition of Acropora palmata and A. cervicornis coral formations in the Tayrone National Natural Park (Colombia). *Bol. Invernar.* 2004, 33: 117-136.

[113] Aronson RB, Precht WF. White-band disease and the changing face of Caribbean coral reefs. *Hidrobiologia* 2001, 460: 25-38.

[114] Formica JV. Crown gall neoplasms. In: The Pathobiology of Neoplasia. Edited by Sirica AE. New York: Plenum Press; 1989: 497-512.

[115] Bayer MH, Kaiser HE, Micozzi MS. Abnormal growth processes in plants and animals: a comparison. *In Vivo* 1994, 8: 3-16.

[116] Brencic A, Angert ER, Winans SC. Unwounded plants elicit Agobacterium *vir* induction and T-DNA transfer: transformed plant cells produce opines yet are tumor free. *Molecular Microbiology* 2005, 57: 1522-1531.

[117] Zhu J, Oger PM, Schrammeijer B, Houykaas PJJ, Farrand SK, Winans SC. The bases of crown gall tumorigenesis. *J. Bacteriology* 2000, 182: 3885-3895.

[118] Ullrich CI, Aloni R. Vascularization is a general requirement for growth of plant and animal tumors. *J. Experimental Botany* 2000, 51: 1951-1960.

[119] Galmarini CM, Galmarini FC. Multidrug resistance in cancer therapy: role of the microenvironment. *Curr. Opin. Investig. Drugs* 2003, 4: 1416-1421.

[120] Cranmer LD, Trevor KT, Hersh EM. Clinical applications of dendritic cell vaccination in the treatment of cancer. Cancer Immunol. *Immunother.* 2004, 53: 275-306.

[121] Tallman MS, Nabhan C, Feusner JH, Rowe JM. Acute promyelocytic leukemia: evolving therapeutic strategies. *Blood* 2002, 99: 759-767.

[122] Schipper H, Turley EA, Braun M. A new biological framework for cancer research. *Lancet* 1996, 348: 1149-1151.

[123] Leclerc E, Corlu A, Griscom L, Baudoin R, Legallais C. Guidance of liver and kidney organotypic cultures inside rectangular silicone microchannels. *Biomaterials* 2006, 27: 4109-4119.

[124] Sirica AE. Chronology of significant events in the study of neoplasia. In: the Pathobiology of neoplasia. Edited by: AE Sirica.New York and London, Plenum Press 1989: 1-24.

[125] Bustuoabad OD, Correa JE. Injury and death in the embryo development process: hypothesis of biological self-organization. *Frontier Perspectives* 2004, 13: 34-38.

In: Cell Growth Processes: New Research
Editor: Daiki Kimura, pp. 123-137
ISBN: 978-1-60456-132-6

Chapter IV

Implication of Asymmetric Dimethylarginine (ADMA) in Endothelial Cell Growth Process in Atherosclerosis

***Seiji Ueda*[1]*, Sho-ichi Yamagishi*[2]*, Kei Fukami*[1]*, Ayako Hayashida*[1]*, Yusuke Kaida*[1]*, Yoshimi Takamiya*[1]*, and Seiya Okuda*[1]**

Divisions of Nephrology[1] and Cardiovascular Medicine[2], Departments of Medicine, Kurume University School of Medicine, Kurume 830-0011, Japan

Abstract

Atherosclerosis is a chronic inflammatory-proliferative disease of the arterial wall characterized by progressive accumulation of lipids and inflammatory cells and enhanced proliferation and migration of smooth muscle cells. Nitric oxide (NO) is well recognized as the anti-atherogenic regulator by inhibiting the inflammatory-proliferative processes. Endothelial dysfunction with reduced nitric oxide (NO) production and/or bioavailability is common in patients with apparent coronary atherosclerosis or patients with risk factors and may contribute to the initiation and the progression of atherosclerosis. An increase of endogenous NO synthase inhibitors such as asymmetric dimethylarginine (ADMA) is now believed to be a major causative factor for the NO deficiency. Plasma ADMA levels are elevated in patients with atherosclerosis and predict future cardiovascular events, thus being a novel biomarker for atherosclerotic vascular diseases. In this review, we focus current knowledge on the role of ADMA in endothelial cell growth in atherosclerosis.

Keywords: asymmetric dimethylarginine, CVD, dimethylarginine dimethylaminohydro -lase, endothelial dysfunction, nitric oxide.

Introduction

Atherosclerosis is a chronic inflammatory-proliferative disease of the arterial wall characterized by progressive accumulation of lipids and inflammatory cells and enhanced proliferation and migration of smooth muscle cells (SMCs) [1]. Recruited SMCs or inflammatory cells to the atherosclerotic lesion produce various types of chemokines, further augmenting the inflammatory responses in arterial wall cells [2,3]. In normal vascular physiology, endothelium-derived nitric oxide (NO) is the most potent endogenous vasodilator and, by the role of its anti-inflammatory, anti-proliferation, and anti-thrombotic effects, it is widely recognized as an endogenous anti-atherogenic factor [4,5]. Endothelial dysfunction with reduced NO production and/or bioavailability is a common feature in patients with apparent coronary atherosclerosis or patients with risk factors and may contribute to the development and progression of atherosclerosis [4,5].

NO is synthesized by stereospecific oxidation of the terminal guanidine nitrogen of L-arginine by the action of the NO synthase (NOS). The synthesis of NO can be blocked by inhibition of the NOS active site with guanidino-substituted analogues of L-arginine, such as asymmetric dimethylarginine (ADMA) [6-8]. We, along with others, have demonstrated that plasma concentrations of ADMA are elevated in patients with hypertension [9], diabetes [10], hyperlipidemia [11] and renal failure [12], being one of the useful biomarkers for future cardiovascular events in these patients [10,13-15]. These observations suggest that the elevation of ADMA may account for endothelial dysfunction, thus contributing to the development and progression of CVD. In this review, we discuss current knowledge on the roles of ADMA in CVD.

Endothelial Function and ADMA

Endothelium is a flat monolayer of cells that cover vascular lumina throughout the body. It is well recognized as the principal regulator of vascular function such as vascular tone, permeability, platelets aggregation, inflammation, and SMC proliferation [4,5]. Endothelial dysfunction, defined as alterations in the normal properties of the endothelium that are inappropriate for preservation of organ function, is characterized by loss of protective endothelial characteristics [4,5]. Endothelial dysfunction is widely recognized as a key factor for the initiation and the progression of atherosclerosis and a common denominator for conditions associated with elevated cardiovascular risks, including hypertension, diabetes mellitus, dyslipidemia, obesity, smoking, and/or chronic kidney disease (CKD) [4,5]. In all of these conditions, endothelial dysfunction is characterized by decreased production and/or bioavailability of NO [4,5,16]. Furthermore, it has been shown that atherosclerosis by itself decreases biological NO activity, leading to a vicious cycle [17,18].

NO deficiency could be induced by increased reactive oxygen species (ROS) generation [17], decreased tetrahydrobiopterin (BH4), an important cofactor for NOS [18], reduced L-arginine bioavailability [19,20], and/or increased endogenous NOS inhibitors such as ADMA [6-8]. Among them, ADMA is thought to play an important role for endothelial dysfunction [6-8]. In fact, an inverse correlation between endothelial function and ADMA levels were observed in patients with essential hypertension [21] and hypercholesterolemia [11]. Further, exogenously administrated ADMA impairs endothelial function *in vivo* [12]. In addition, short-term reduction of circulating ADMA by hemodialysis is associated with amelioration of endothelial dysfunction in patients with end stage renal disease (ESRD) [22]. These observations suggest that ADMA may have an active participation in endothelial dysfunction.

Biochemical Characteristics of ADMA

There are three kinds of methylated arginines (ADMA, N^G-monomethyl-L-arginine (L-NMMA), and symmetric dimethylarginine (SDMA), an inert isomer of ADMA) *in vivo* [6-8]. The biological action of methylated arginine was first deonstrated by Hibbs *et al.* They reported in their paper that L-NMMA inhibited macrophage activation *in vitro* [23]. Further, Vallance *et al.* reported in 1992 that ADMA and L-NMMA were found in human plasma and urine, and acted as endogenous competitive inhibitors of NOS [12]. Since only minor amounts of L-NMMA are found in human plasma and that SDMA has no direct effect on NOS activity, ADMA is now thought to be a major type of endogenously generated methylated arginines that possess the inhibitory activity of NOS [6-8].

ADMA inhibits all 3 isoforms of NOS, which is reversed by excess L-arginine supplementation, and its inhibitory activity of the 3 isoforms of NOS is approximately equipotent with that of L-NMMA [6,12]. The IC_{50} is dependent on the prevailing L-arginine concentration [24] and is approximately 2-10 μM [25]. *In vitro*-studies have demonstrated that physiologically relevant levels of ADMA significantly inhibits NOS and subsequently reduces NO generation in cultured ECs and in isolated human blood vessels [26-28]. Administration of ADMA to normal rats causes an increase in renal vascular resistance and blood pressure (BP) [29,30]. Further, plasma from patients with renal failure, in whom plasma ADMA concentrations are markedly elevated, significantly inhibits NO production in cultured EC [31]. These observations suggest a biologically relevant action of ADMA *in vivo*. However, since it has been reported that circulating L-arginine levels in human plasma are in the range of 40-100 μM and that L-arginine concentration as low as 3 μM is sufficient to induce half-maximal activity of NOS [32,33], it is still questioned that 1-10 μM of circulating ADMA could actually compete with L-arginine for the NOS activity. However, Cardounel *et al.* recently demonstrated that intracellular level of ADMA was greatly higher than extracellular one, which was sufficient to inhibit endothelial NOS in cultured ECs [34]. In addition, they showed that ADMA levels in balloon injured artery were elevated by 4-folds, which caused prominent impairment of vascular relaxation. Taken together, these observations suggest endogenous ADMA is pathophysiologically relevant in CVD. Recently, chronic infusion of ADMA induces vascular lesions in eNOS knockout mice [35], thus suggesting some additional unrecognized action (NO/NOS-independent action) of ADMA *in vivo*. In addition, all three methylarginines are reported to interfere with L-arginine transport mediated by plasma membrane cationic amino acid transporter (y^+) [36]. Therefore, it is also possible that SDMA could inhibit NO generation by limiting intracellular L-arginine availability [36], although it has no direct effects on NOS activity. Indeed, Bode-Boger *et al.* recently demonstrated that SDMA dose-dependently blocked NO synthesis by cultured ECs and that plasma levels of SDMA were positively associated with extent of coronary artery disease (CAD) in patients with CAD [37].

ADMA Synthesis –Role of PRMT

Humans generate approximately 300 μmol of ADMA per day [38]. Various types of cells can generate ADMA, L-NMMA, and SDMA [6-8]. These dimethylarginines are derived from the degradation products of methylated proteins, which have been produced by protein methyltranferases (PRMTs) [39]. S-adenosylmethionine works as a methyl donor in the reactions mediated by PRMTs [39]. After proteolysis of arginine-methylated proteins, free dimethylarginines are released from the cells. To date nine PRMT genes, PRMT1-9, are reported [40-42]. There are 2 types of PRMTs: type-I (PRMT1-4, 6 and 8) catalyzes asymmetrical dimethylation and monomethylation of arginine residues and produces ADMA and L-NMMA, whereas type-II (PRMT5, 7 and 9) catalyzes symmetrical dimethylation and monomethylation of arginine residues and produces SDMA and L-NMMA [40]. Thus, type-I PRMT is thought to be mainly involved in the generation of ADMA *in vivo*. Type-I PRMTs are found in various organs including vessels, heart, and kidney [43-45]. Recently, oxidized low-density lipoprotein (LDL) was reported to increase the generation of ADMA by ECs via up-regulation of type-I PRMT genes [39]. Further, ADMA secretion from ECs is also enhanced by shear stress through the induction of PRMT genes [46]. These observations suggest that generation of ADMA may be regulated partly by type-I PRMT *in vivo*.

ADMA Degradation –Role of DDAH

It is known that more than 90% of the circulating ADMA is metabolized by the action of dimethylarginine dimethylaminohydrolase (DDAH) in rats [47]. ADMA and L-NMMA, but not SDMA, are metabolized into citrulline and dimethylamine by the action of DDAH [6-8]. DDAH is widely distributed throughout the body [48-51] and there are two isoforms of DDAH (DDAH-I and DDAH-II) [52]. DDAH-I is located in tissues expressing neuronal NOS, whereas DDAH-II predominates in tissues containing the endothelial NOS [52,53]. It has been demonstrated that 4124W, an inhibitor of DDAH, inhibits methylarginine metabolism and increases ADMA sufficiently enough to inhibit endothelial NOS [54]. In addition, we have previously reported that DDAH overexpression decreases the production of ADMA by cultured vascular smooth muscle cells, thus resulting in the enhancement of NO generation via inducible NOS [55]. In DDAH-I$^{+/-}$ mice (DDAH activity is about 50% of that in control mice), endothelial dysfunction associated with increases in plasma ADMA and blood pressure (BP) levels was observed [53], thus suggesting the active participation of the ADMA-DDAH axis in endothelial function. Furthermore, it has been reported that the impaired metabolism of ADMA by DDAH is associated with elevated ADMA levels in hypercholesterolemia and diabetes [48,56]. Taken together, these observations suggest that decreased DDAH action may be one of the main mechanisms for the elevation of ADMA levels in patients with risk factors for CVD.

ADMA and ECs

NO not only inhibits apoptotic cell death, but also increases growth and migration of ECs [57-62], thus contributing to maintain vascular homeostasis. Prolonged and/or repeated NO deficiency by cardiovascular risk factors can ultimately exhaust the protective effect of endogenous anti-inflammatory systems within ECs [5,16]. As a consequence, the endothelium not only becomes dysfunctional, but ECs can also lose integrity, progress to senescence, and detach into the circulation [63]. Endothelial integrity depends on the extent of injury as well as the endogenous capacity for repair. Two mechanisms by which this process of repair occurs have been recently identified. Adjacent mature endothelial cells can replicate locally, and replace the lost and damaged cells. A recent study suggested that, although local ECs would be sufficient to maintain vascular integrity throughout life in healthy circumstances, in the presence of risk factors, loss of endothelial integrity would rapidly develop if local replication were the only repair mechanism [64]. More recently, it has been demonstrated that circulating endothelial progenitor cells (EPCs) are an alternative mechanism for maintenance and repair of the endothelium [65]. These cells are recruited from the bone marrow, circulate in the peripheral blood, and can differentiate into mature cells with endothelial characteristics. Mobilization of these cells is in part NO-dependent, and may thus be impaired in patients with cardiovascular risk factors [66]. Since ADMA is a pivotal endogenous NOS inhibitor and its plasma levels are elevated in patients with cardiovascular risk factors, it is conceivable that endogenously accumulated ADMA in CVD patients may play a role in the development of vascular injury via decreasing endothelial repair. Indeed, an inverse correlation between plasma levels of ADMA and circulating EPCs was observed in patients with coronary artery disease [67]. In addition, transfection of DDAH into cultured ECs enhances VEGF mRNA expression and stimulates tube formation of ECs [68]. DDAH overexpression promotes tumor angiogenesis and glioma growth *in vivo* [69]. Furthermore, in a murine model of hindlimb ischemia, enhanced neovascularization and limb perfusion were observed in DDAH transgenic mice, which were associated with reduced plasma levels of ADMA [70]. When the femoral arteries were injured by a straight spring wire, enhanced regeneration of the EC monolayer was also observed in DDAH transgenic mice [71]. These observations suggest that the ADMA-DDAH axis may play a role in angiogenic responses and/or endothelial repair, thus being involved in the pathogenesis of vascular injury in atherosclerosis.

Biological Effects of ADMA

ADMA elevation has been reported in a wide range of cardiovascular disorders [6-12]. In this section, we discuss the biological effects of ADMA.

(1) CVD

We, along with others, have demonstrated that high levels of ADMA are observed in patients with hypertension [9], hypercholesterolemia [10], diabetes [11], and CKD [12,72]. Furthermore, we have previously shown that there was a close association between plasma ADMA levels and carotid intima-media thickness, one of the surrogate markers of atherosclerosis in general population [9]. In addition, ADMA is a strong prognostic marker of cardiovascular events in patients with coronary artery disease [13,14], diabetes [10] and ESRD [15]. These observations suggest that the elevation of ADMA could be a novel biomarker for atherosclerosis and contribute to accelerated atherosclerosis by inactivating NOS in patients with coronary risk factors.

(2) Hypertension

There is a growing body of evidence to show that NO plays an important role in the regulation of vascular tonus and BP [73,74]. There are two possible mechanisms by which endogenous ADMA is involved in the pathogenesis of hypertension; 1) ADMA may exert vasoconstrictor/pressor effects by inhibiting endothelial NOS activity [38,75]; and 2) ADMA may inhibit renal sodium excretion by reducing NO synthesis in the kidney [76-78]. Indeed, increased urinary levels of ADMA were observed in Dahl salt sensitive rats, which was associated with an increase in BP [79] Further, BP levels were significantly lower in DDAH-transgenic mice than those in wild type mice [80], whereas the levels of BP were higher in DDAH-$I^{+/-}$ mice, compared with control mice [53]. In human, we have recently found that the plasma levels of ADMA are associated with mean BP levels in healthy subjects [9]. Further, we have recently reported that plasma levels of ADMA are strongly correlated to BP levels in a rat remnant kidney model, a CKD rat model [81]. In addition, adenovirus-mediated DDAH gene transfer was found to decrease plasma levels of ADMA and subsequently prevent the elevation of BP levels in this model [81]. These observations further support the pathological relevance of the increased plasma ADMA levels in hypertension.

(3) Diabetes

CVD is responsible for about 70 % of all deaths in patients with type 2 diabetes [82]. In the Framingham study, the incidence of CVD was 2-4 times greater in diabetic patients than in the general population [83]. The levels of conventional risk factors for atherosclerosis were certainly increased in diabetic patients, but not enough to explain the exaggerated risk for macrovascular complications in this population [84]. Plasma ADMA levels were elevated in animal models of diabetes [48,85] and in patients with impaired glucose tolerance [9], insulin resistance [86] or diabetes [10,87]. It has been also reported that elevated plasma levels of ADMA are associated with increased risks of nonfatal stroke and myocardial infarction in patients with early diabetic nephropathy [10]. These findings suggest that the elevated ADMA in diabetes could contribute to accelerated atherosclerosis in this population.

(4) Hypercholesterolemia

Plasma ADMA levels were also elevated in hyperlipidemic animals and patients [11,88]. *In vitro*-experiments have suggested the involvement of ROS in the elevation of ADMA levels in hypercholesterolemia. Indeed, oxidized LDL was shown not only to stimulate the expression of PRMTs but also suppress the activity of DDAH through ROS generation [39,56]. Therefore, although it is still controversial, lipid lowing agents, 3-hydroxy-3-methylglutaryl coenzyme A reductase inhibitors (statins), could decrease the ADMA levels in hypercholesterolemic patients, at least in part, through its anti-oxidative properties [89-93].

(5) Heart failure

We, along with others, have previously shown that ADMA levels were elevated in patients or animal models with heart failure [94-96]. ADMA levels were associated with left ventricular hypertrophy and dysfunction in patients with ESRD as well [97]. Since an intravenous injection of ADMA significantly reduced heart rates and cardiac output in healthy subjects [38], ADMA may be one of the causative factors for heart failure in humans.

Whether heart failure itself may affect synthesis or metabolism of ADMA remains to be clarified. However, the reduction in renal clearance or metabolism of ADMA may account for the elevation of ADMA in patients with heart failure because renal excretory function deteriorates as heart failure progress [98]. Moreover, recently, DDAH activity was shown to be reduced in coronary endothelium, smooth muscles, and cardiac myocytes in an animal model of heart failure [99]. A positive feedback loop may exist between heart failure and ADMA levels.

(6) CKD

Plasma levels of ADMA are elevated in patients with ESRD [12] as well as mild to moderate CKD [72]. Elevation of plasma ADMA levels may be a reason for the high prevalence of CVD in patients with CKD [7,8]. In fact, it has been reported that plasma ADMA was strongly associated with carotid intima-media thickness [100], left ventricular htpertrophy [97], cardiovascular complication [15], and mortality [15] in patients with ESRD. These observations suggest that ADMA may have an active participation in the development and progression of atherosclerosis in patients with ESRD.

Both hypertension and proteinuria are well-recognized risk factors for the progression of CKD [101]. As mentioned above, ADMA plays an important role in the pathogenesis of hypertension. In addition, a positive correlation between ADMA and proteinuria was observed in patients with CKD [102]. Therefore, it is conceivable that ADMA may be involved in CKD progression by BP- and/or proteinuria-dependent mechanisms. In fact, plasma level of ADMA is a strong predictor for the progression of renal dysfunction in patients with CKD [103,104]. NOS inhibition may accelerate the progression of renal injury by impairing the angiogenic responses and subsequent loss of peritubular capillaries via suppression of NO [105,106]. In addition, we have recently found that plasma levels of ADMA are associated with decreased number of peritubular capillaries, enhanced tubulointerstitial fibrosis and progressive loss of renal function in the remnant kidney model, all of which were prevented by DDAH overexpression-elicited ADMA reduction [107]. These results suggest that accumulated ADMA may play important roles in not only accelerated atherosclerosis but also renal damage progression in patients with CKD.

Conclusion

ADMA is a circulating endogenous NOS inhibitor and is now considered one of the strongest biomarkers of CVD. Suppression of ADMA by enhancement of DDAH, a key limiting enzyme for ADMA, or inhibition of PRMT may be a novel strategy for preventing the accelerated atherosclerosis in patients with coronary risk factors, at least in part by modulating the impaired endothelial cell growth process in this devastating disorder.

References

[1] Ross R. Atherosclerosis--an inflammatory disease. *N Engl J Med.* 1999; 340: 115-26.

[2] Braun M, Pietsch P, Schror K, Baumann G, Felix SB. Cellular adhesion molecules on vascular smooth muscle cells. *Cardiovasc. Res.* 1999; 41: 395-401.

[3] Marumo T, Schini-Kerth VB, Fisslthaler B, Busse R. Platelet-derived growth factor-stimulated superoxide anion production modulates activation of transcription factor NF-kappaB and expression of monocyte chemoattractant protein 1 in human aortic smooth muscle cells. *Circulation* 1997: 96; 2361-7.

[4] Davignon J, Ganz P. Role of endothelial dysfunction in atherosclerosis. *Circulation.* 2004; 109: III27-32.

[5] Deanfield JE, Halcox JP, Rabelink TJ. Endothelial function and dysfunction: testing and clinical relevance. *Circulation.* 2007; 115: 1285-95.

[6] Cooke JP. Does ADMA cause endothelial dysfunction? *Arterioscler. Thromb. Vasc. Biol.* 2000; 20: 2032-7.

[7] Ueda S, Yamagishi S, Matsumoto Y, Fukami K, Okuda S. Asymmetric dimethylarginine (ADMA) is a novel emerging risk factor for cardiovascular disease and the development of renal injury in chronic kidney disease. *Clin. Exp. Nephrol.* 2007 ; 11: 115-21.

[8] Ueda S, Yamagishi S, Kaida Y, Okuda S. Asymmetric dimethylarginine (ADMA) may be a missing link between cardiovascular disease and chronic kidney disease. *Nephrology (Carlton)* 2007; 12: 582-90.

[9] Miyazaki H, Matsuoka H, Cooke JP, Usui M, Ueda S, et al. Endogenous nitric oxide synthase inhibitor: a novel marker of atherosclerosis. *Circulation* 1999; 99:1141-6.

[10] Tarnow L, Hovind P, Teerlink T, Stehouwer CD, Parving HH. Elevated plasma asymmetric dimethylarginine as a marker of cardiovascular morbidity in early diabetic nephropathy in type 1 diabetes. *Diabetes Care* 2004; 27: 765-9.

[11] Boger RH, Bode-Boger SM, Szuba A, Tsao PS, Chan JR, et al. Asymmetric dimethylarginine (ADMA): a novel risk factor for endothelial dysfunction: its role in hypercholesterolemia. *Circulation* 1998; 98: 1842-7.

[12] Vallance P, Leone A, Calver A, Collier J, Moncada S. Accumulation of an endogenous inhibitor of nitric oxide synthesis in chronic renal failure. *Lancet* 1992; 339: 572-5.

[13] Valkonen VP, Paiva H, Salonen JT, Lakka TA, Lehtimaki T, et al. Risk of acute coronary event and serum concentration of asymmetrical dimethylarginine. *Lancet* 2001; 358: 2127-8.

[14] Schnabel R, Blankenberg S, Lubos E, Lackner KJ, Rupprecht HJ, et al. Asymmetric dimethylarginine and the risk of cardiovascular events and death in patients with coronary artery disease: results from the AtheroGene Study. *Circ Res.* 2005; 97: e53-9.

[15] Zoccali C, Bode-Boger S, Mallamaci F, Benedetto F, Tripepi G, et al. Plasma concentration of asymmetrical dimethylarginine and mortality in patients with end-stage renal disease: a prospective study. *Lancet* 2001; 358: 2113-7.

[16] Matsuoka H. Endothelial dysfunction associated with oxidative stress in human. *Diabetes Res. Clin .Pract.* 2001; 54: S65-72.

[17] Cooke JP, Dzau VJ. Nitric oxide synthase: role in the genesis of vascular disease. *Annu. Rev. Med.*, 1997; 48: 489–509.

[18] Ueda S, Matsuoka H, Miyazaki H, Usui M, Okuda S, Imaizumi T. Tetrahydrobiopterin restores endothelial function in long-term smokers. *J. Am. Coll. Cardiol.* 2000; 35: 71-5.

[19] Schwartz IF, Ayalon R, Chernichovski T, Reshef R, Chernin G, et al. Arginine uptake is attenuated through modulation of cationic amino-acid transporter-1, in uremic rats. *Kidney Int.* 2006; 69: 298-303.

[20] Boger RH. The pharmacodynamics of L-arginine. *J. Nutr.* 2007; 137: 1650S-5.

[21] Perticone F, Sciacqua A, Maio R, Perticone M, Maas R, et al. Asymmetric dimethylarginine, L-arginine, and endothelial dysfunction in essential hypertension. *J. Am. Coll. Cardiol.* 2005; 46: 518-23.

[22] Cross JM, Donald A, Vallance P, Deanfield JE, Woolfson RG., MacAllister RJ. Hemodialysis improves endothelial function in humans. *Nephrol. Dial. Transplant.* 2001; 16: 1823-9.

[23] Hibbs JB Jr, Vavrin Z, Taintor RR. L-arginine is required for expression of the activated macrophage effector mechanism causing selective metabolic inhibition in target cells. *J. Immunol.* 1987; 138: 550-65.

[24] MacAllister RJ, Whitley GS, Vallance P. Effects of guanidino and uremic compounds on nitric oxide pathways. *Kidney Int.* 1994; 45: 737-42.

[25] Boger RH. The emerging role of asymmetric dimethylarginine as a novel cardiovascular risk factor. *Cardiovasc. Res.* 2003; 59: 824-33.

[26] Faraci FM, Brian JE Jr, Heistad DD. Response of cerebral blood vessels to an endogenous inhibitor of nitric oxide synthase. *Am. J. Physiol.* 1995; 269 :H1522-7.

[27] Segarra G, Medina P, Ballester RM, Lluch P, Aldasoro M, et al. Effects of some guanidino compounds on human cerebral arteries. *Stroke* 1999; 30: 2206-10.

[28] Segarra G, Medina P, Vila JM, Chuan P, Domenech C, et al. Inhibition of nitric oxide activity by arginine analogs in human renal arteries. *Am. J. Hypertens*. 2001; 14: 1142-8.

[29] Gardiner SM, Kemp PA, Bennett T, Palmer RM, Moncada S. Regional and cardiac haemodynamic effects of NG, NG,dimethyl-L-arginine and their reversibility by vasodilators in conscious rats. *Br. J. Pharmacol.* 1993; 110: 1457-64.

[30] Jin JS, D'Alecy LG. Central and peripheral effects of asymmetric dimethylarginine, an endogenous nitric oxide synthetase inhibitor. *J. Cardiovasc. Pharmacol.* 1996; 28: 439-46.

[31] Xiao S, Wagner L, Schmidt RJ, Baylis C. Circulating endothelial nitric oxide synthase inhibitory factor in some patients with chronic renal disease. *Kidney Int.* 2001; 59: 1466-72.

[32] Pollock JS, Forstermann U, Mitchell JA, Warner TD, Schmidt HH, et al. Purification and characterization of particulate endothelium-derived relaxing factor synthase from cultured and native bovine aortic endothelial cells. *Proc. Natl. Acad. Sci. U S A.* 1991; 88: 10480-4.

[33] Jeserich M, Munzel T, Just H, Drexler H. Reduced plasma L-arginine in hypercholesterolaemia. *Lancet.* 1992; 339: 561.

[34] Cardounel AJ, Cui H, Samouilov A, Johnson W, Kearns P, et al. Evidence for the pathophysiological role of endogenous methylarginines in regulation of endothelial NO production and vascular function. *J. Biol. Chem.* 2007 ; 282: 879-87

[35] Suda O, Tsutsui M, Morishita T, Tasaki H, Ueno S, et al. Asymmetric dimethylarginine produces vascular lesions in endothelial nitric oxide synthase-deficient mice: involvement of renin-angiotensin system and oxidative stress. Arterioscler *Thromb. Vasc. Biol.* 2004; 24: 1682-8.

[36] Closs EI, Basha FZ, Habermeier A, Forstermann U. Interference of L-arginine analogues with L-arginine transport mediated by the y+ carrier hCAT-2B. *Nitric. Oxide.* 1997; 1: 65-73.

[37] Bode-Boger SM, Scalera F, Kielstein JT, Martens-Lobenhoffer J, Breithardt G, et al. Symmetrical dimethylarginine: a new combined parameter for renal function and extent of coronary artery disease. *J. Am. Soc. Nephrol.* 2006; 17: 1128-34.

[38] Achan V, Broadhead M, Malaki M, Whitley G, Leiper J, et al. Asymmetric dimethylarginine causes hypertension and cardiac dysfunction in humans and is actively metabolized by dimethylarginine dimethylaminohydrolase. *Arterioscler. Thromb. Vasc. Biol.* 2003; 23: 1455–9.

[39] Boger RH, Sydow K, Borlak J, Thum T, Lenzen H, et al. LDL cholesterol upregulates synthesis of asymmetrical dimethylarginine in human endothelial cells: involvement of S-adenosylmethionine-dependent methyltransferases. *Circ. Res.* 2000; 87: 99-105.

[40] Anthony S, Leiper J, Vallance P. Endogenous production of nitric oxide synthase inhibitors. *Vascular Medicine* 2005; 10: S3-9.

[41] Lee J, Sayegh J, Daniel J, Clarke S, Bedford MT. PRMT8, a new membrane-bound tissue-specific member of the protein arginine methyltransferase family. *J. Biol. Chem.* 2005; 280: 32890-6.

[42] Cook JR, Lee JH, Yang ZH, Krause CD, Herth N, et al. FBXO11/PRMT9, a new protein arginine methyltransferase, symmetrically dimethylates arginine residues. *Biochem. Biophys. Res. Commun.* 2006; 342: 472-81.

[43] Rawal N, Rajpurohit R, Paik WK, Kim S. Purification and characterization of S-adenosylmethionine-protein-arginine N-methyltransferase from rat liver. *Biochem. J.* 1994; 300: 483–9.

[44] Ghosh SK, Paik WK, Kim S. Purification and molecular identification of two protein methylase I from calf brain: myelin basic protein- and histone-specific enzyme. *J. Biol. Chem.* 1988; 263: 19024-33.

[45] Gary JD, Clarke S. RNA and protein interactions modulated by protein arginine methylation. P*rog. Nucl. Acid. Res. Mol. Biol.* 1998; 61: 65-131.

[46] Osanai T, Saitoh M, Sasaki S, Tomita H, Matsunaga T, Okumura K. Effect of shear stress on asymmetric dimethylarginine release from vascular endothelial cells. *Hypertension* 2003; 42: 985–90.

[47] Ogawa T, Kimoto M, Sasaoka K. Occurrence of a new enzyme catalyzing the direct conversion of NG, NG-dimethyl-L-arginine to L-citrulline in rats. *Biochem. Biophys. Res. Commun.* 1987; 148: 671–7.

[48] Lin KY, Ito A, Asagami T, Tsao PS, Adimoolam S, et al. Impaired nitric oxide synthase pathway in diabetes mellitus role of asymmetric dimethylarginine and dimethylarginine dimethylaminohydrolase. *Circulation* 2002; 106: 987-92.

[49] MacAllister RJ, Fickling SA, Whitley GS, Vallance P. Metabolism of methylarginines by human vasculature; implications for the regulation of nitric oxide synthesis. *Br. J. Pharmacol.* 1994; 112: 43-8.

[50] Kimoto M, Whitley GS, Tsuji H, Ogawa T. Detection of N^G,N^G-dimethylarginine dimethylaminohydrolase in human tissues using a monoclonal antibody. *J. Biochem.* Tokyo 1995; 117: 237-8.

[51] Kimoto M, Miyatake S, Sasagawa T, Yamashita H, Okita M, et al. Purification, cDNA cloning and expression of human N^G,N^G-dimethylarginine dimethylaminohydrolase. Eur *J. Biochem.* 1998; 258: 863-8.

[52] Leiper JM, Santa Maria J, Chubb A, MacAllister RJ, Charles IG, et al. Identification of two human dimethylarginine dimethylaminohydrolases with distinct tissue distributions and homology with microbial arginine deaminases. *Biochem. J.* 1999; 343: 209-14.

[53] Leiper J, Nandi M, Torondel B, Murray-Rust J, Malaki M, et al. Disruption of methylarginine metabolism impairs vascular homeostasis. *Nat. Med.* 2007; 13: 198-203.

[54] MacAllister RJ, Parry H, Kimoto M, Ogawa T, Russell RJ, et al. Regulation of nitric oxide synthesis by dimethylarginine dimethylaminohydrolase. *Br. J. Pharmacol.* 1996; 119: 1533-40.

[55] Ueda S, Kato S, Matsuoka H, Kimoto M, Okuda S, et al. Regulation of cytokine-induced nitric oxide synthesis by asymmetric dimethylarginine: role of dimethylarginine dimethylaminohydrolase. *Circ. Res.* 2003; 92: 226-33.

[56] Ito A, Tsao PS, Adimoolam S, Kimoto M, Ogawa T, Cooke JP. Novel mechanism for endothelial dysfunction. Dysregulation of dimethylarginine dimethylaminohydrolase. *Circulation* 1999; 99: 3092-5.

[57] Dimmeler S, Hermann C, Galle J, Zeiher AM. Upregulation of superoxide dismutase and nitric oxide synthase mediates the apoptosis-suppressive effects of shear stress on endothelial cells. *Arterioscler. Thromb. Vasc. Biol.* 1999; 19: 656–64.

[58] Rossig L, Fichtlscherer B, Breitschopf K, Haendeler J, Zeiher AM, et al. Nitric oxide inhibits caspase-3 by S-nitrosation in vivo. *J. Biol. Chem.* 1999;274:6823– 6.

[59] Morbidelli L, Chang CH, Douglas JG, Granger HJ, Ledda F, Ziche M. Nitric oxide mediates mitogenic effect of VEGF on coronary venular endothelium. *Am. J. Physiol.* 1996; 270: H411–5.

[60] Ziche M, Parenti A, Ledda F, Dell'Era P, Granger HJ, et al. Nitric oxide promotes proliferation and plasminogen activator production by coronary venular endothelium through endogenous bFGF. *Circ. Res.* 1997; 80: 845–52.

[61] Ziche M, Morbidelli L, Masini E, Amerini S, Granger HJ, et al. Nitric oxide mediates angiogenesis in vivo and endothelial cell growth and migration in vitro promoted by substance P. *J. Clin. Invest.* 1994; 94: 2036 –44.

[62] Murohara T, Witzenbichler B, Spyridopoulos I, Asahara T, Ding B, et al. Role of endothelial nitric oxide synthase in endothelial cell migration. *Arterioscler. Thromb. Vasc. Biol.* 1999;19:1156 –61.

[63] Woywodt A, Bahlmann FH, De Groot K, Haller H, Haubitz M. Circulating lating endothelial cells: life, death, detachment and repair of the endothelial cell layer. *Nephrol. Dial. Transplant.* 2002; 17: 1728 –30.

[64] Op den Buijs J, Musters M, Verrips T, Post JA, Braam B, Van Riel N. Mathematical modeling of vascular endothelial layer maintenance: the role of endothelial cell division, progenitor cell homing and telomere shortening. *Am. J. Physiol. Heart Circ. Physiol.* 2004; 287: H2651–8.

[65] Asahara T, Murohara T, Sullivan A, Silver M, van der Zee R, et al. Isolation of putative progenitor endothelial cells for angiogenesis. *Science.* 1997; 275: 964–7.

[66] Aicher A, Heeschen C, Mildner-Rihm C, Urbich C, Ihling C, et al. Essential role of endothelial nitric oxide synthase for mobilization of stem and progenitor cells. *Nat. Med.* 2003; 9: 1370–6.

[67] Thum T, Tsikas D, Stein S, Schultheiss M, Eigenthaler M, et al. Suppression of endothelial progenitor cells in human coronary artery disease by the endogenous nitric oxide synthase inhibitor asymmetric dimethylargine. *J. Am. Coll. Cardiol.* 2005; 46: 1693-701.

[68] Smith CL, Birdsey GM, Anthony S, Arrigoni FI, Leiper JM, Vallance P. Dimethylarginine dimethylaminohydrolase activity modulates ADMA levels, VEGF expression, and cell phenotype. Biochem Biophys Res Commun. 2003; 308: 984-9.

[69] Kostourou V, Robinson SP, Cartwright JE, Whitley GS. Dimethylarginine dimethylaminohydrolase I enhances tumour growth and angiogenesis. *Br. J. Cancer.* 2002; 87: 673-80.

[70] Jacobi J, Sydow K, von Degenfeld G, Zhang Y, Dayoub H, et al. Overexpression of dimethylarginine dimethylaminohydrolase reduces tissue asymmetric dimethylarginine levels and enhances angiogenesis. *Circulation*. 2005; 111:1431-8.

[71] Konishi H, Sydow K, Cooke JP. Dimethylarginine dimethylaminohydrolase promotes endothelial repair after vascular injury. J Am Coll Cardiol. 2007; 49: 1099-105.

[72] Kielstein JT, Boger RH, Bode-Boger SM, Frolich JC, Haller H, et al. Marked increase of asymmetric dimethylarginine in patients with incipient primary chronic renal disease. *J. Am. Soc. Nephrol.* 2002; 13: 170-6.

[73] Dominiczak AF, Bohr DF. Nitric oxide and its putative role in hypertension. *Hypertension* 1995; 25: 1202-11.

[74] Arnal J-F, Michel J-B, Harrison DG. Nitric oxide in the pathogenesis of hypertension. *Curr. Opin. Nephrol. Hypertens* 1995; 4: 182-8.

[75] MacAllister RJ, Vallance P. Nitric oxide in essential and renal hypertension. *J. Am. Soc. Nephrol.* 1994; 5: 1057-65.

[76] Bech JN, Nielsen CB, Pedersen EB. Effects of systemic NO synthesis inhibition on RPF, GFR, Una, and vasoactive hormones in healthy humans. *Am. J. Physiol.* 1996; 270: F845–51.

[77] Ruilope LM, Lahera VK Rodicio JL, Romero JC. Participation of nitric oxide in the regulation of renal function: possible role in the genesis of arterial hypertension. *J. Hypertens.* 1994; 12: 625–31.

[78] Kielstein JT, Simmel S, Bode-Boger SM, Roth HJ, Schmidt-Gayk H, et al. Subpressor dose asymmetric dimethylarginine modulates renal function in humans through nitric oxide synthase inhibition. *Kidney Blood Press Res.* 2004; 27: 143-7.

[79] Matsuoka H, Itoh S, Kimoto M, Kohno K, Tamai O, et al. Asymmetrical dimethylarginine, an endogenous nitric oxide synthase inhibitor, in experimental hypertension. *Hypertension* 1997; 29: 242–7.

[80] Dayoub H, Achan V, Adimoolam S, Jacobi J, Stuehlinger MC, et al. Dimethylarginine dimethylaminohydrolase regulates nitric oxide synthesis: genetic and physiological evidence. *Circulation* 2003; 108: 3042–7.

[81] Matsuguma K, Ueda S, Yamagishi S, Matsumoto Y, Kaneyuki U, et al. Molecular mechanism for elevation of asymmetric dimethylarginine and its role for hypertension in chronic kidney disease. *J. Am. Soc. Nephrol.* 2006; 17: 2176-83.

[82] Laakso M. Hyperglycemia and cardiovascular disease in type 2 diabetes. *Diabetes* 1999; 48: 937-42.

[83] Haffner SM, Lehto S, Ronnemaa T, Pyorala K, Laakso M. Mortality from coronary heart disease in subjects with type 2 diabetes and in nondiabetic subjects with and without prior myocardial infarction. *N. Engl. J. Med.* 1998; 339: 229-34.

[84] Standl E, Balletshofer B, Dahl B, Weichenhain B, Stiegler H, et al. Predictors of 10-year macrovascular and overall mortality in patients with NIDDM: the Munich General Practitioner Project. *Diabetologia* 1996; 39: 1540-5.

[85] Xiong Y, Fu YF, Fu SH, Zhou HH. Elevated levels of the serum endogenous inhibitor of nitric oxide synthase and metabolic control in rats with streptozotocin-induced diabetes. *J. Cardiovasc. Pharmacol.* 2003; 42:191-6.

[86] Stuhlinger MC, Abbasi F, Chu JW, et al. Relationship between insulin resistance and an endogenous nitric oxide synthase inhibitor. *JAMA* 2002; 287:1420-6.

[87] Paiva H, Lehtimaki T, Laakso J, Ruokonen I, Rantalaiho V, et al. Plasma concentrations of asymmetric-dimethyl-arginine in type 2 diabetes associate with glycemic control and glomerular filtration rate but not with risk factors of vasculopathy. *Metabolism* 2003; 52: 303-7.

[88] Boger RH, Bode-Boger SM, Sydow K, Heistad DD, Lentz SR. Plasma concentration of asymmetric dimethylarginine, an endogenous inhibitor of nitric oxide synthase, is elevated in monkeys with hyperhomocyst(e)inemia or hypercholesterolemia. Arterioscler. *Thromb. Vasc. Biol.* 2000; 20: 1557–64.

[89] Tsiara S, Elisaf M, Mikhailidis DP. Early vascular benefits of statin therapy. *Curr. Med. Res. Opin.* 2003; 19: 540-56.

[90] Tsubouchi H, Inoguchi T, Sonta T, Sato N, Sekiguchi N, et al. Statin attenuates high glucose-induced and diabetes-induced oxidative stress in vitro and in vivo evaluated by electron spin resonance measurement. *Free Radic. Biol. Med.* 2005; 39: 444-52.

[91] Jiang JL, Jiang DJ, Tang YH, Li NS, Deng HW, Li YJ. Effect of simvastatin on endothelium-dependent vasorelaxation and endogenous nitric oxide synthase inhibitor. *Acta Pharmacol. Sin.* 2004 ; 25: 893-901.

[92] Paiva H, Laakso J, Lehtimaki T, Isomustajarvi M, Ruokonen I, Laaksonen R. Effect of high-dose statin treatment on plasma concentrations of endogenous nitric oxide synthase inhibitors. *J. Cardiovasc. Pharmacol.* 2003; 41: 219-22.

[93] Lu TM, Ding YA, Leu HB, Yin WH, Sheu WH, Chu KM. Effect of rosuvastatin on plasma levels of asymmetric dimethylarginine in patients with hypercholesterolemia. *Am. J. Cardiol.* 2004; 94: 157-61.

[94] Usui M, Matsuoka H, Miyazaki H, Ueda S, Okuda S, et al. Increased endogenous nitric oxide synthase inhibitor in patients with congestive heart failure. *Life Sci.* 1998; 62: 2425–30.

[95] Feng Q, Lu X, Fortin AJ, Pettersson A, Hedner T, et al. Elevation of an *endogenous inhibitor of nitric oxide synthesis in experimental congestive heart failure. Cardiovasc. Res.* 1998; 37: 667–75.

[96] Saitoh M, Osanai T, Kamada T, Matsunaga T, Ishizaka H, et al. High plasma level of asymmetric dimethylarginine in patients with acutely exacerbated congestive heart failure: role in reduction of plasma nitric oxide level. *Heart Vessels* 2003; 18: 177–82.

[97] Zoccali C, Mallamaci F, Maas R, Benedetto FA, Tripepi G, et al. Left ventricular hypertrophy, cardiac remodeling and asymmetric dimethylarginine (ADMA) in hemodialysis patients. *Kidney Int.* 2002; 62: 339-45.

[98] Harris P. Congestive cardiac failure: central role of the arterial blood pressure. *Br. Heart J.* 1987; 58: 190-203.

[99] Chen Y, Li Y, Zhang P, Traverse JH, Hou M, et al. Dimethylarginine dimethylaminohydrolase and endothelial dysfunction in failing hearts. *Am. J. Physiol. Heart Circ. Physiol*.2005; 289: H2212-9.

[100] Zoccali C, Benedetto FA, Maas R, Mallamaci F, Tripepi G, et al. Asymmetric dimethylarginine, C-reactive protein, and carotid intima-media thickness in end-stage renal disease. *J. Am.Soc. Nephrol.* 2002; 13: 490-6.

[101] Jafar TH, Stark PC, Schmid CH, Landa M, Maschio G, et al. Progression of chronic kidney disease: the role of blood pressure control, proteinuria, and angiotensin-converting enzyme inhibition: a patient-level meta-analysis. *Ann. Intern. Med.* 2003; 139: 244-52.

[102] Caglar K, Yilmaz MI, Sonmez A, Cakir E, Kaya A, et al. ADMA, proteinuria, and insulin resistance in non-diabetic stage I chronic kidney disease. *Kidney Int.* 2006; 70: 781-7.

[103] Fliser D, Kronenberg F, Kielstein JT, Morath C, Bode-Boger SM, et al. Asymmetric dimethylarginine and progression of chronic kidney disease: the mild to moderate kidney disease study. *J. Am. Soc. Nephrol.* 2005; 16: 2456-61.

[104] Ravani P, Tripepi G, Malberti F, Testa S, Mallamaci F, Zoccali C. Asymmetrical dimethylarginine predicts progression to dialysis and death in patients with chronic kidney disease: a competing risks modeling approach. *J. Am. Soc. Nephrol.* 2005; 16: 2449-55.

[105] Kang DH, Nakagawa T, Feng L, Johnson RJ. Nitric oxide modulates vascular disease in the remnant kidney model. *Am. J. Pathol.* 2002; 161: 239-48.

[106] Erdely A, Wagner L, Muller V, Szabo A, Baylis C. Protection of Wistar Furth rats from chronic renal disease is associated with maintained renal nitric oxide synthesis. *J. Am. Soc. Nephrol.* 2003; 14: 2526-33.

[107] Matsumoto Y, Ueda S, Yamagishi S, Matsuguma K, Shibata R, et al. Dimethylarginine dimethylaminohydrolase prevents progression of renal dysfunction by inhibiting loss of peritubular capillaries and tubulointerstitial fibrosis in a rat model of chronic kidney disease. *J. Am. Soc. Nephrol.* 2007; 18: 1525-33.

In: Cell Growth Processes: New Research
Editor: Daiki Kimura, pp. 139-157
ISBN: 978-1-60456-132-6

Chapter V

Epidermal Growth Factor Receptor Dependent Mitogenic Signaling in Normal and Malignant Mammary Epithelial Cells

Paul W. Sylvester, Sunitha V. Bachawal, Vikram B. Wali, and Amit B. Shirode
College of Pharmacy, University of Louisiana at Monroe,
700 University Ave, Monroe, LA 71209-0470, USA

Abstract

The EGF-receptor belongs to the HER (human) or ErbB (mouse) family of receptor tyrosine kinases that includes four members (HER1-4/ErbB1-4). Activation of these receptors by their appropriate ligands will induce the formation of homodimers and/or heterodimers. Heterodimer formation between different EGF-receptor family members has been shown to provide greater signal duration, diversification and amplification, as compared to their corresponding homodimer receptor complex. This is particularly true for the HER2/ErbB2 and the HER3/ErbB3 receptors because the HER2/ErbB2 receptor has no known ligand, whereas the HER3/ErbB3 receptor has no intrinsic tyrosine kinase activity, but heterodimer formation thereby allows these EGF-receptor family members to participate in mitogen-dependent signal transduction. Studies have also shown that the most potent EGF-receptor receptor complex for stimulating cell growth and transformation is composed of HER3/ErbB3 with either HER1/ErbB1 or HER2/ErbB2. Following ligand binding and homo- or heterodimerization, the EGF-R becomes transphosphorylated by its dimer partner, and the tyrosine phosphorylated EGF-receptor dimer is now able to interact and activate several important intracellular signaling proteins involved in the activation of several mitogenic signaling pathways including the mitogen-activated protein kinase (MAPK) cascade or Ras/Raf/MEK/ERK; the phosphatidylinositol 3-kinase (PI3K)/PI3K-dependent kinase (PDK)/Akt pathway; and the cAMP-dependent protein kinase (PKA) pathway. For example, studies have shown that HER3/ErbB3 heterodimers are the most important and potent activators of the

PI3K/PDK/Akt mitogenic pathway in neoplastic mammary epithelial cells. Excessive Akt signaling promotes oncongenic expression by stimulating tumor cell proliferation, transformation and metastasis, and overexpression of constitutively active Akt has been shown to reverse the proliferative block induced by treatments targeted at inhibiting EGF-receptor mitogenic signaling in neoplastic mammary epithelial cells. In addition, these distinct mitogenic signaling pathways can greatly influence the activation and intensity of the other signaling pathways through the process of cross-talk. This review will describe the role of EGF-receptor family members in the activation and modulation of various mitogenic signaling pathways in normal mammary epithelial cells, and how alterations in the expression or activation of these EGF-receptors can lead to neoplastic transformation and enhance the development, growth and malignant progression of breast cancer in women.

Introduction

Breast cancer is the most common malignancy in women in the United States and until only recently, the incidence of breast cancer has steadily increased over the past 70 years [1]. Studies investigating the etiology of breast cancer have concluded that ovarian steroids are required for the onset and progression of breast cancer in pre- and post menopausal women [2]. Estrogen and progesterone stimulation of normal and neoplastic mammary gland development and growth can be clearly demonstrated in both women and experimental animals [3]. However, the exact mechanism(s) involved in mediating the effects of these steroids remains unclear. Estrogen and progesterone do not exert direct mitogenic effects on normal murine mammary epithelial cells, but instead induces normal cell proliferation indirectly by stimulating paracrine hormone and growth factor production, and by enhancing the responsiveness of mammary epithelial cells to these locally produced mitogens [3, 4]. Since growth factors produced locally in the mammary gland appear to be the primary mitogens stimulating normal mammary epithelial cell proliferation, defects and alterations in their signal transduction pathways may be responsible for causing unregulated or abnormal cell growth that characteristically occurs during the initial stages of carcinogenesis [5, 6].

The sensitivity of mammary epithelial cells to various hormone and growth factor induced proliferation is dependent on the parous state of the gland [3]. Mammary epithelial cells isolated from nulliparous mice consist mainly of ductal epithelium [3]. These cells are particularly sensitive to prolactin or the combination of prolactin and progesterone stimulated proliferation when grown in primary culture [7]. In contrast, estrogen has no mitogenic effect on these same isolated cells *in vitro* [7]. Epidermal growth factor (EGF) stimulates cell growth to the same extent as the combination of prolactin and progesterone, but the combination of EGF, prolactin and progesterone does not further enhance the growth rate of these cells as compared to EGF alone [3, 7]. Although corticosterone and aldosterone are not mitogenic when given alone, these steroids significantly enhance prolactin-dependent cell growth by upregulating prolactin receptors in these cells [3, 7].

In contrast to the nulliparous mammary gland, mammary epithelial cells isolated from midpregnant mice are derived predominately from alveolar epithelium [3]. These cells are very responsive to EGF- and transforming growth factor α (TGFα) induced mitogenesis [3].

However, these cells respond with only a modest increase in cell number when treated with prolactin or the combination of prolactin and progesterone [3]. The striking differences in mitogenic sensitivities of mammary epithelial cells isolated from a nulliparous and parous mice to EGF, prolactin, and steroid hormones suggest that ductal epithelial cells are equally responsive to prolactin- and EGF-induced mitogenesis, whereas alveolar cells are primarily dependent on EGF/TGFα as a stimulus for growth [3].

The role of EGF in breast cancer development, growth, and progression has been extensively investigated. Previous studies have established that in normal mammary epithelial cells, ovarian hormones enhance cell proliferation by stimulating paracrine hormone and growth factor production and by increasing the responsiveness of these cells to these mitogens by upregulating mitogen receptor levels [8]. However, excessive overexpression of oncogene products such as EGF-like ligands and their receptors is associated with malignant transformation [9]. Advanced malignant or metastatic breast cancer growth is also associated with overexpression of other EGF-receptor superfamily oncogene products [10]. In addition, these different EGF receptors can form heterodimers, which can result in enhanced mitogenic signaling in both normal and neoplastic mammary epithelial cells [11]. Therefore, although development of breast cancer is dependent on ovarian hormones, the overexpression of growth EGF receptor ligands is associated with breast carcinogenesis. It is now well established that alterations in EGF-dependent signal transduction pathways play a vital role in the establishment, growth and progression of breast cancer.

EGF Receptor Superfamily

Receptor tyrosine kinases are a class of membrane receptors with intrinsic tyrosine kinase activity that enables them to catalyze the phosphorylation of proteins at specific tyrosine residues (Figure 1). Members of the EGF superfamily of receptors are typical receptor tyrosine kinases that are widely distributed throughout the animal kingdom [12]. In humans, these receptors are termed as human epidermal growth factor receptors (HERs) and in rodents, they are referred to as ErbB receptors. The EGF receptor superfamily consists of four transmembrane receptors named: 1) EGF receptor (HER1, ErbB1); 2) HER2 (ErbB2, neu); 3) HER3 (ErbB3) and HER4 (ErbB4). All EGF receptors have a glycosylated extracellular ligand binding domain where ligands can bind and activate the receptor, a transmembrane domain embedded within the plasma membrane responsible for the cooperation between different ErbB receptors, and an intracellular tyrosine kinase domain that catalyzes the tyrosine phosphorylation of intracellular substrates involved in the second messenger signal transduction that is responsible for the biological responses associated with EGF receptor activation [13, 14]. However, individual EGF receptors differ from one another in their ligand binding affinity and tyrosine kinase activity [15, 16].

EGF receptors function to regulate growth, differentiation, migration, and apoptosis in normal cells in various tissues throughout the body, but the importance of these receptors in the embryonic development was clearly demonstrated with knockout studies in mice. Results from these studies clearly showed that loss of ErbB1 gene activation during the embryonic stage of development leads to abnormalities in a wide range of organs and tissues including the brain, gastrointestinal tract, and lungs [17-19]. These EGF receptor knockout mice also displayed a high rate of embryonic death and many of the animals that survived birth characteristically displayed progressive neurodegenerative disorders [17-19].

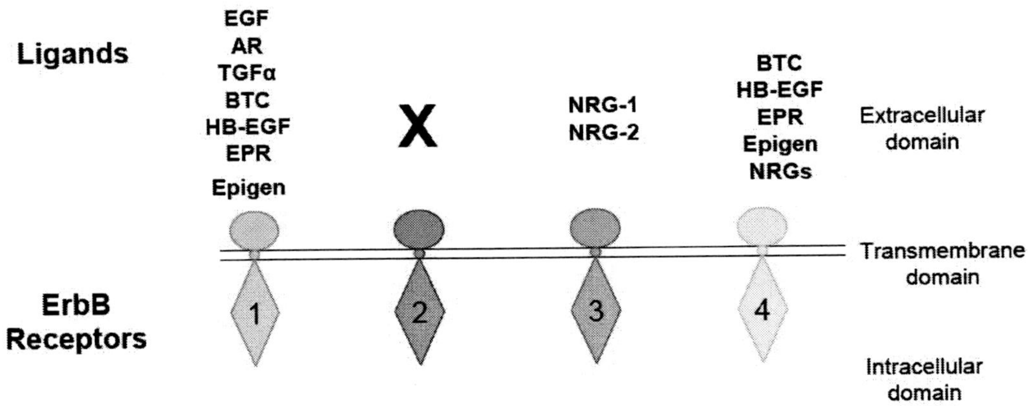

Figure 1. EGF receptors and their associated ligands. The different ligands that bind to the various members of the EGF family of receptors are classified into different groups. EGF, AR, TGFα exclusively bind to HER1/ErbB1 receptor. BTC, HB-EGF, EPR and epigen bind to HER1/ErbB1 and HER4/ErbB4. NRGs are divided into subgroups out of which NRG-1 and NRG-2 bind to both HER3/ErbB3 and HER4/ErbB4 whereas NRG-3 and NRG-4 bind to HER4/ErbB4, but not HER3/ErbB3.

In contrast, EGF ligand knockout mice showed only minor alterations in growth and development as compared to normal mice [17-19]. However, EGF ligand knockouts commonly displayed abnormalities in skin and eye development [17-19]. Knockout of the ErbB2 receptor resulted in embryonal fatality before the eleventh day *in utero*, due to cardiac defects such as trabecular malformation. In other studies, ErbB2 null mutant embryos were genetically rescued from cardiac defects in order to study the effects of ErbB2 knockout on neuronal development [20]. These genetically rescued mutant mice displayed defects in the peripheral nervous system, demonstrating the importance of ErbB2 receptors in the normal peripheral nervous system development [20]. Subsequent studies have shown that functional ErbB3 receptors are required for the normal development of Schwann cell myelination of axons, and severe neuropathies occur in mice with targeted mutations of ErbB3 [21, 22]. ErbB3 knock out studies also showed that embryonal death in mice was due to defective valve formation in the heart, and lack of functional ErbB4 receptors results in aberrant neural and cardiac development [23].

All members of the EGF receptors are also involved in the normal mammary gland development. The female mammary gland undergoes extensive postnatal development and all the four EGF receptors are present in mammary glands [24]. Studies have shown that different ErbB receptors have different roles in the mammary gland development. Specifically, the ErbB1 receptor is responsible for promoting the ductal growth, where as the ErbB2 and ErbB4 receptors are responsible for lobular differentiation and lactation. At present, the functional role of the ErbB3 receptor in normal mammary gland development, differentiation and lactation remains unknown [25-27].

Although, the ErbB receptors expressed in various tissues play an important role at different stages of development, the loss of regulation of normal ErbB receptor function can result in cellular transformation and subsequent tumor growth and development. In cancer cells, abnormal activation of ErbB receptors results in enhanced cell proliferation, survival, and metastasis. The abnormal activation of ErbB receptors may be due to several reasons like overproduction of ligands or receptors, mutations in either extracellular or intracellular kinase domain of the receptor, or constitutive activation of the receptor [28]. The correlation between the ligand expression and tumorogenicity is complex. Among all the ligands, the expression of TGFα is best studied and is co-expressed with ErbB1 receptor in human carcinomas like lung, colon, and ovarian cancers [28]. Overexpression of ErbB2 receptors is found in breast, colon, ovarian, endometrial, and pancreatic cancers and is associated with the poor prognosis of the disease. Deletion mutations within ErbB1 were observed in brain tumors and duplication in the kinase domain of ErbB2 receptor was observed in lung cancer [28].

EGF Receptor Ligands

Ligands that bind to and activate the various EGF receptors belong to EGF family of growth factors (Figure 1). The EGF family ligands are characterized by the presence of an EGF-like domain with three disulfide bonds, responsible for binding with the receptor, and also other structural motifs such as immunoglobulin like domains, heparin binding sites and glycosylation sites [24]. These growth factors are formed by the proteolytic cleavage of the soluble transmembrane precursors that are biologically active and can activate the adjacent receptors through juxtacrine stimulation [14, 29]. The EGF related peptide growth factors can be divided into different groups according to their binding affinity to the receptor. Epidermal growth factor (EGF), transforming growth factor alpha (TGFα) and amphiregulin (AR) specifically bind to ErbB1 (Figure 1). Betacellulin (BTC), heparin-binding EGF (HB-EGF), epiregulin (EPR) and epigen are the ligands that exhibit nonspecific binding. EPR and epigen are the ligands with low binding affinity. BTC, HB-EGF, EPR and epigen can bind to both HER1/ErbB1 and HER4/ErbB4 receptors (Figure 1). Neuregulins (NRGs) are a class of ligands which are further divided into four subclasses. NRG-1 and NRG-2 bind to HER3/ErbB3 and HER4/ErbB4 receptors, whereas NRG-3 and NRG-4 exclusively bind to HER4/ErbB4 receptor [30-34]. NRG-1, also called as heregulin or Neu differentiation factor (NDF), is a polypeptide with both immunoglobulin and EGF-like domain (Figure 1). Tomoregulin and epiglycan-C have been suggested as the genes that encode NRG-5 and NRG-6, respectively, but further studies are needed to ascertain it [15]. A large number of ligands have been identified that can activate HER1/ErbB1, HER3/ErbB3, and HER4/ErbB4 receptors, yet no known ligand has been identified that can bind HER2/ErbB2 directly. In contrast, HER3/ErbB3 receptor can bind to ligands like NRGs but lacks the tyrosine kinase domain where the phosphorylation of tyrosine residues occurs [35]. However, both HER2/ErbB2 and HER3/ErbB3 receptors can participate in the signal transduction by acting as co-receptors in the formation of heterodimers (Figure 1). Dimerization is an important step in the activation of EGF receptors and the intensity of activation of the downstream signaling pathway depends on the type of the dimer formed.

EGF Receptor Dimerization

Ligand binding to the extracellular domain of EGF receptor results in the receptor homo- or heterodimerization. Dimers formed by the pairing of two similar receptors are called homodimers while those involving two different receptors are called heterodimers (Figure 2). This dimerization of EGF receptors is favored by the bivalent nature of the ligand. EGF related peptide growth factors bind to a specific receptor with greater affinity and recruit the other EGF receptor monomer which acts as co-receptor for the formation of dimer.

Structural studies of the EGF receptor extracellular domain have suggested a possible mechanism for the ligand induced dimerization of the receptor (36-38). These structural studies have revealed that the extracellular domain is further divided into subdomains I to IV (39, 40). Subdomain I and III together are responsible for ligand binding and subdomain II is composed of a dimerization loop responsible for interaction with other receptors. In an inactive conformation, the dimerization site is occluded, due to interaction of subdomain II with subdomain IV, and holds subdomain I and III far apart such that it does not allow the ligand to bind to the receptor (Figure 2). Thus, in this tethered state, the receptor is closed for ligand binding and cannot undergo dimerization. Ligands like EGF and other related peptides induce conformational changes in the receptor such that the EGF receptor exposes the dimerization loop and makes the ligand binding site available for the ligand to bind (28, 36, 41, 42). Juxtapositioning of subdomains I and III is necessary for ligand binding, which in an inactive tethered confirmation, is prevented due to the intramolecular attraction between subdomain II and IV (Figure 2). The ErbB receptor has to undergo various conformational changes to reach a stable expanded conformation in which the dimerization loop is exposed and is ready to dimerize (Figure 2). The exact mechanism by which these conformational changes are induced is unclear, but the tethered and expanded confirmations are always in equilibrium and the ligand binding traps the EGF receptor in expanded confirmation and drives the equilibrium towards receptor dimer. This drive of equilibrium towards receptor dimer occurs because of greater stability of receptor dimer than their corresponding monomer. Stabilizing this tethered conformation might be a new approach to inhibit the EGF receptor signaling, which is often lost in breast cancer cells.

The structure of the extracellular domain of the HER2/ErbB2 receptor is different from other EGF receptors. The HER2/ErbB2 receptor does not have tethered conformation, it always appears in expanded conformation. This is because the intramolecular attraction between the subdomain I and III in the HER2/ErbB2 receptor is very strong and these domains are held very close to each other and hence, do not favor any ligand binding to the receptor explaining why the HER2/ErbB2 receptor has no known ligand. Even in the absence of the ligand, the HER2/ErbB2 receptor is present in a favorable dimerization conformation and hence, is a preferred dimerization partner for all other EGF receptors (43, 44). Although HER3/ErbB3 receptor lacks intrinsic tyrosine kinase activity, ligand binding and dimerization with other EGF receptors results in the cross phosphorylation of the intracellular domain of HER3/ErbB3 receptor, which can then activate the p85 regulatory subunit of PI3K and initiate PI3K/Akt mitogenic signaling (45).

Heterodimers formed by HER2/ErbB2 and HER/3ErbB3 receptors are potent activators of mitogenic signaling and can function as an oncogenic unit to drive the tumor cell proliferation (46, 47). The HER2/ErbB2 and HER3/ErbB3 receptor heterodimer complex has been shown to enhance the binding affinity for receptor ligands and increase the spectrum of ligands binding to the receptor. Ligand induced receptor endocytosis is a major pathway for the inactivation of the EGF receptor signaling (48). This process is highly regulated in HER1/ErbB1 receptors by an E3 ubiquitin ligase, whereas ligand activation of the other EGF receptors occurs at a slower rate (47, 48). Other studies have shown that excessive activation of HER3/ErbB3 receptors is able to overcome endocytosis mediated degradation by its ability to undergo enhanced recycling to the plasma membrane. In addition, HER2/ErbB2 and HER3/ErbB3 heterodimers activate major signaling pathways like PI3K/Akt, MAPK, and Janus kinase, while at the same time suppress the recruitment of proteins that negatively regulate mitogenic signals (47-49).

EGF Receptor-Dependent Mitogenic Signaling

EGF receptor mitogenic signaling pathways are highly conserved and commonly found in lower and higher organisms. *Caenorhabditis elegans*, a nematode, contains only a single ErbB receptor homologue that is activated by an EGF-like ligand called Lin-3 [50, 51]. Functional activation of this pathway is required for vulva development [50, 51]. In the fruit fly, *Drosophila melanogaster*, there also exist a single EGF receptor homologue, but four specific EGF receptor ligands have been identified that activate this single receptor [52]. EGF receptor signaling in the *Drosophila* is crucial during several developmental stages of this insect. As discussed earlier, the four EGF receptor subtypes can bind multiple EGF-related ligands and consequently provide a high degree of signal diversity in mammals.

EGF receptors are activated by binding to specific ligands resulting in receptor dimerization, tyrosine kinase activation, and transphosphorylation at receptor tyrosine residues. These phosphotyrosine residues located in the cytoplasmic region of the EGF receptor serve as docking sites for various proteins containing Src homology (SH2 and SH3) and phosphotyrosine binding (PTB) domains [13]. Recruitment of specific intracellular substrates and adaptor proteins to these SH2, SH3 and PTB docking sites is required for receptor second messenger signaling necessary to produce specific biological effects. The recruited adaptor proteins include Shc, Grb7, Grb2, Crk, Nck, PLCγ intracellular kinases such as Src and PI3K; protein tyrosine phosphatases SHP1 and SHP2; and Cbl E3 ubiquitin ligase [13]. Depending upon the type of ligand and the receptor dimer formed, specific patterns of phosphotyrosine sites are associated with specific substrate and adaptor protein recruitment. The specificity of the intracellular signals is mainly determined by the binding pattern of these proteins to the intracellular sites of activated receptor homodimers and heterodimers. EGF receptor signaling is then subsequently inactivated by internalization and destruction of the activated receptor/ligand complex [53]. Major signaling pathways activated by EGF receptors include the MAPK, PI3K/Akt, PLCγ protein kinase C (PKC), and Janus kinase cascades [54-58].

EGF Receptors are G Protein-Coupled Receptors:

Previous studies have shown that EGF receptors are G protein-coupled receptors (GPCRs) and ligand-induced activation of these receptors results in an increase in intracellular cAMP levels and activation of cAMP-dependent protein kinases in both normal and neoplastic mammary epithelial cells [59-61]. Pharmacological agents that either enhance G protein function (cholera and pertussis toxin) or increase intracellular cAMP levels (forskolin and 8-Br-cAMP) have been shown to stimulate, whereas treatments that decrease intracellular levels of cAMP have been found to inhibit EGF receptor dependent normal and neoplastic mammary epithelial cell proliferation [62]. Activation of cAMP-dependent protein kinases is a key event that initiates additional mitogenic cascades including the PI3K/Akt and MAPK pathways [63].

Activation of EGF receptors results in a conformational change in their associated G-protein structure that dramatically enhances its interactions with GTP-binding proteins. G-proteins are heterotrimeric complexes consisting of α, β and γ-subunits. G proteins undergo a cycle of activation and deactivation that transmits signal from the receptor to effector molecules [64, 65]. In the inactive state, the three subunits form a complex with a GDP bound to the α subunit. EGF receptor activation triggers the exchange of GDP for GTP on the α-subunit, resulting in dissociation of the complex into receptor, GTP-bound Gα, and Gβγ subunits. The dissociated Gα-GTP and Gβγ subunits are now able to interact with intracellular effector molecules such as adenylyl cyclase, PI3K, PLC, etc, to initiate second messenger signal transduction [64, 65]. GPCRs can also transmit cytosolic signals to the nucleus and inducing changes in gene transcription processes that regulate cell growth.

Recent findings indicate that G proteins act as proto-oncogene by appropriate gain-of-function mutations, altering the activity of mitogenic signaling pathways [66]. Similar to Ras protein, G protein α-subunits undergo GTPase-inhibiting mutations and behave as oncogenes [66]. Mutations in different α-subunits can induce a transformed phenotype in specific fibroblast cell lines. GPCRs can also undergo mutations resulting in constitutively active receptors, contributing to increased cell proliferation and tumorigenesis. GPCRs contribute to some human tumors by stimulating growth through autocrine and paracrine activation. In these systems, specific cell types express peptides that can positively regulate cell growth by binding to and activating their GPCRs [67-70].

EGF Receptor Activation of MAPK

MAPK signaling cascades are essential for the normal cellular metabolism, cell survival, migration, differentiation, and cell death [54, 55, 59]. Each of these parallel cascades are comprised of three protein kinases identified as: 1) MAPK kinase kinase (MAPKKK); 2) MAPK kinase (MAPKK); and 3) MAPK. Activation of this pathway results in the sequential serine/threonine phosphorylation of each of these three kinases [60]. In mammals, four such signaling cascades have been identified as Erk1 and 2, JNK, p38, and Erk5, respectively [55]. The Erk1 and 2 play a major role in cell proliferation and survival. All four EGF receptors can initiate Erk1 and 2 activation [71]. Activation of the EGF receptor, causes the adaptor protein, Grb2, to directly bind to a specific phosphotyrosine site or indirectly bind through the interaction with another adaptor protein called Shc [72]. Grb2 is associated with Sos, a guanine nucleotide exchange factor specific for Ras, and Sos activates Ras by exchanging GDP for GTP. In the active GTPase-bound state, Ras interacts with several effector proteins like Raf and stimulates numerous intracellular processes [72]. Activated Raf stimulates the MAPKK, MEK, by phosphorylating a serine residue in the activation loop [73, 74]. MEK then activates Erk1 and 2 by phosphorylating serine and threonine residues located in the activation loop of these particular kinases. After Erk1 and 2 are activated, then interact with and phosphorylate a wide range of cytoplasmic and membrane bound substrates. Erk 1 and 2 can also translocate to the nucleus and activate a number of transcription factors such as sp1, PEA3, E2F, ElK1, and AP1, to modulate the expression of various genes [73, 74]. Activated Erk 1 and 2 have been found to phosphorylate and regulate the activity of over 160 cellular proteins [75].

The Ras/Raf/MEK/Erk pathway has also been shown to play a vital role in oncogenesis [55]. Overexpression or mutation of the HER1/ErbB1 or HER2/ErbB2 receptors causes hyperactivation of Ras and a large increase of MAPK signaling in many types of human cancers [76-78]. Furthermore, Ras mutations and subsequent hyperactivation of Ras have found to be present in 30% of all cancers [79]. The mutant Ras genes in human cancers translate into mutant proteins with single amino-acid substitutions. Mutant Ras proteins are insensitive to GTPase-activating protein (GAP) resulting in constitutive activation of the downstream signal. In addition, it has been recently shown that a variety of human cancers have been found to contain the B-Raf mutant, but not a concurrent Ras mutation in melanomas, colorectal carcinomas, thyroid carcinomas and lung cancer [80, 81]. These findings suggest that Ras function is mediated mainly through Raf activation [80, 81]. However, studies have also demonstrated that additional effectors such as p110 subunits (p110α, β, γ, and δ) of PI3K, Tiam1, RalGDS, RGL, RGL2, RGL3 and PLCε are utilized by Ras to promote tumorigenesis [82, 83].

EGF Receptor Activation of PI3K/Akt

One class of phosphoinositol-3-kinases or PI3Ks is a heterodimer composed of a regulatory subunit p85 and a catalytic subunit p110. EGF receptors activate PI3K by recruiting its p85 regulatory subunit to the activated receptors, resulting in the allosteric activation of the p110 catalytic subunit [84, 85]. HER3/ErbB3 and HER4/ErbB4 receptors contain 6 and 1 putative p85 binding sites, respectively, while HER1/ErbB1 and HER2/ErbB2 receptors bind p85 indirectly through adapter proteins. Activated PI3K phosphorylates phosphatidylinositol-4-phosphate [PtdIns[4]P] and phosphatidylinositol-4,5-diphosphate [PtdIns(4,5)P2] to form phosphatidylinositol-3,4-diphosphate [PtdIns[3,4]P2] and phosphatidylinositol-3,4,5-triphosphate [PtdIns[3,4,5]P3], respectively. These newly formed signaling phospholipids then act to mediate the membrane translocation of many Pleckstrin homology (PH) domain containing signaling proteins like PDK1 and Akt [84, 85]. Unlike Akt, PDK1 is present as a constitutively active enzyme in cells. When in close proximity with Akt at membrane, PDK1 activates Akt by phosphorylating it generally at Thr308, followed by second phosphorylation at Ser473 by PDK2 [86]. Activated Akt detaches from the membrane and phosphorylates an array of proteins in the cytoplasm and nucleus, thus regulating cell growth and apoptosis [86].

Mutation or amplification of genes encoding HER2 or PTEN, results in the hyperactivation and signaling of the PI3K/Akt pathway and this has been associated as a significant risk factor in the development of breast cancer in humans [87]. Akt1, Akt2, and Akt3 were activated to different degrees by EGF stimulation through the PI3K signaling pathway in human esophageal cancer cells. These studies showed that activation of Akt was indicated by increased Akt-phosphorylation and higher kinase activity both of which were abolished by pre-incubating the cells with the PI3K inhibitors [88]. In addition, EGF receptor activation of PI3K/Akt signaling contributes to tumor cell growth in other cancers such as prostate and non-small cell lung cancer [89, 90].

Cross-Talk between EGF Receptors and other Growth Factor Receptors

Receptor tyrosine kinases and cytokine receptors have been shown to display significant cross-talk that modulates receptor signaling. Specifically, there exists a complex interaction between HER1/ErbB1 and prolactin receptor in a mammary gland. Prolactin receptor activation has a negative impact on HER1/ErbB1 induced signaling. Continuous prolactin receptor activation by prolactin in mammary epithelial cells results in the phosphorylation of specific threonine residues on the HER1/ErbB1 receptor and diminished kinase activity [31, 91]. In contrast, earlier studies reported that excessive activation of EGF receptors had a negative influence upon prolactin receptor signaling [92, 93]. Prolonged exposure to EGF reduces the capacity of mammary epithelial cells to respond to prolactin-induced signaling, due to EGF receptor dependent upregulation of protein tyrosine phosphatases, which inversely regulates the prolactin receptor associated Janus kinase autophosphorylation and Jak2 activation [31]. This interaction appears to play an important role in maintaining a balance between cell proliferation and differentiation during the developmental phases of mammary gland. Prolactin receptor activation and signaling is required for mammary epithelial cell differentiation and milk synthesis, whereas EGF receptor activation and signaling is associated with mammary epithelial cell proliferation and mammary gland growth. Activation of EGF receptors facilitates the proliferation of ductal epithelial cells in proliferative phase of mammary gland development, when prolactin receptor activity is reduced [31]. In breast cancer cells that overexpress HER2/ErbB2 receptors, prolactin-mediated Jak2 activation has been shown to enhance tyrosine phosphorylation of HER2/ErbB2 receptors and stimulate downstream MAPK activity [94]. This effect can be blocked by inhibiting autocrine prolactin production and suggests a possible growth mechanism during breast tumor development [94]. Similarly, growth hormone receptor induced Jak2 activation leads to HER1/ErbB1 phosphorylation, followed by Grb2 association and stimulation of MAPK pathway [95]. Furthermore, G-protein coupled receptor (GPCR) agonists stimulate MAPK by transactivating HER1/ErbB1 [96]. GPCRs for lysophosphatidic acid (LPA), thrombin and endothelin has been shown to promote EGF receptor signaling through multiple mechanisms including the activation of matrix metalloproteinases which cleave membrane bound EGF receptor ligands, such as HB-EGF, making them available for activation of EGF receptors. GPCRs can also indirectly activate Src which phosphorylates and enhances the signaling of EGF receptors. In addition, steroid hormones activate genes encoding EGF receptor ligands and EGF receptor dependent activation of the MAPK cascade can exert a positive feed back effect by stimulating the transcription of genes involved with the synthesis of EGF receptor ligands [24].

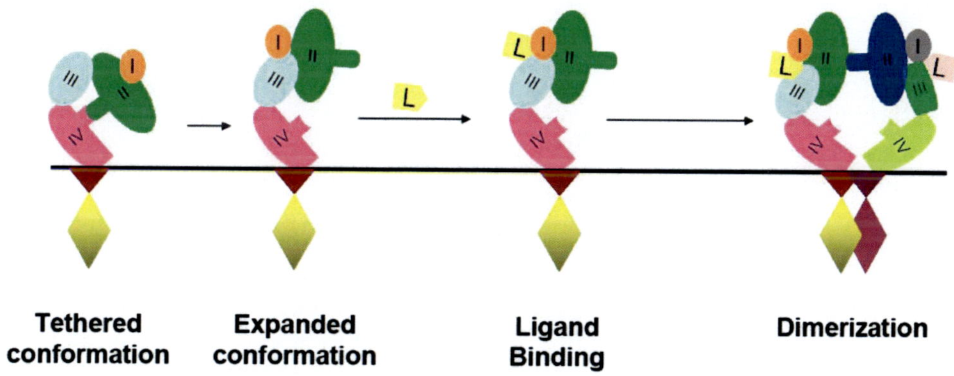

Figure 2. Schematic representation of ligand induced activation and dimerization of EGF receptors. The extracellular domain of EGF receptor is divided into subdomains. A tethered conformation exists before ligand binding to domain II (comprising the dimerization loop) and domain IV, and an intramolecular bond prevents the formation of high affinity ligand binding site in domain I and III. The expanded conformation exists when dimerization loop becomes exposed and ligand binding is possible. After ligand binding, the EGF receptor monomer undergoes dimerization with other EGF receptor and the intracellular catalytic domain is activated by transphosphorylation.

Conclusions

The EGF family of receptors plays a pivotal role in normal mammary epithelial cell growth, viability and differentiation. However, overexpression or excessive EGF receptor signaling has been linked to cellular transformation and promotion of mammary tumor development, growth and progression. A great deal of progress has recently been made in identifying not only those receptors and ligands that are risk factors for breast cancer, but also the intracellular signaling pathways that are stimulated by the activation of these receptors. These findings are especially important because although EGF receptors and their ligands vary considerably in their structure among family members, they all tend to activate common mechanisms and second messenger signal transduction pathways involved in initiating cell growth and function. This knowledge of EGF receptor ligand activation of EGF receptor dependent mitogenic signaling has provided an excellent opportunity for improving cancer diagnosis and therapy.

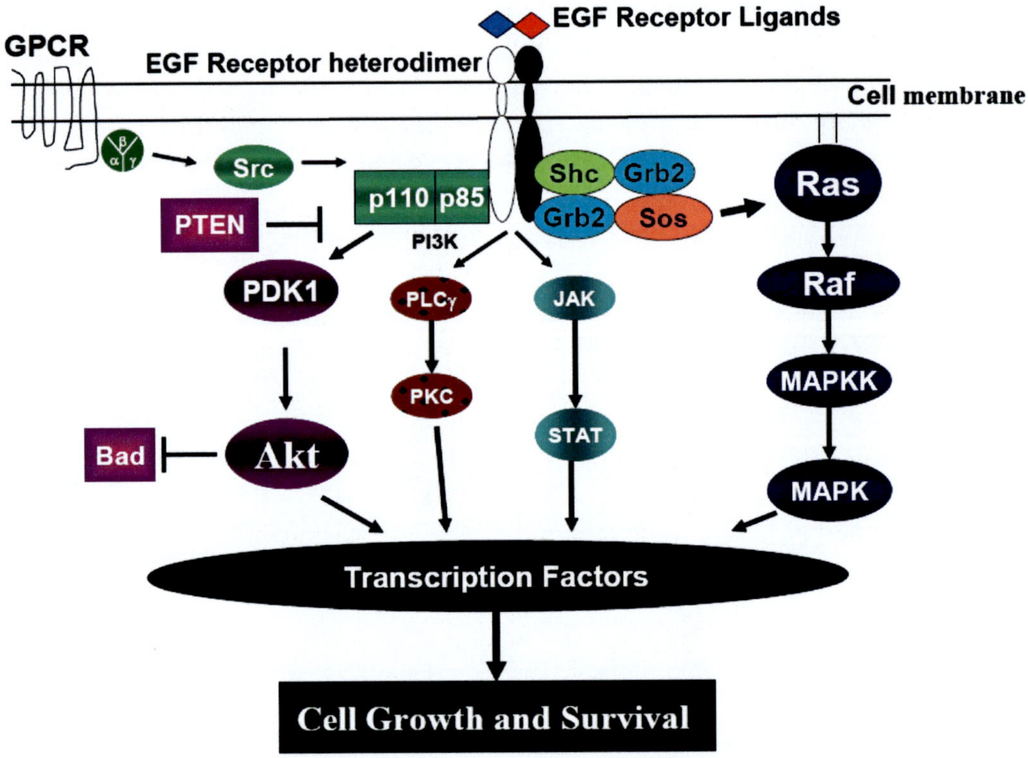

Figure 3. Major EGF mitogenic signaling pathways. EGF receptors are potent stimulators of MAPK and PI3K/Akt signaling pathways. Ligand activated EGF receptor dimers bind adaptor protein, Grb2 directly or through interaction with Shc. Grb2 is associated with Sos, which activates Ras. In the active state, Ras interacts with several downstream effector proteins like Raf, which in turn leads to the activation of the MAPK cascade. Activated MAPK can also translocate to the nucleus and activate a number of transcription factors regulating cell growth. G-protein coupled receptor (GPCR) agonists also stimulate MAPK by transactivating HER1/ErbB1. GPCRs can indirectly activate Src which phosphorylates EGF receptors. EGF receptor activation of PI3K results from the recruitment of the p85 regulatory subunit to the activated receptors and subsequent allosteric activation of the p110 catalytic subunit on PI3K. Activated PI3K mediate the membrane translocation and activation of many signaling proteins like PDK1 and Akt. Activated Akt phosphorylates many proteins in cytoplasm and nucleus to regulate cell growth and apoptosis. PTEN is a phosphatase which inhibits PI3K activity while Bad is a proapoptotic protein whose activity is inhibited by Akt. The activated β/γ subunits of G-proteins can also activate PI3K. In addition, activation of PLCγ, PKC, and Jak-STAT pathways can also be initiated through EGF ErbB receptor signaling, which then can enhance gene transcription and cell proliferation.

It is now understood that various EGF receptor activities are commonly amplified in breast cancer, and the extent of amplification displays a direct correlation with the stage of tumor progression and prognosis. Diagnostic assays are currently being used to measure normal and abnormal EGF receptor signaling in order to optimizing the effectiveness of therapeutic protocols in the treatment and cure of breast cancer. Newly developed therapies are also currently being used to block tumor formation and progression by interfering with receptor activation and downstream signal transduction. This can be done in a number of different ways such as the use of EGF receptor antagonist or antibodies that bind with high affinity and specificity to individual receptors [97, 98]. Therapeutic agents have also been developed and are being used clinically to block EGF receptor tyrosine kinase activity by preventing the interaction of ATP with the cytoplasmic catalytic sites of the receptor.

Inhibiting EGF receptor signal transduction by developing therapies that target intracellular proteins and enzymes involved in mediating second messenger signal transduction is another approach that is being extensively investigated (Figure 3). Tyrophostins are one such group of reagents that act as soluble chemical inhibitors of protein tyrosine phosphorylation [99]. By blocking tyrosine phosphorylation, tyrophostins block the autophosphorylation of SH2 and SH3 sites on the intracellular catalytic region of the EGF receptor and prevent the interaction of substrates and adaptor proteins with the activated receptor, and thereby nullify any mitogenic signaling that would result from ligand activation of the receptor. Other agents specifically target components within the signaling pathways downstream of the EGF receptor such as specific inhibitors of PI3K, Akt, MAPK, PKA, PKC and PLC (Figure 3). It is clearly evident that the more information obtained related to EGF receptor structure and downstream signaling, the greater the opportunity to use this information to develop rational and efficient approaches to inhibit breast cancer growth and progression.

References

[1] Marshall, E. Epidemiology. Search for a killer: focus shifts from fat to hormones [news] [see comments]. *Science, 259*: 618-621, 1993.

[2] Vogel, V. G., Yeomans, A., and Higginbotham, E. Clinical management of women at increased risk for breast cancer. *Breast Cancer Res. Treat, 28*: 195-210, 1993.

[3] Imagawa, W., Bandyopadhyay, G. K., and Nandi, S. Regulation of mammary epithelial cell growth in mice and rats. *Endocr. Rev., 11*: 494-523, 1990.

[4] Haslam, S. Z. Stromal-epithelial interactions in normal and neoplastic mammary gland. *Cancer Treat Res., 53*: 401-420, 1991.

[5] Cross, D. A., Alessi, D. R., Cohen, P., Andjelkovich, M., and Hemmings, B. A. Inhibition of glycogen synthase kinase-3 by insulin mediated by protein kinase B. *Nature, 378*: 785-789, 1995.

[6] McIntyre, B. S., Briski, K. P., Hosick, H. L., and Sylvester, P. W. Effects of protein tyrosine phosphatase inhibitors on EGF- and insulin- dependent mammary epithelial cell growth. *Proc. Soc. Exp. Biol. Med., 217*: 180-187, 1998.

[7] Imagawa, W., Tomooka, Y., Hamamoto, S., and Nandi, S. Stimulation of mammary epithelial cell growth in vitro: interaction of epidermal growth factor and mammogenic hormones. *Endocrinology, 116*: 1514-1524, 1985.

[8] Haslam, S. Z. and Nummy, K. A. The ontogeny and cellular distribution of estrogen receptors in normal mouse mammary gland. *J. Steroid Biochem. Mol. Biol., 42*: 589-595, 1992.

[9] Nandi, S., Guzman, R. C., and Yang, J. Hormones and mammary carcinogenesis in mice, rats, and humans: a unifying hypothesis. *Proc. Natl. Acad. Sci. U S A, 92*: 3650-3657, 1995.

[10] Dougall, W. C., Qian, X., and Greene, M. I. Interaction of the neu/p185 and EGF receptor tyrosine kinases: implications for cellular transformation and tumor therapy. *J. Cell Biochem.*, *53*: 61-73, 1993.

[11] Samant, G. V. and Sylvester, P. W. gamma-Tocotrienol inhibits ErbB3-dependent PI3K/Akt mitogenic signalling in neoplastic mammary epithelial cells. *Cell Prolif., 39*: 563-574, 2006.

[12] Stein, R. C. Prospects for phosphoinositide 3-kinase inhibition as a cancer treatment. *Endo. Relat. Can.*, *8*: 237-248, 2001.

[13] Schlessinger, J. Cell signaling by receptor tyrosine kinases. *Cell, 103*: 211-225, 2000.

[14] Normanno, N., Bianco, C., De Luca, A., Maiello, M. R., and Salomon, D. S. Target-based agents against ErbB receptors and their ligands: a novel approach to cancer treatment. *Endocr Relat Cancer, 10*: 1-21, 2003.

[15] Bazley, L. A. and Gullick, W. J. The epidermal growth factor receptor family. *Endocr. Relat. Cancer, 12 Suppl 1*: S17-27, 2005.

[16] McIntyre, B. S., Birkenfeld, H. P., and Sylvester, P. W. Relationship between epidermal growth factor receptor levels, autophosphorylation and mitogenic-responsiveness in normal mouse mammary epithelial cells in vitro. *Cell Prolif., 28*: 45-56, 1995.

[17] Sibilia, M. and Wagner, E. F. Strain-dependent epithelial defects in mice lacking the EGF receptor. *Science, 269*: 234-238, 1995.

[18] Miettinen, P. J., Berger, J. E., Meneses, J., Phung, Y., Pedersen, R. A., Werb, Z., and Derynck, R. Epithelial immaturity and multiorgan failure in mice lacking epidermal growth factor receptor. *Nature, 376*: 337-341, 1995.

[19] Threadgill, D. W., Dlugosz, A. A., Hansen, L. A., Tennenbaum, T., Lichti, U., Yee, D., LaMantia, C., Mourton, T., Herrup, K., Harris, R. C., et al. Targeted disruption of mouse EGF receptor: effect of genetic background on mutant phenotype. *Science, 269*: 230-234, 1995.

[20] Morris, J. K., Lin, W., Hauser, C., Marchuk, Y., Getman, D., and Lee, K. F. Rescue of the cardiac defect in ErbB2 mutant mice reveals essential roles of ErbB2 in peripheral nervous system development. *Neuron, 23*: 273-283, 1999.

[21] Riethmacher, D., Sonnenberg-Riethmacher, E., Brinkmann, V., Yamaai, T., Lewin, G. R., and Birchmeier, C. Severe neuropathies in mice with targeted mutations in the ErbB3 receptor. *Nature, 389*: 725-730, 1997.

[22] Erickson, S. L., O'Shea, K. S., Ghaboosi, N., Loverro, L., Frantz, G., Bauer, M., Lu, L. H., and Moore, M. W. ErbB3 is required for normal cerebellar and cardiac development: a comparison with ErbB2-and heregulin-deficient mice. *Development, 124*: 4999-5011, 1997.

[23] Gassmann, M., Casagranda, F., Orioli, D., Simon, H., Lai, C., Klein, R., and Lemke, G. Aberrant neural and cardiac development in mice lacking the ErbB4 neuregulin receptor. *Nature*, *378*: 390-394, 1995.

[24] Yarden, Y. and Sliwkowski, M. X. Untangling the ErbB signalling network. *Nat. Rev. Mol. Cell Biol.*, *2*: 127-137, 2001.

[25] Hennighausen, L. and Robinson, G. W. Think globally, act locally: the making of a mouse mammary gland. *Genes.Dev.*, *12*: 449-455, 1998.

[26] Jones, F. E. and Stern, D. F. Expression of dominant-negative ErbB2 in the mammary gland of transgenic mice reveals a role in lobuloalveolar development and lactation. *Oncogene, 18*: 3481-3490, 1999.

[27] Jones, F. E., Welte, T., Fu, X. Y., and Stern, D. F. ErbB4 signaling in the mammary gland is required for lobuloalveolar development and Stat5 activation during lactation. *J. Cell. Biol., 147*: 77-88, 1999.

[28] Citri, A. and Yarden, Y. EGF-ERBB signalling: towards the systems level. *Nat Rev. Mol. Cell Biol., 7*: 505-516, 2006.

[29] Massague, J. and Pandiella, A. Membrane-anchored growth factors. *Annu. Rev. Biochem., 62*: 515-541, 1993.

[30] Carpenter, G. and Cohen, S. Epidermal growth factor. *J. Biol. Chem., 265*: 7709-7712, 1990.

[31] Hynes, N. E., Horsch, K., Olayioye, M. A., and Badache, A. The ErbB receptor tyrosine family as signal integrators. *Endocr. Relat. Cancer, 8*: 151-159, 2001.

[32] Pinkas-Kramarski, R., Shelly, M., Guarino, B. C., Wang, L. M., Lyass, L., Alroy, I., Alimandi, M., Kuo, A., Moyer, J. D., Lavi, S., Eisenstein, M., Ratzkin, B. J., Seger, R., Bacus, S. S., Pierce, J. H., Andrews, G. C., and Yarden, Y. ErbB tyrosine kinases and the two neuregulin families constitute a ligand-receptor network. *Mol. Cell Biol., 18*: 6090-6101, 1998.

[33] Zhang, D., Sliwkowski, M. X., Mark, M., Frantz, G., Akita, R., Sun, Y., Hillan, K., Crowley, C., Brush, J., and Godowski, P. J. Neuregulin-3 (NRG3): a novel neural tissue-enriched protein that binds and activates ErbB4. *Proc. Natl. Acad. Sci. U S A, 94*: 9562-9567, 1997.

[34] Harari, D., Tzahar, E., Romano, J., Shelly, M., Pierce, J. H., Andrews, G. C., and Yarden, Y. Neuregulin-4: a novel growth factor that acts through the ErbB-4 receptor tyrosine kinase. *Oncogene, 18*: 2681-2689, 1999.

[35] Guy, P. M., Platko, J. V., Cantley, L. C., Cerione, R. A., and Carraway, K. L., 3rd. Insect cell-expressed p180erbB3 possesses an impaired tyrosine kinase activity. *Proc .Natl .Acad .Sci .U S A, 91*: 8132-8136, 1994.

[36] Ferguson, K. M. Active and inactive conformations of the epidermal growth factor receptor. *Biochem. Soc. Trans., 32*: 742-745, 2004.

[37] Zhang, X., Gureasko, J., Shen, K., Cole, P. A., and Kuriyan, J. An allosteric mechanism for activation of the kinase domain of epidermal growth factor receptor. *Cell, 125*: 1137-1149, 2006.

[38] Hubbard, S. R. EGF receptor activation: push comes to shove. *Cell, 125*: 1029-1031, 2006.

[39] Garrett, T. P., McKern, N. M., Lou, M., Elleman, T. C., Adams, T. E., Lovrecz, G. O., Zhu, H. J., Walker, F., Frenkel, M. J., Hoyne, P. A., Jorissen, R. N., Nice, E. C., Burgess, A. W., and Ward, C. W. Crystal structure of a truncated epidermal growth factor receptor extracellular domain bound to transforming growth factor alpha. *Cell, 110*: 763-773, 2002.

[40] Ogiso, H., Ishitani, R., Nureki, O., Fukai, S., Yamanaka, M., Kim, J. H., Saito, K., Sakamoto, A., Inoue, M., Shirouzu, M., and Yokoyama, S. Crystal structure of the

complex of human epidermal growth factor and receptor extracellular domains. *Cell, 110*: 775-787, 2002.

[41] Normanno, N., Bianco, C., Strizzi, L., Mancino, M., Maiello, M. R., De Luca, A., Caponigro, F., and Salomon, D. S. The ErbB receptors and their ligands in cancer: an overview. *Curr Drug Targets, 6*: 243-257, 2005.

[42] Bublil, E. M. and Yarden, Y. The EGF receptor family: spearheading a merger of signaling and therapeutics. *Curr .Opin. Cell. Biol., 19*: 124-134, 2007.

[43] Karunagaran, D., Tzahar, E., Beerli, R. R., Chen, X., Graus-Porta, D., Ratzkin, B. J., Seger, R., Hynes, N. E., and Yarden, Y. ErbB-2 is a common auxiliary subunit of NDF and EGF receptors: implications for breast cancer. *Embo. J., 15*: 254-264, 1996.

[44] Tzahar, E., Pinkas-Kramarski, R., Moyer, J. D., Klapper, L. N., Alroy, I., Levkowitz, G., Shelly, M., Henis, S., Eisenstein, M., Ratzkin, B. J., Sela, M., Andrews, G. C., and Yarden, Y. Bivalence of EGF-like ligands drives the ErbB signaling network. *Embo J., 16*: 4938-4950, 1997.

[45] Kim, H. H., Sierke, S. L., and Koland, J. G. Epidermal growth factor-dependent association of phosphatidylinositol 3-kinase with the erbB3 gene product. *J. Biol. Chem., 269*: 24747-24755, 1994.

[46] Alimandi, M., Romano, A., Curia, M. C., Muraro, R., Fedi, P., Aaronson, S. A., Di Fiore, P. P., and Kraus, M. H. Cooperative signaling of ErbB3 and ErbB2 in neoplastic transformation and human mammary carcinomas. *Oncogene, 10*: 1813-1821, 1995.

[47] Holbro, T., Beerli, R. R., Maurer, F., Koziczak, M., Barbas, C. F., 3rd, and Hynes, N. E. The ErbB2/ErbB3 heterodimer functions as an oncogenic unit: ErbB2 requires ErbB3 to drive breast tumor cell proliferation. *Proc. Natl. Acad. Sci. U S A, 100*: 8933-8938, 2003.

[48] Holbro, T., Civenni, G., and Hynes, N. E. The ErbB receptors and their role in cancer progression. *Exp. Cell Res., 284*: 99-110, 2003.

[49] McIntyre, B. S. and Sylvester, P. W. Genistein and erbstatin inhibition of normal mammary epithelial cell proliferation is associated with EGF-receptor down-regulation. *Cell Prolif., 31*: 35-46, 1998.

[50] Hill, R. J. and Sternberg, P. W. The gene lin-3 encodes an inductive signal for vulval development in C. elegans. *Nature, 358*: 470-476, 1992.

[51] Sternberg, P. W., Hill, R. J., Jongeward, G., Huang, L. S., and Carta, L. Intercellular signaling during Caenorhabditis elegans vulval induction. *Cold Spring Harb. Symp. Quant. Biol., 57*: 353-362, 1992.

[52] Shvartsman, S. Y., Muratov, C. B., and Lauffenburger, D. A. Modeling and computational analysis of EGF receptor-mediated cell communication in Drosophila oogenesis. *Development, 129*: 2577-2589, 2002.

[53] Waterman, H. and Yarden, Y. Molecular mechanisms underlying endocytosis and sorting of ErbB receptor tyrosine kinases. *FEBS Lett, 490*: 142-152, 2001.

[54] Davis, R. J. Signal transduction by the JNK group of MAP kinases. *Cell, 103*: 239-252, 2000.

[55] Roberts, P. J. and Der, C. J. Targeting the Raf-MEK-ERK mitogen-activated protein kinase cascade for the treatment of cancer. *Oncogene, 26*: 3291-3310, 2007.

[56] Sylvester, P. W., Birkenfeld, H. P., Hosick, H. L., and Briski, K. P. Fatty acid modulation of epidermal growth factor-induced mouse mammary epithelial cell proliferation in vitro. *Exp. Cell Res., 214*: 145-153, 1994.

[57] Birkenfeld, H. P., McIntyre, B. S., Briski, K. P., and Sylvester, P. W. Protein kinase C isoenzyme expression in normal mouse mammary epithelial cells grown in primary culture. *Proc .Soc. Exp. Biol .Med., 213*: 65-70, 1996.

[58] Birkenfeld, H. P., McInntyre, B. S., Briski, K. P., and Sylvester, P. W. Role of protein kinase C in modulating epidermal growth factor- and phorbol ester-induced mammary epithelial cell growth in vitro. *Exp. Cell Res., 223*: 183-191, 1996.

[59] Pearson, G., Robinson, F., Beers Gibson, T., Xu, B. E., Karandikar, M., Berman, K., and Cobb, M. H. Mitogen-activated protein (MAP) kinase pathways: regulation and physiological functions. *Endocr. Rev., 22*: 153-183, 2001.

[60] Johnson, G. L. and Lapadat, R. Mitogen-activated protein kinase pathways mediated by ERK, JNK, and p38 protein kinases. *Science, 298*: 1911-1912, 2002.

[61] Jorissen, R. N., Walker, F., Pouliot, N., Garrett, T. P., Ward, C. W., and Burgess, A. W. Epidermal growth factor receptor: mechanisms of activation and signalling. *Exp. Cell Res., 284*: 31-53, 2003.

[62] Sylvester, P. W., Nachnani, A., Shah, S., and Briski, K. P. Role of GTP-binding proteins in reversing the antiproliferative effects of tocotrienols in preneoplastic mammary epithelial cells. *Asia Pacific J. Clin. Nutr., 11*: S452-S459, 2002.

[63] Carpenter, G. ErbB-4: mechanism of action and biology. *Exp. Cell Res., 284*: 66-77, 2003.

[64] Neer, E. J. Heterotrimeric G proteins: organizers of transmembrane signals. *Cell, 80*: 249-257, 1995.

[65] Treisman, R. Regulation of transcription by MAP kinase cascades. *Curr. Opin. Cell Biol., 8*: 205-215, 1996.

[66] Gupta, S. K., Gallego, C., and Johnson, G. L. Mitogenic pathways regulated by G protein oncogenes. *Mol. Biol. Cell, 3*: 123-128, 1992.

[67] Schuller, H. M. Receptor-mediated mitogenic signals and lung cancer. *Cancer Cells, 3*: 496-503, 1991.

[68] Cuttitta, F., Carney, D. N., Mulshine, J., Moody, T. W., Fedorko, J., Fischler, A., and Minna, J. D. Bombesin-like peptides can function as autocrine growth factors in human small-cell lung cancer. *Nature, 316*: 823-826, 1985.

[69] Sethi, T., Langdon, S., Smyth, J., and Rozengurt, E. Growth of small cell lung cancer cells: stimulation by multiple neuropeptides and inhibition by broad spectrum antagonists in vitro and in vivo. *Cancer Res., 52*: 2737s-2742s, 1992.

[70] Hoosein, N. M., Kiener, P. A., Curry, R. C., Rovati, L. C., McGilbra, D. K., and Brattain, M. G. Antiproliferative effects of gastrin receptor antagonists and antibodies to gastrin on human colon carcinoma cell lines. *Cancer Res., 48*: 7179-7183, 1988.

[71] Citri, A., Skaria, K. B., and Yarden, Y. The deaf and the dumb: the biology of ErbB-2 and ErbB-3. *Exp. Cell Res., 284*: 54-65, 2003.

[72] Marshall, C. J. MAP kinase kinase kinase, MAP kinase kinase and MAP kinase. *Curr. Opin. Genet. Dev., 4*: 82-89, 1994.

[73] Zuber, J., Tchernitsa, O. I., Hinzmann, B., Schmitz, A. C., Grips, M., Hellriegel, M., Sers, C., Rosenthal, A., and Schafer, R. A genome-wide survey of RAS transformation targets. *Nat. Genet., 24*: 144-152, 2000.

[74] Schulze, A., Nicke, B., Warne, P. H., Tomlinson, S., and Downward, J. The transcriptional response to Raf activation is almost completely dependent on Mitogen-activated Protein Kinase Kinase activity and shows a major autocrine component. *Mol.Biol .Cell, 15*: 3450-3463, 2004.

[75] Yoon, S. and Seger, R. The extracellular signal-regulated kinase: multiple substrates regulate diverse cellular functions. *Growth Factors, 24*: 21-44, 2006.

[76] Lynch, T. J., Bell, D. W., Sordella, R., Gurubhagavatula, S., Okimoto, R. A., Brannigan, B. W., Harris, P. L., Haserlat, S. M., Supko, J. G., Haluska, F. G., Louis, D. N., Christiani, D. C., Settleman, J., and Haber, D. A. Activating mutations in the epidermal growth factor receptor underlying responsiveness of non-small-cell lung cancer to gefitinib. *N. Engl. J. Med., 350*: 2129-2139, 2004.

[77] Hynes, N. E. and Lane, H. A. ERBB receptors and cancer: the complexity of targeted inhibitors. *Nat. Rev. Cancer, 5*: 341-354, 2005.

[78] Grandis, J. R. and Sok, J. C. Signaling through the epidermal growth factor receptor during the development of malignancy. *Pharmacol. Ther., 102*: 37-46, 2004.

[79] Malumbres, M. and Barbacid, M. RAS oncogenes: the first 30 years. *Nat. Rev. Cancer, 3*: 459-465, 2003.

[80] Rajagopalan, H., Bardelli, A., Lengauer, C., Kinzler, K. W., Vogelstein, B., and Velculescu, V. E. Tumorigenesis: RAF/RAS oncogenes and mismatch-repair status. *Nature, 418*: 934, 2002.

[81] Sieben, N. L., Macropoulos, P., Roemen, G. M., Kolkman-Uljee, S. M., Jan Fleuren, G., Houmadi, R., Diss, T., Warren, B., Al Adnani, M., De Goeij, A. P., Krausz, T., and Flanagan, A. M. In ovarian neoplasms, BRAF, but not KRAS, mutations are restricted to low-grade serous tumours. *J. Pathol., 202*: 336-340, 2004.

[82] Repasky, G. A., Chenette, E. J., and Der, C. J. Renewing the conspiracy theory debate: does Raf function alone to mediate Ras oncogenesis? *Trends Cell Biol., 14*: 639-647, 2004.

[83] Shaw, R. J. and Cantley, L. C. Ras, PI(3)K and mTOR signalling controls tumour cell growth. *Nature, 441*: 424-430, 2006.

[84] Soltoff, S. P. and Cantley, L. C. p120cbl is a cytosolic adapter protein that associates with phosphoinositide 3-kinase in response to epidermal growth factor in PC12 and other cells. *J. Biol. Chem., 271*: 563-567, 1996.

[85] Normanno, N., De Luca, A., Bianco, C., Strizzi, L., Mancino, M., Maiello, M. R., Carotenuto, A., De Feo, G., Caponigro, F., and Salomon, D. S. Epidermal growth factor receptor (EGFR) signaling in cancer. *Gene, 366*: 2-16, 2006.

[86] Brazil, D. P., Yang, Z. Z., and Hemmings, B. A. Advances in protein kinase B signalling: AKTion on multiple fronts. *Trends Biochem Sci, 29*: 233-242, 2004.

[87] Navolanic, P. M., Steelman, L. S., and McCubrey, J. A. EGFR family signaling and its association with breast cancer development and resistance to chemotherapy (Review). *Int. J. Oncol., 22*: 237-252, 2003.

[88] Okano, J., Gaslightwala, I., Birnbaum, M. J., Rustgi, A. K., and Nakagawa, H. Akt/protein kinase B isoforms are differentially regulated by epidermal growth factor stimulation. *J. Biol. Chem., 275*: 30934-30942, 2000.

[89] El Sheikh, S. S., Domin, J., Abel, P., Stamp, G., and Lalani el, N. Phosphorylation of both EGFR and ErbB2 is a reliable predictor of prostate cancer cell proliferation in response to EGF. *Neoplasia, 6*: 846-853, 2004.

[90] Engelman, J. A., Janne, P. A., Mermel, C., Pearlberg, J., Mukohara, T., Fleet, C., Cichowski, K., Johnson, B. E., and Cantley, L. C. ErbB-3 mediates phosphoinositide 3-kinase activity in gefitinib-sensitive non-small cell lung cancer cell lines. *Proc. Natl. Acad. Sci. U S A, 102*: 3788-3793, 2005.

[91] Quijano, V. J., Jr. and Sheffield, L. G. Prolactin decreases epidermal growth factor receptor kinase activity via a phosphorylation-dependent mechanism. *J. Biol. Chem., 273*: 1200-1207, 1998.

[92] Hynes, N. E., Taverna, D., Harwerth, I. M., Ciardiello, F., Salomon, D. S., Yamamoto, T., and Groner, B. Epidermal growth factor receptor, but not c-erbB-2, activation prevents lactogenic hormone induction of the beta-casein gene in mouse mammary epithelial cells. *Mol. Cell Biol., 10*: 4027-4034, 1990.

[93] Marte, B. M., Jeschke, M., Graus-Porta, D., Taverna, D., Hofer, P., Groner, B., Yarden, Y., and Hynes, N. E. Neu differentiation factor/heregulin modulates growth and differentiation of HC11 mammary epithelial cells. *Mol. Endocrinol., 9*: 14-23, 1995.

[94] Yamauchi, T., Yamauchi, N., Ueki, K., Sugiyama, T., Waki, H., Miki, H., Tobe, K., Matsuda, S., Tsushima, T., Yamamoto, T., Fujita, T., Taketani, Y., Fukayama, M., Kimura, S., Yazaki, Y., Nagai, R., and Kadowaki, T. Constitutive tyrosine phosphorylation of ErbB-2 via Jak2 by autocrine secretion of prolactin in human breast cancer. *J. Biol .Chem., 275*: 33937-33944, 2000.

[95] Yamauchi, T., Ueki, K., Tobe, K., Tamemoto, H., Sekine, N., Wada, M., Honjo, M., Takahashi, M., Takahashi, T., Hirai, H., Tushima, T., Akanuma, Y., Fujita, T., Komuro, I., Yazaki, Y., and Kadowaki, T. Tyrosine phosphorylation of the EGF receptor by the kinase Jak2 is induced by growth hormone. *Nature, 390*: 91-96, 1997.

[96] Luttrell, L. M., Daaka, Y., and Lefkowitz, R. J. Regulation of tyrosine kinase cascades by G-protein-coupled receptors. *Curr. Opin. Cell Biol., 11*: 177-183, 1999.

[97] Normanno, N., Campiglio, M., De, L. A., Somenzi, G., Maiello, M., Ciardiello, F., Gianni, L., Salomon, D. S., and Menard, S. Cooperative inhibitory effect of ZD1839 (Iressa) in combination with trastuzumab (Herceptin) on human breast cancer cell growth. *Ann. Oncol., 13*: 65-72, 2002.

[98] Britten, C. D. Targeting ErbB receptor signaling: a pan-ErbB approach to cancer. *Mol. Cancer Ther., 3:* 1335-1342, 2004.

[99] Yaish, P., Gazit, A., Gilon, C., and Levitzki, A. Blocking of EGF-dependent cell proliferation by EGF receptor kinase inhibitors. *Science, 242*: 933-935, 1988.

In: Cell Growth Processes: New Research
Editor: Daiki Kimura, pp. 159-182

ISBN: 978-1-60456-132-6

Chapter VI

Regulation of Endothelial Cell Proliferation by Estrogen in Reproductive Organs

Keiichi Matsubara [*1], ***Yuko Matsubara*** [*1], ***Adam G. King*** [*2], ***Jing Zheng*** [*3], ***Emiko Abe*** [*1], ***Masaharu Ito*** [*1], ***and Ronald R. Magness*** [*3]

[*1]Department of Obstetrics and Gynecology, Ehime University School of Medicine,
[*2]Internal Medicine/Pediatrics Residency Program, Medical College of Wisconson,
[*3]Department of Obstetrics and Gynecology/ Perinatal Research Laboratories, University of Wisconsin Madison, USA

Endothelial Cell Proliferation

Estrogen and Endothelial Cell

In the female reproductive system, angiogenesis in the uterus, corpus luteum, and placenta is essential for embryo implantation and to support the dramatic elevation of uterine blood flow that occurs during pregnancy [1, 2]. Disruption of uteroplacental vascular development is associated with pregnancy loss, preeclampsia, and intrauterine fetal growth restriction [3]. Angiogenesis in the uterine endometrium [4, 5] is partly regulated by periodic changes in ovarian hormones such as estrogen, levels of which are markedly increased during the periovulatory period and throughout several important stages of pregnancy [6-8]. Estrogen Receptors (ERs) are expressed in uterine arterial endothelial cells (ECs) [9] and elsewhere in other ECs [10], suggesting that estrogen can act directly on ECs to alter uterine vascular function and mediate endogenous and exogenous estrogen-induced rises in uterine perfusion [11]. Estrogen can enhance angiogenesis [12] in systemic lupus erythematosus [13], ovarian vascular network [14], and ischemic lesion [15] and metastasis of breast cancer [16]. Thus, estrogen is thought to be important in fertility.

[1] Address all correspondence to: Keiichi Matsubara, Department of Obstetrics and Gynecology, Ehime University School of Medicine, Shitsukawa, Toon, Ehime 791-0295, Japan Telephone: +81-89-960-5379, Fax: +81-89-960-5381, E-mail: keiichi@m.ehime-u.ac.jp.

Estrogen has been reported to exert a protective, anti-atherosclerosis effect on the cardiovascular system [17, 18]. Consistent with these observations, the incidence of cardiovascular diseases in women is lower than in men but increases after menopause [19, 20] to that seen in male. Estrogen replacement therapy reduces the frequency of angina pectoris in postmenopausal women with syndrome X [21]. Moreover, estrogen reduces vascular neointimal proliferation involved in atherosclerosis repair in experimental animal models of vascular injury [22]. However, in the presence of progestins, estrogen may not contribute as much to cardiac protection and may even increase the risk of coronary heart disease (CHD) among otherwise healthy postmenopausal women [23]; although this may be age dependent. It is therefore unclear whether hormone replacement therapy is suited to the prevention of cardiovascular disease in older menopausal women. Among 50 to 59 years old women, who had undergone hysterectomy, the calcified-plaque burden in the coronary arteries with the use of electron-beam or multi detector-row computed tomography was lower in women assigned to estrogen than in those assigned to placebo [24].

Furthermore, randomized trials revealed that estrogen has been shown to ameliorate CHD among women 50-59 years, but not in those women aged 70–79 years [25]. Other randomized studies failed to demonstrate an effect of estrogen on CHD [26], possibly due to use of a broad age range.

Estrogen Receptors in Endothelial Cells

ERs, which mediate the effects of estrogen, were shown to be expressed in endothelium [27]. The study also showed that estradiol led to increased proliferation of human coronary artery and umbilical vein ECs, while the ER antagonist ICI 182,780 completely inhibited this increase. ER was shown to be nuclearly localized in ECs by immunostaining, and ER mRNA expression was increased by estrogen itself. However, no sex-differences in ER expression were detected in the endothelium. Interestingly, aortic tissue isolated from both female and male mice exhibit similar estrogen binding affinity [28]. However, estrogen-induced endothelium-dependent vascular relaxation is known to contribute to gender-differences in vascular tone. For example, norepinephrine causes less forearm vasoconstriction in women than in men [29]. Thus sex-specific difference in the effects of estrogen on ECs may not be caused by differences in ER expression but by cellular reactivity or other mechanisms.

Two subtypes of ER, ERα and ERβ, are encoded by separate genes, and might be responsible for the different vascular actions of estrogen, depending on the type of response element in target gene promoters and other cell specific coregulators [30]. According to the classical model, ligand-bound ER translocated to the nucleus to regulate gene transcription.

ERα expression and activity have been proposed to influence the prognosis of female cancers through regulation of tumor growth [31]. ERβ is localized to cell nuclei, and is expressed constitutively in both large and small vascular endothelial cells [32]. ERβ, which is almost ubiquitously expressed, may play a major role in mediating the effects of estrogen, but has not been detected in human uterine endometrial glandular epithelial cells, suggesting that ER expression in the uterus may differ from other organs. Although ERβ expression does not appear to be linked to ERα expression, both receptors are expressed in vascular ECs [32].

The uterine artery plays an important role in supplying nutrients and oxygen to the uterus and in sustaining the growing fetus. ERα and ERβ mRNA and protein are both expressed in uterine arterial endothelial cells (UAECs) [9, 33] and crosstalk between ERα and ERβ could influence endothelial function. ERα is uniformly expressed in UAECs, whereas ERβ is localized mainly in the nucleus [9]. Functionally, ERα is reported to mediate protective effects against vascular injury. Treatment with 17β-estradiol significantly increases expression of VEGF receptor (VEGFR) type 2 via ERα in myometrial microvascular endothelial cells [34]. 17β-estradiol-dependent angiogenesis [35] and re-endothelialization is mainly mediated by ERα [36], and lead to vascular protective effect. During the follicular phase of menstrual cycle, the expression of ERα is increased in UAECs, compared to the luteal phase [9]. ERα expression may be increased by estrogen itself since estrogen levels are similarly altered during different phases of menstrual cycle. Moreover, the expression levels of both ERα and estrogen are similar during pregnancy. ERα in UAECs is not altered by ovariectomy but is significantly increased by E2β. Estrogen must therefore play a key role in regulating the expression of ERα in UAECs. ERβ is also increased during follicular phase and in pregnancy compared with luteal phase [9]. ERβ expression levels are reduced to nearly half of the levels found in the luteal phase, and expression is significantly increased by E2β. ERβ is also important in vascular response to injury [37], and the effects of estrogen are dependent on complex interactions between ERα and ERβ.

Estrogen-Induced Angiogenesis

Angiogenesis occurs regularly in association with cyclic changes in uterine endometrium [38]. Estrogen can modulate angiogenesis under normal and pathological conditions, especially in the reproductive organs [39]. Endothelial abnormalities appear to be central to the pathogenesis of atherosclerosis and interfere with antithrombotic responses and endothelium-dependent vasodilation through endothelial nitric oxide (NO) availability [40]. After vascular injury, estrogen accelerates functional endothelial recovery through VEGF production [41]. The finding that estradiol inhibits vascular medial enlargement in the injured carotid artery of ERβ-deficient mice [42] indicates that ERα, but not ERβ, mediates the beneficial effect of estrogen on reendothelialization and potentially the prevention of atherosclerosis [36]. In the uterine endometrium, estrogen also regulates neovascularization by regulating expression and secretion of VEGF by glandular epithelial cells of the endometrium [43] or by stimulating basic fibroblast growth factor production through a paracrine mechanism during the menstrual cycle [44]. Estrogen is thus an important factor in promoting angiogenesis in reproductive organs.

Estrogen-Induced Uterine Arterial Angiogenesis

Uterine arterial angiogenesis is important for ensuring much blood flow to the uterus and the maintenance of pregnancy. Several methods are useful for investigating the effects of ovarian hormones (estrogen and progesterone) on angiogenesis (vascular differentiation and endothelial cellular proliferation), including a three-dimensional basement membrane assay (capillary tube formation assay) and [^{3}H] thymidine incorporation assay (DNA synthesis assay). We showed that E2β significantly increased capillary tube formation and DNA synthesis in UAECs (Figure 1).

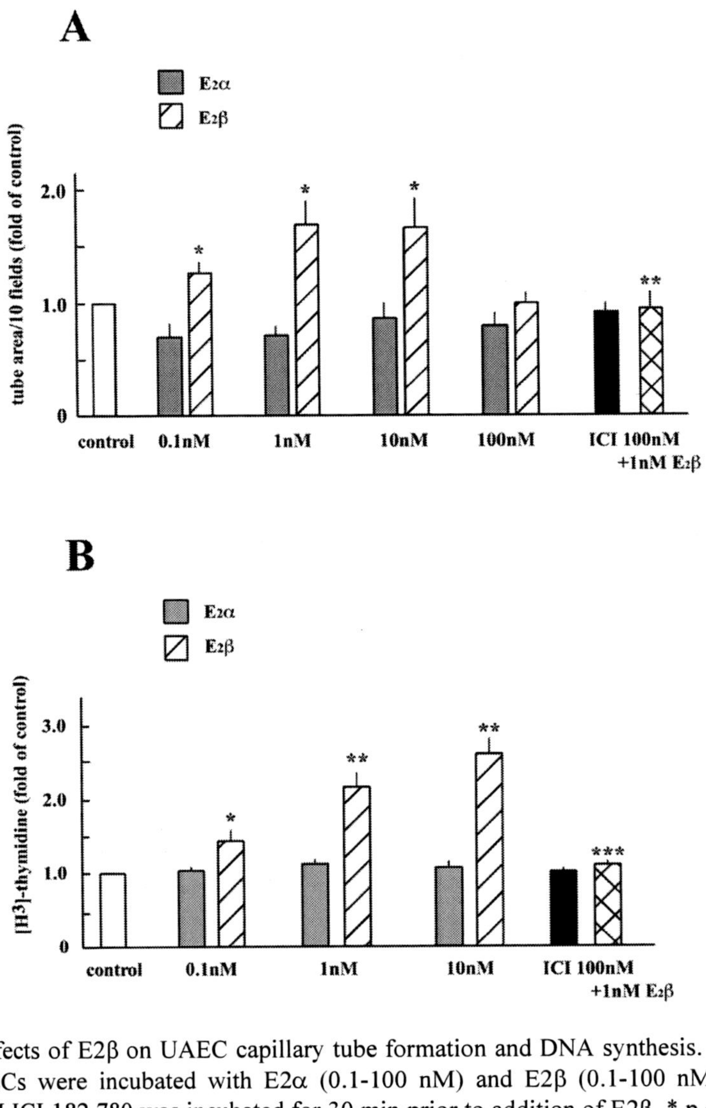

Figure 1. The effects of E2β on UAEC capillary tube formation and DNA synthesis. A) Capillary tube formation. UAECs were incubated with E2α (0.1-100 nM) and E2β (0.1-100 nM) for 6 hours on Matrigel. 100nM ICI 182.780 was incubated for 30 min prior to addition of E2β. * $p < 0.05$ vs. control, ** $p < 0.0005$ vs. 1nM E2β. B) DNA synthesis. UAECs were incubated with E2α (0.1-10 nM) or E2β (0.1-10 nM) for 16 hours. * $p < 0.05$, ** $p < 0.0001$ vs. control, *** $p < 0.0001$ vs. 1nM E2β. 100 nM ICI 182.780 was incubated for 30 min prior to E2β addition. Data are expressed as means ± SE.

E2β enhanced capillary tube formation on Matrigel in a dose-dependent fashion (Figure 1A). A maximal response was observed at low concentration (1nM). E2β also significantly stimulated DNA synthesis in UAECs in a dose-dependent manner (Figure 1B).

This observation suggests that E2β promotes UAEC angiogenesis not only by promoting cell proliferation, but also by influencing cell differentiation at some concentrations. In contrast, 17α-estradiol (E2α), which exhibits a low affinity for ERs, did not significantly alter capillary tube formation or DNA synthesis in UAECs. This observation further supports the proposal that the effects on UAECs were specific to E2β, mediated by ER, and not due to a nonspecific effect of steroid hormones. The ERα agonist 16-α-iodo-17β-estradiol (16αIE2) stimulated capillary tube formation and DNA synthesis in a dose-dependent manner (Figure 2A, B).

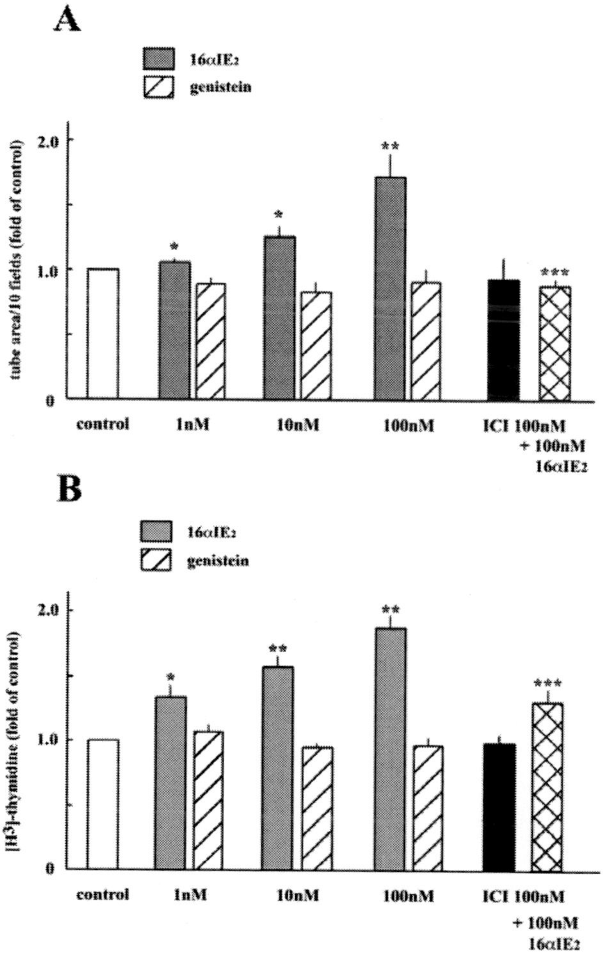

Figure 2. Effects of ERα and ERβ agonists on UAECs capillary tube formation and DNA synthesis. A) Capillary tube formation. UAECs were incubated with 16αIE2 (1-100 nM, ERα agonist) or genistein (1-100 nM, ERβ agonist) for 6 hours on Matrigel. 100 nM ICI 182.780 was incubated for 30 min prior to 100 nM E2β addition. * $p < 0.05$, ** $p < 0.005$ vs. control, *** $p < 0.005$ vs. 100 nM 16αIE2. B) DNA synthesis. UAECs were incubated with 16αIE2 (1-100 nM) or genistein (1-100 nM) for 16 hours on Matrigel. * $p < 0.005$, ** $p < 0.0001$ vs. control, *** $p < 0.01$ vs. 1 nM 16αIE2. 100 nM ICI 182.780 was incubated for 30 min prior to 100 nM 16αIE2 addition. Data are expressed as means ± SE.

In contrast to ERα stimulation, ERβ activated with genistein (ERβ agonist) had no effect on either capillary tube formation or DNA synthesis (Figure 2A, B). 16αIE2 has a high affinity for ERα (ERα:ERβ = about 30:1, Km=34.2nM) [45], whereas, genistein, which acts as a specific ERβ agonist at concentrations less than 1μM, has an affinity of ERβ: ERα of about

30:1 [46, 47]. These studies demonstrate that E2β-induced angiogenic effects are mediated by ERα but not ERβ in UAECs.

Estrogen-Induced Angiogenesis and Nitric Oxide

ERα is involved in E2β-mediated activation of NO synthase (NOS), resulting in the release of the relaxing factor NO from ECs. We showed that E2β-induced capillary tube formation was inhibited by Nω-nitro-L-arginine-methyl-ester (L-NAME) but not by an inactive stereoisomer, Nω-nitro-D-arginine-methyl-ester (D-NAME) (Figure 3A).

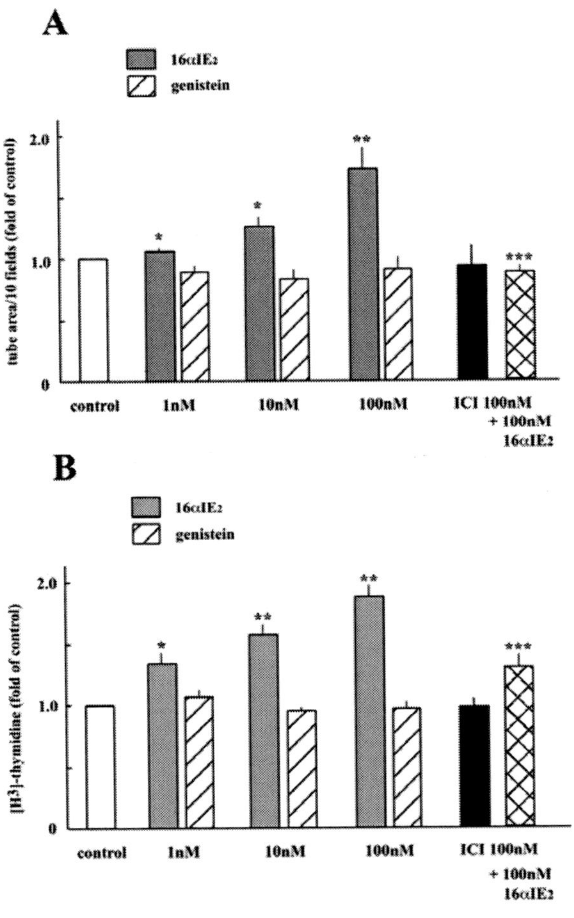

Figure 3. Effects of NO inhibition on E2β-induced UAEC angiogenesis. A) Capillary tube formation. 1 nM E2β-induced capillary tube formation was inhibited by 1 μM L-NAME on Matrigel. * $p < 0.005$ vs. control, ** $p < 0.05$ vs. 1 nM E2β. B) DNA synthesis. 1 nM E2β-induced DNA synthesis was not inhibited by L-NAME on Matrigel. Data are expressed as means ± SE.

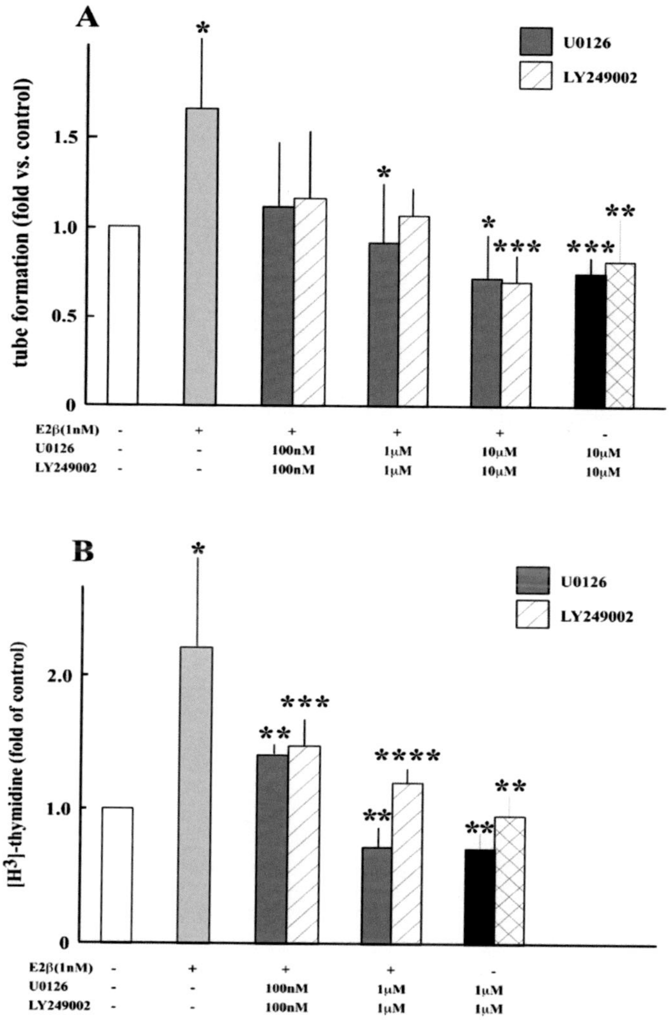

Figure 4. Signaling during E2β-induced UAEC angiogenesis. A) Capillary tube formation. 1 nM E2β-induced capillary tube formation was inhibited by 1 μM and 10 μM U0126 (MEK inhibitor) or 10 μM LY249002 (PI3-kinase inhibitor) on Matrigel. Inhibitors alone had no effect. * $p < 0.05$ vs. control, ** $p < 0.05$ vs. 1 nM E2β, *** $p < 0.01$ vs. 1 nM E2β. B) DNA synthesis. 1 nM E2β-induced DNA synthesis was also inhibited by 100 nM and 1 μM U0126 or 100 nM and 1 μM LY249002 on Matrigel. * $p < 0.005$ vs. control, ** $p < 0.0001$ vs. 1 nM E2β, *** $p < 0.05$ vs. 1 nM E2β, **** $p < 0.0005$ vs. 1 nM E2β. Data are expressed as means ± SE.

Furthermore, 10 μM sodium nitroprusside (SNP) and 10 μM cyclic guanosine monophosphate (c-GMP) enhanced capillary tube formation. It is possible that estrogen induces capillary tube formation through c-GMP and NO production increased by ERα activation. Capillary tube formation is rapid and might be mediated by activation of estrogen-binding sites on the UAEC membrane. An alternatively spliced, truncated ERα isoform exhibits a unique ability to mediate acute but not transcriptional responses to estrogen in human ECs. Truncated ERα localizes to the EC membrane where it mediates activation of estrogen-induced endothelial NOS (eNOS) [48, 49]. E2β is known to rapidly induce phosphorylation and activation of eNOS through the phosphatidylinositol 3 (PI3)-kinase-Akt pathway [50] and mitogen-activated protein kinase (MAPK) [51]. Since E2β-induced UAEC capillary tube formation and DNA synthesis were inhibited by treatment with U0126 (MAPK inhibitor) or LY249002 (PI3-kinase inhibitor) E2β-induced UAEC angiogenesis could be mediated by NO, which is produced in ECs through PI3-kinase and MAPK signaling (Figure 4A, B).

Given that ERα predominates in the uterus [45], ERα may play an important role in angiogenesis in female reproductive organs. In EC nuclei in most spiral arteries, immunoreactive ERβ is detectable, but not ERα is not [52]. Also, ERβ, but not ERα, mRNA is detectable in uterine endometrial ECs [53]. Using immunocytochemistry, we observed the expression of both ERα and ERβ in UAECs during capillary tube formation (Figure 5). In E2β-induced capillary tubes, ERα was detected in both nucleus and cytoplasm (Figure 5A), whereas, ERβ was primarily localized to nuclei only (Figure 5C). It has been recently shown that ERα is present in endothelial caveolae [54]. Nongenomic and short-term effects of E2β on NO release occurs in the caveolae through ERα activation. For instance, E2β treatment of MCF-7 cells triggers association of ERα with Src and p85, the regulatory subunit of PI3-kinase, leading to DNA synthesis [55]. E2β-induced rapid non-genomic actions through binding to ERα might be important for cell proliferation [56] and angiogenesis in reproductive organs.

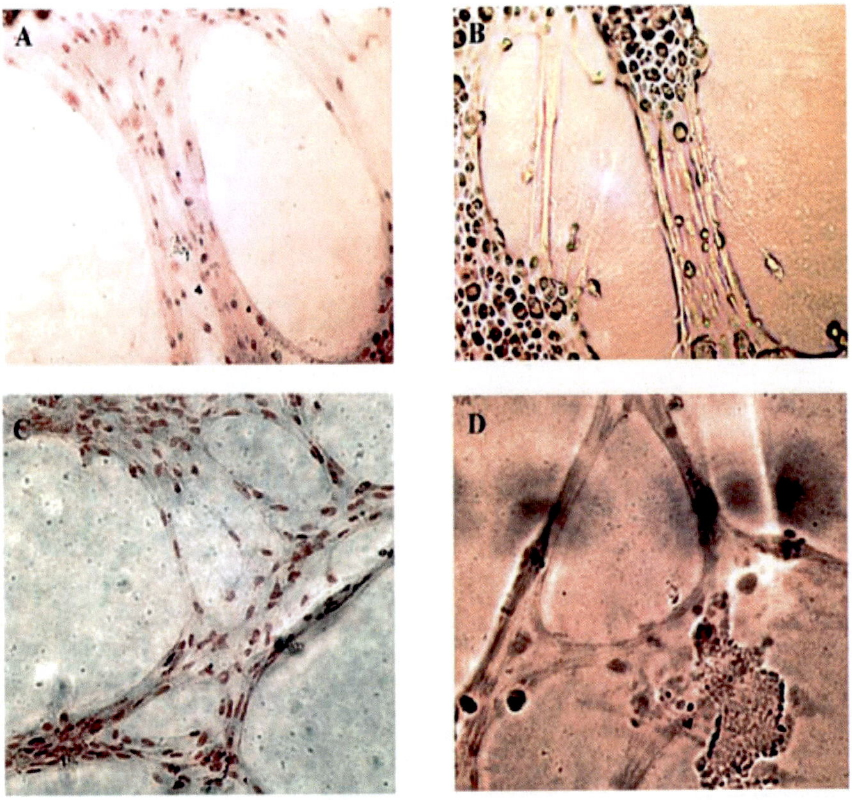

Figure 5. Immunostaining of capillary tubes with ERα and ERβ antibodies. A) Nucleus and cytoplasm were stained with ERα antibody. B) No immunoreactivity was observed with control IgG. C) Nucleus only was stained with ERβ antibody. D) No immunoreactivity was observed with control IgG.

Progesterone Effect on Estrogen-Induced Angiogenesis

Late in pregnancy, P4 increases to 30 times the levels detected during the luteal phase. P4 is a potent inhibitor of E2β-induced cell proliferation. In the uterus, estrogen cooperates with P4 in endometrial proliferation and differentiation prior to embryo implantation by regulating P4 receptor (PR) gene expression [57]. In addition, the preovulatory surge in estrogen levels upregulates uterine expression of ER and PR genes, and endometrial levels of ER and PR mRNAs increased 500% and 75%, respectively [58]. We showed that P4 could inhibit E2β-induced angiogenic response (Figure 6). Physiological levels of P4 reportedly arrest the endothelial cell cycle and inhibit proliferation through a nuclear receptor-mediated mechanism [59]. Furthermore, P4 reduces ER and PR expression in the human uterine vascular endothelium [5], as well as E2β-induced cell proliferation as the uterine epithelium changes from the proliferative phase to the secretory phase [22]. P4 might reduce E2β-induced angiogenesis in part by decreasing ER and PR expression. P4 also reduces E2β-induced VEGF binding to myometrial microvascular endothelial cells [34]. Negative cross-talk between estrogen and P4 during angiogenesis could be regulated by several mechanisms.

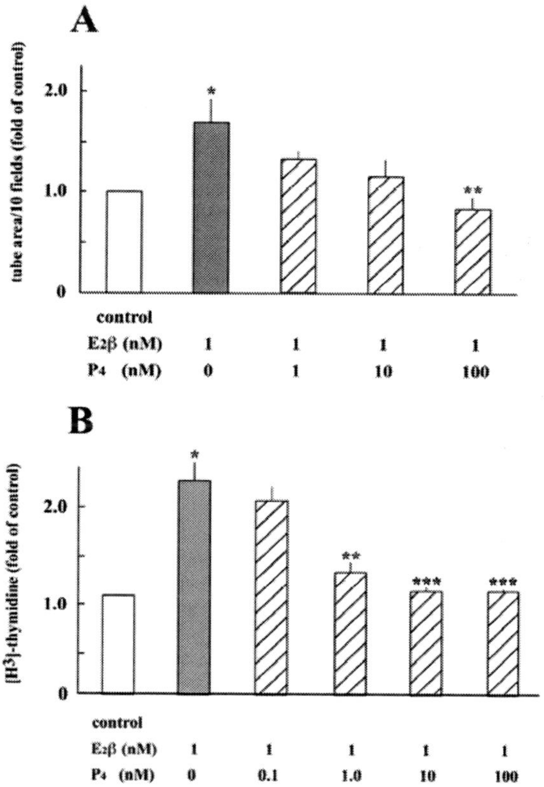

Figure 6.The effects of P4 on E2β-induced UAEC angiogenesis. A) Capillary tube formation. 1 nM E2β-induced capillary tube formation was reduced by 100 nM P4 on Matrigel. * $p < 0.05$ vs. control, ** $p < 0.05$ vs. 1 nM E2β. B) DNA synthesis. 1 nM E2β-induced DNA synthesis was also reduced by 1.0, 10, and 100 nM P4 on Matrigel. * $p < 0.0001$ vs. control, ** $p < 0.001$, *** $p < 0.0001$ vs. 1 nM E2β. Data are expressed as means ± SE.

Uterine angiogenesis during pregnancy could be strongly mediated by steroid hormones, including estrogen and P4. However, these hormonal effects are complex and controversial. Estrogen and P4 independently stimulate DNA synthesis and capillary tube formation, but P4 reduces estrogen-induced angiogenesis. P4 is known to influence mitotic rate associated with the endometrial vasculature and is implicated as an anti-angiogenic steroid in the suppression of neovascularization [60, 61]. On the contrary, Heryanto *et al.* reported that P4 stimulated further E2β-induced endometrial ECs' proliferation at high dose [61]. P4 may increase endothelial cellular proliferation synergistically at high dose of estrogen as in pregnancy. The balance of estrogen and P4 could play an important role in the regulation of angiogenesis in the uterine endometrium and corpus luteum during the menstrual cycle.

Estrogen-Induced Angiogenesis and Co-Factors

Estrogen-induced endometrial epithelial cell proliferation is mediated in part through paracrine signaling by stromal cell-secreted insulin-like growth factor I (IGF-I). This mechanism is thought to be involved in decidualization during embryo implantation. Indeed, 17ß-estradiol induces IGF-I release from endometrial stromal cells in culture [62]. IGF-1, probably in addition to other factors, induces proliferation of epithelial cells including ECs. In our hands, complete-Matrigel was essential for estrogen-mediated effects on proliferation and differentiation of cultured UAECs. However, these effects were not observed when growth factor-reduced (GFR) Matrigel was used (Figure 7). Thus factors present in Matrigel are likely to contribute to induction of EC differentiation (capillary tube formation) in the absence of stromal cells. To find the essential growth factors on E2β-induced UAECs' capillary tube formation we examined the effects of epidermal growth factor (EGF), platelet derived growth factor (PDGF) and IGF-1, which was significantly reduced in GFR-Matrigel, compared to complete-Matrigel.

EGF consists of 53 amino acids and the receptor is identified in embryo and uterine endometrium [63]. Since serum concentrations are decreased with advancing age, EGF is thought to be regulated by ovarian hormones. EGF could reduce estrogen production and stimulate P4 production by ovarian garanulosa cells. Also, EGF affects ovarian proliferation and differentiation [64]. In contrast, estrogen stimulates EGF production in the uterine endometrium. Interaction between estrogen and EGF is thought to play an important role in uterus. Estrogen-induced uterine endometrial proliferation and angiogenesis during the proliferative phase is partly mediated by EGF, but EGF production is not altered during menstrual cycle. However, expression of the EGF receptor fluctuates periodically, peaking before ovulation [65]. EGF, which is important for embryonic attachment to the uterine endometrium, may be regulated by paracrine interactions between the endometrial epithelium and the stroma. Also, EGF might be involved in the cellular proliferation and differentiation. However, EGF did not affect of capillary tube formation or DNA synthesis on Matrigel. Furthermore, in GFR-Matrigel, EGF content (<0.5ng/ml) is lower than complete-Matrigel (0.5-1.3 ng/ml). Addition of EGF GFR-Matrigel, however, did not improve E2β-induced capillary tube formation. E2β and P4 also did not influence EGF-induced capillary tube formation in our study. Thus EGF is not essential for E2β-induced angiogenesis.

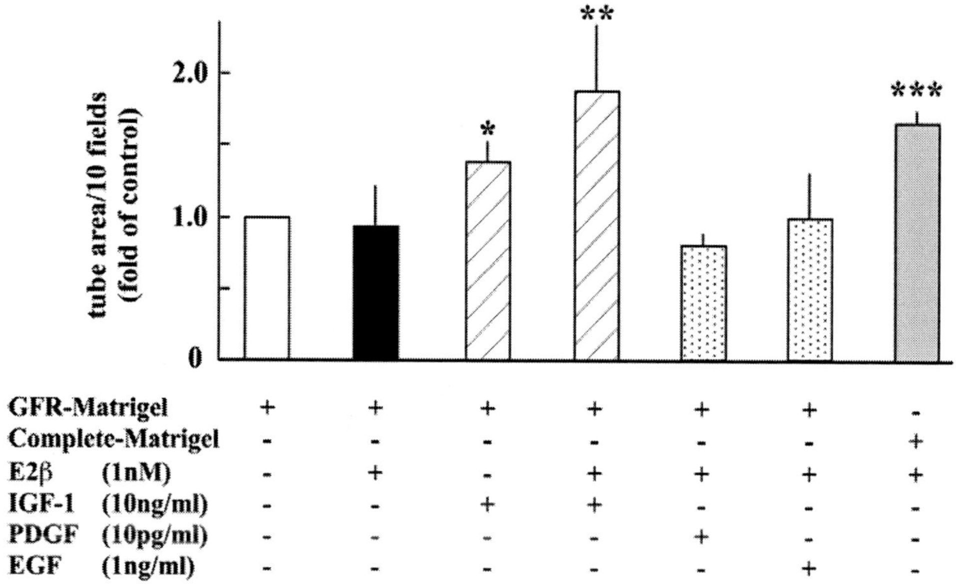

Figure 7. E2β-induced UAEC angiogenesis on GFR-Matrigel. 1 nM E2β did not induce capillary tube formation on GFR-Matrigel. IGF-1 induced capillary formation on GFR-Matrigel and 1 nM E2βincreased the angiogenic effect synergistically to levels induced by 1 nM E2β on complete-Matrigel. * $p < 0.005$ vs. 1 nM E2β on GFR-Matrigel, ** $p < 0.001$ vs. 1 nM E2β on GFR-Matrigel, *** $p < 0.005$ vs. 1 nM E2β on GFR-Matrigel . Data are expressed as means ± SE.

PDGF is produced by stromal cells and induces EC proliferation. In addition, release of PDGF is stimulated by estrogen. However, the amount of PDGF is not significantly altered during the menstrual cycle. Our data suggest that PDGF did not influence capillary tube formation or DNA synthesis on Matrigel. Furthermore, although the PDGF content in GFR-Matrigel (<0.5 pg/ml) was significantly decreased compared to complete-Matrigel (12 pg/ml), addition of PDGF to GFR-Matrigel did not improve E2β-induced capillary tube formation. We therefore conclude that PDGF is also not essential for E2β-induced angiogenesis.

Competence factors such as basic fibroblast growth factor (bFGF), PDGF, and estrogen stimulate transition from G0 to G1 phase of the cell cycle. On the other hand, progression factors such as IGF-1 stimulate transition from G1 to S. These factors are expected to produce a synergistic effect. In our study, EGF and PDGF had no stimulative effect on capillary tube formation or DNA synthesis in UAECs on complete-Matrigel. In contrast, IGF-1 significantly increased capillary tube formation and DNA synthesis in UAECs on complete-Matrigel as initially expected. VEGF and bFGF induced capillary tube formation in UAECs on Matrigel; however, this effect was reduced by E2β or P4 (data not shown). Interestingly, IGF-1-induced capillary tube formation was accelerated by E2β but DNA synthesis was not. Since E2β-induced capillary tube formation was completely inhibited by anti-IGF-1 receptor antibody (data not shown), the synergistic effect is likely to be specific to IGF-1. In GFR-Matrigel, IGF-1 content (5 ng/ml) was significantly lower than complete-Matrigel (15.6 ng/ml). Since addition of IGF-1 improved E2β-induced capillary tube formation in GFR-Matrigel, we propose that IGF-1 is essential for E2β-induced angiogenesis (Figure 7).

IGF-1 mRNA and protein are expressed during the luteal phase of the menstrual cycle and in the first month of pregnancy in bovine corpus luteum [66]. VEGF is a key angiogenesis-regulating factor and is associated with ovarian luteinization and formation of corpus luteum [67]. IGF-1, a 70-amino acid, indirectly affects angiogenesis in the luteinized corpus luteum by stimulating production of VEGF. IGF-1 signaling is mediated by IGF-1 receptor (IGF-1R), which consists of α and β subunits. The α subunit of IGF-1R is similar to insulin receptor and β subunit is also regarded as homologous with 60% of insulin receptor. IGF-1 and IGF-1R conjugation could phosphorylate IRS-1 and the phosphorylation activates MAPK and PI3-kinase through Ras activation.

Estrogen-induced uterine endometrial epithelial cell proliferation is partly mediated and enhanced by stromal cell-derived IGF-1 [62]. The serum IGF-1 concentration is increased by 1.6-folds late in pregnancy and decreased after delivery. Serum IGF-1 levels may therefore be regulated by placental factors, including estrogen. IGF-1, which is widely but selectively distributed in all developing tissues, is mainly produced in liver and placenta [68], where it could stimulate proliferation of many types of cells such as fibroblasts, smooth muscle cells, and epithelial cells [69]. IGF-1 is normally expressed abundantly in decidual cells at the implantation site, where it functions as a proliferative factor, in autocrine and paracrine manners [70]. IGF-1 is regulated by estrogen and P4. Surprisingly, growth hormone-induced hepatic IGF-1 gene expression and serum IGF-1 concentrations in the rat uterus are reduced by chronic estrogen administration. On the contrary, estrogen has no effect on hepatic IGF-I expression in acute experiments [71]. Therefore, chronic E2β exposure may inhibit the synergistic effects on estrogen-induced angiogenesis. Also, acute estrogen infusion (2 hours) increased uterine blood flow; however, chronic infusion (several days) decreased substantially almost back to baseline [72, 73].

We showed that IGF-1 could stimulate capillary tube formation in UAECs on Matrigel, but did not affect the DNA synthesis. Though capillary tube formation can be completed within 6 hours, [^{3}H]-thymidine incorporation requires at least 12 hours; thus the results do not conflict with the fact that IGF-1 production is reduced by chronic estrogen administration. However, it is reported that estrogen did not affect the production of IGF-1 in acute experiments [71]. Estrogen does not reduce IGF-1 production in Matrigel and may work synergistically with IGF-1 in capillary tube formation via a short angiogenic reaction.

IGF-1 signaling leads to phosphorylation of IGF-1 receptor and its substrates (IRS-1 and IRS-2), which in turn associate with enzyme PI3-kinase. IRS-1/2 are components of the insulin/IGF-IR signaling cascade and major substrates of IGF-1 receptor tyrosine kinases. Tyrosine-phosphorylated IRS-1 binds PI3-kinase, which may be involved in the translocation of glucose transporters and in activation of the MAP kinase cascade [74]. IGF-1 stimulates MAP kinase and extracellular signal regulated kinase (ERK). IGF-1 and activated ERK are thought to play an important role for EC proliferation and differentiation [75].

Estrogen increases phosphorylation of ERK and Akt in rat, and estrogen and IGF-1 synergistically activate of PI3-kinase/Akt signaling [33, 76, 77]. These two signaling cascades are coupled to the IGF-1 receptor [78]. Furthermore, it is known that ERα is involved in IGF-1-dependent dendritic growth [78] and neuronal survival [77]. The normal angiogenic function of estrogen may be partly mediated by IGF-1 in female reproductive organs (Figure 8).

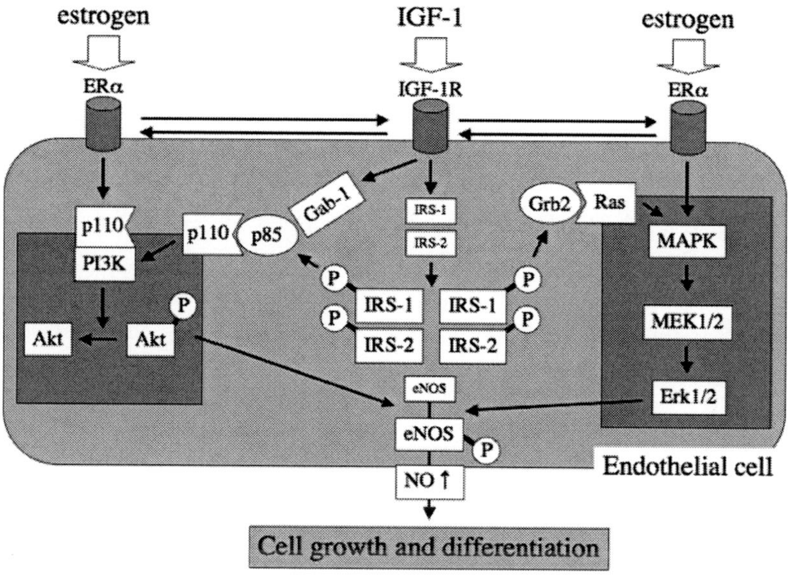

Figure 8. Model of estrogen signaling transduction and IGF-1 in angiogenesis. IGF-1 receptor binding leads to phosphorylation of the IRS family. Subsequent interactions with signaling molecules result in activation of PI3 and MAP kinases. Estrogen also activates PI3-kinase and MAP kinase through ERα activation at the EC membrane. These signaling pathways interact with and coordinate EC proliferation and differentiation in part through NOS activation.

Endothelial Progenitor Cell Proliferation

Generation of vessels can be classified into two general categories: angiogenesis and vasculogenesis. Angiogenesis is defined as the sprouting of capillaries from preexisting vessels, for example (e.g. tumor angiogenesis). Vasculogenesis is defined as the development of blood vessels from differentiating ECs (e.g. fetal vascularization) [79]. Early in pregnancy, fetal vascular system is comprised of circulating hematopoietic cells and ECs that line the blood vessels. Hematopoietic cells are derived from endothelial progenitor cells (EPCs) on the periphery of blood islands [80].

CD34-positive EPCs have been detected in peripheral blood from adults [81], and vasculogenesis has also been found to occur postnatally. EPCs, which are derived from bone marrow [82], are incorporated into areas of myocardial infarction and involved in neovascularization of ischemic lesions [83]. EPCs in the peripheral blood serve as a reserve for EC maintenance and contribute up to 25% of ECs in newly formed vessels [84]. In female reproductive organs, EPCs appear to play a crucial role in vascularization of the uterine endometrium at the site of embryo implantation and placentation [83, 85, 86], consistent with the proposal that angiogenesis, and probably vasculogenesis, occur in the uterine endometrium during the luteal phase through early pregnancy.

During menstrual cycle, the number of circulating EPCs in peripheral blood exhibits a cyclic change, markedly increasing during the luteal phase and menstruation, and decreasing in the follicular phase (Figure 9). EPCs have been identified in corpus luteum and the uterine endometrium following ovulation, and are involved in neovascularization [83]. Circulating bone marrow-derived EPCs are thought to serve as EC reservoirs in female reproductive organs. The menstrual cycle was associated with increased numbers of EPCs in the peripheral blood, and estrogen and P4 could be involved in their proliferation and mobilization. However, the number of EPCs has been shown to be decreased during the course of pregnancy [87] (Figure 9). During pregnancy, other placental factors may reduce EPC proliferation and mobilization. Estrogen stimulates extensive placental neovascularization through upregulation of angiogenic factors in the uterus, especially during the early stages of pregnancy [88]. Thus circulating EPCs may be involved in neovascularization of female reproductive organs.

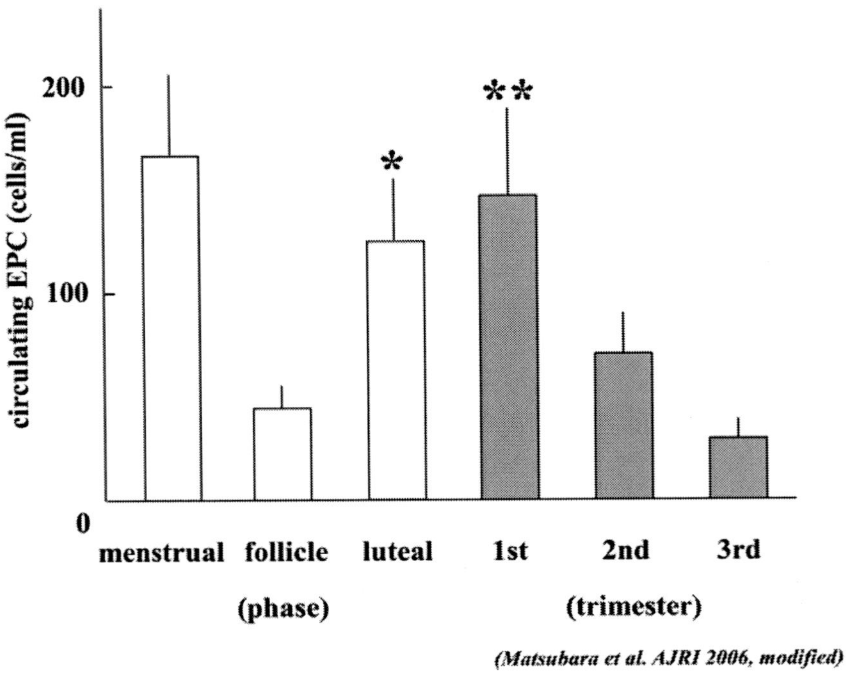

Figure 9. Circulating EPCs in the peripheral blood. The number of circulating EPCs is significantly increased in the luteal phase compared with the follicular phase of the menstrual cycle. This number decreased throughout pregnancy. * $p < 0.05$ vs. follicle phase, ** $p < 0.001$ vs. the 3rd trimester. Data are expressed as means ± SE.

Estrogen and Endothelial Progenitor Cell Proliferation

Interestingly, the increase in EPC in menstrual blood was not significantly changed after 7 days in culture compared to the number of EPCs in luteal blood. The discrepancy between menstrual and luteal EPC number may be due to different serum concentrations of estrogen and progesterone. We showed that E2β stimulated menstrual EPC proliferation in a dose-dependent manner. However, E2β did not influence luteal EPC proliferation after 7 days in culture (Figure 10). These observations suggest that EPC proliferation may require priming by estrogen. While luteal EPCs might be already primed by estrogen, the dose may be limited. Estrogens reportedly stimulate uterine endometrial proliferation through increased expression of ERs during the follicular phase and the differentiation of uterine endometrial glands through PR activation in the luteal phase. The uterine endometrium expresses ERα and ERβ, although higher levels of ERα are generally detected. We showed that in luteal EPCs more ERβ than ERα was detected in the peripheral blood by real-time PCR (data not shown). In contrast, more ERα was detected in menstrual EPCs than ERβ (data not shown). Increased expression of ERα may at least partially explain the increase in menstrual EPC proliferation compared with luteal EPCs.

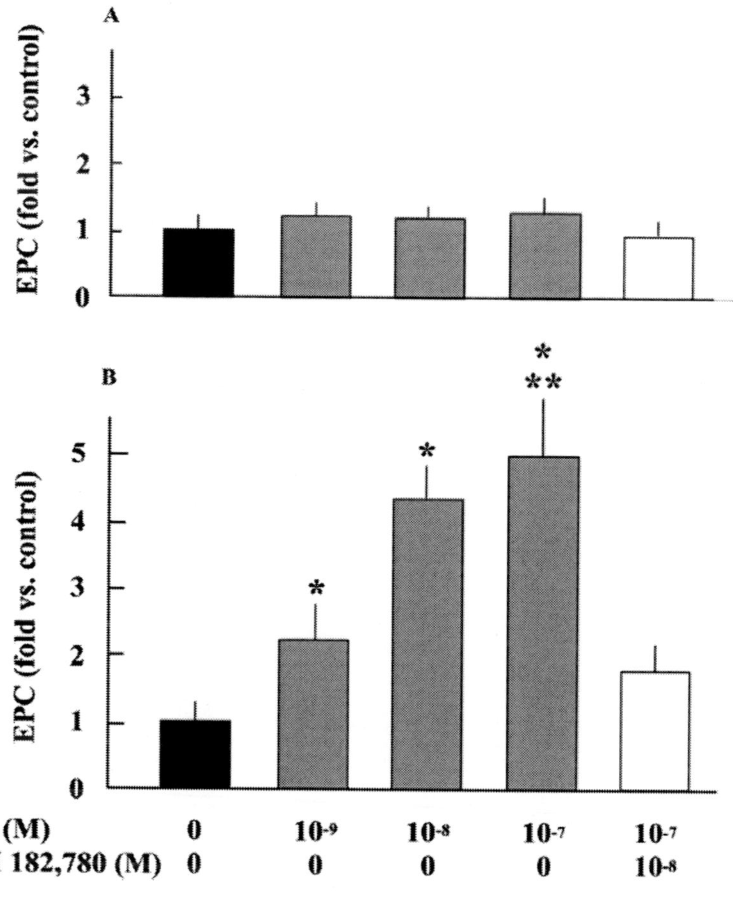

Figure 10. EPC proliferation after 7 days of culture with E2β. A) Proliferation of EPCs derived from the peripheral blood during the luteal phase. Treatment with E2β did not affect EPC proliferation. B) Proliferation of EPCs derived from peripheral blood during menstruation. E2β treatment significantly increased proliferation in a dose-dependent manner. * $p < 0.05$ vs. 0 M E2β, ** $p < 0.05$ vs. 10^{-8} M ICI 182.780. Data are expressed as means ± SE.

EPCs accelerate vascular repair by restoring blood flow [89], and reduce atherosclerosis, and improve cardiac function after myocardial infarction, ameliorating injured coronary arteries [90]. Estrogens may increase the number of EPCs by inhibiting caspase-8-dependent anti-apoptotic effects, thereby inducing vascular repair [91]. Activation of both ERα and ERβ contribute to E2β-induced EPC proliferation, since estrogen could not efficiently induce gene transcription in EPCs obtained from ERα or ERβ-knockout models [92]. Furthermore, estrogen increases ERα mRNA levels in EPCs, leading to VEGF expression and proliferation in EPCs [92]. An increase in EPCs could be beneficial for vascular injury, and estrogen might contribute to recruitment of EPCs into neovascularization foci [93], stimulation of EPC production and survival, and an increase in the number of circulating EPCs. Estrogen accelerates re-endothelialization of injured arteries in part through mobilization of bone marrow-derived EPCs through a PI3-kinase–dependent pathway [83, 94], which promotes repair of human atherosclerosis [91] and perfusion salvage enhancement [89]. Estrogen preserves ischemic heart function during myocardial infarction by augmenting mobilization and incorporation of bone marrow-derived EPCs into sites of neovascularization. The effect is mediated by eNOS-induced augmentation of MMP-9 expression in bone marrow, since estrogen-induced EPC mobilization was reduced it in eNOS-knockout or MMP-9-knockout mice [93]. A decreased number of healthy circulating progenitor cells might be contribute to coronary artery disease risk [95]. Estrogen also stimulates extensive placental neovascularization through upregulation of angiogenic factors in the uterus during pregnancy [96] and the number of circulating EPCs was increased throughout the 1st trimester [87]. This finding is consistent with the fact that vasculogenesis peaks during embryogenesis [79, 80]. Estrogen must be a key to modulating EPC proliferation and mobilization during the menstrual cycle and in pregnancy by stimulating growth of the uterine endometrial vasculature.

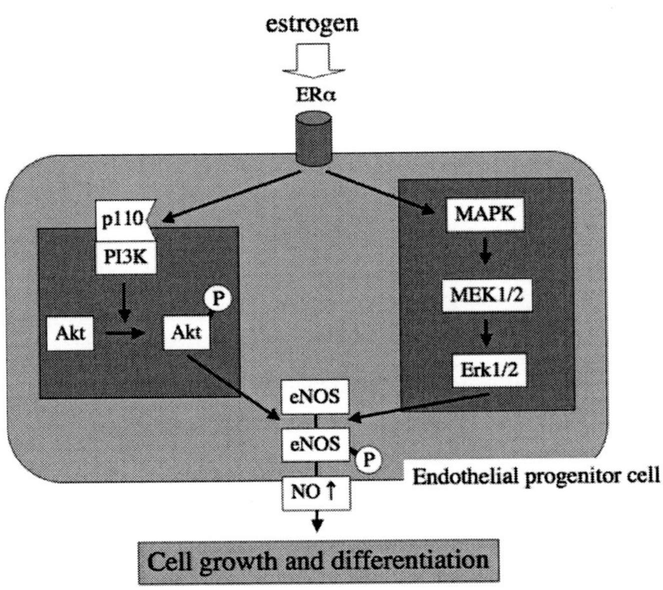

Figure 11.Model of estrogen signaling during EPC proliferation. Estrogen activates PI3 and MAP kinases through ERα activation at the EPC membrane. This signaling pathway coordinates EPC proliferation in part through NOS activation.

After menopause, the estrogen concentration in the peripheral blood is significantly reduced. Depletion of EPCs for uterine vascularization prior to menopause could lead to depletion of EPCs. The number and availability of EPCs is reduced in females after menopause, compared to males [93]. Estrogen can also delay the onset of senescence in bone marrow-derived EPCs, leading to increased cell survival [97].

PI3-kinase/Akt and eNOS/NO signaling pathways are critical for estrogen-induced, ERα-dependent proliferation and differentiation of ECs and EPCs. Estrogen induces not only reendothelialization of vascular injury, but also vasculogenesis and angiogenesis in female reproductive organs by stimulating endothelial cell proliferation and differentiation, EPC proliferation and mobilization during implantation and pregnancy. This review supports the hypothesis that estrogen is critical in promoting angiogenesis and vasculogenesis in female reproductive organs (Figure 11).

Acknowledgement

This work is supported in part by the following grants: National Institutes of Health Grants HL38843, HL49210, and HL87144 (Professor Magness), and a Grant-in-Aid for Japan Scientific Research from the Ministry of Education, Culture, Sports, Science and Technology (18591803).

References

[1] Magness R, Zheng J. Maternal Cardiovascular Alterations During Pregnancy. In: Pediatrics and Perinatology: The Scientific Basis 2nd edition, Editors: PD Gluckman and MA Heymann, Arnold Publishing, London 1996: 762-772.

[2] Magness RR. Maternal cardiovascular and other physiologic responses to the endocrinology of pregnancy. In: Endocrinology of Pregnancy Ed: Fuller Bazer, Humana Press Inc, Totowa NJ, Chapter 1998; 18: 507-539.

[3] Perrot-Applanat M, Deng M, Fernandez H, Lelaidier C, Meduri G, Bouchard P. Immunohistochemical localization of estradiol and progesterone receptors in human uterus throughout pregnancy: expression in endometrial blood vessels. *J. Clin. Endocrinol. Metab.* 1994; 78: 216-224.

[4] Hyder S, Stancel G. Regulation of angiogenic growth factors in the female reproductive tract by estrogens and progestins. *Mol. Endocrinol.* 1999; 13: 806-811.

[5] Perrot-Applanat M, Ancelin M, Buteau-Lozano H, Meduri G, Bausero P. Ovarian steroids in endometrial angiogenesis. Steroids 2000; 65: 599-603.

[6] Carnegie J, Robertson H. Conjugated and unconjugated estrogens in fetal and maternal fluids of the pregnant ewe: a possible role for estrone sulfate during early pregnancy. *Biol. Reprod.* 1978; 19: 202-211.

[7] Magness R, Rosenfeld C, Carr B. Protein kinase C in uterine and systemic arteries during the ovarian cycle and pregnancy. *Am. J. Physiol.* 1991; 260: E464-E470.

[8] Tungsubutra G, France J. Steroid levels in pregnancy. *Aust. NZ J. Obstet. Gynecol.* 1978; 18: 97.

[9] Byers M, Zangl A, Phernetton T, Lopez G, Chen D, Magness R. Endothelial vasodilator production by ovine uterine and systemic arteries: ovarian steroid and pregnancy control of ERalpha and ERbeta levels. *J. Physiol.* 2005; 565: 85-99.

[10] Chen Z, Yuhanna I, Galcheva-Gargova Z, Karas R, Mendelsohn M, Shaul P. Estrogen receptor alpha mediates the nongenomic activation of endothelial nitric oxide synthase by estrogen. *J. Clin. Invest.* 1999; 103: 401-406.

[11] Magness R, Phernetton T, Gibson T, Chen D. Uterine blood flow responses to ICI 182 780 in ovariectomized oestradiol-17beta-treated, intact follicular and pregnant sheep. *J. Physiol.* 2005; 565: 71-83.

[12] Morales D, McGowan K, Grant D, Maheshwari S, Bhartiya D, Cid M, Kleinman HK, Schnaper HW. Estrogen promotes angiogenic activity in human umbilical vein endothelial cells in vitro and in a murine model. *Circulation* 1995; 91: 755-763.

[13] Grossman C. Possible underlying mechanisms of sexual dimorphism in the immune response: fact and hypothesis. *J. Steroid. Biochem.* 1989; 34: 241-251.

[14] Grazul-Bilska A, Navanukraw C, Johnson M, Vonnahme K, Ford S, Reynolds L, Redmer DA. Vascularity and expression of angiogenic factors in bovine dominant follicles of the first follicular wave. *J. Anim. Sci.* 2007; 85: 1914-1922.

[15] Cho M, Ziats N, Abdul-Karim F, Pal D, Goldfarb J, Utian W, Gorodeski GI. Effects of estrogen on tight junctional resistance in cultured human umbilical vein endothelial cells. *J. Soc. Gynecol. Investig.* 1998; 5: 260-270.

[16] Dabrosin C, Margetts P, Gauldie J. Estradiol increases extracellular levels of vascular endothelial growth factor in vivo in murine mammary cancer. *Int. J. Cancer* 2003; 107: 535-540.

[17] Clark S, Scofield P, Grace A, Metcalfe J, Kirschenlohr H. Tamoxifen effects on endothelial function and cardiovascular risk factors in men with advanced atherosclerosis. *Circulation* 2001; 103: 1497-1502.

[18] Vischer U. Estrogens and atherosclerosis: a direct protective effect on the vascular wall? *Eur. J. Endocrinol.* 1997; 137: 343-344.

[19] Colditz G, Willett W, Stampfer M, Rosner B, Speizer F, Hennekens C. Menopause and the risk of coronary heart disease in women. N Engl J Med 1987; 316: 1105-1110.

[20] Stevenson J. HRT and the primary prevention of cardiovascular disease. *Maturitas* 2007; 57: 31-34.

[21] Rosano G, Peters N, Lefroy D, Lindsay D, Sarrel P, Collins P, Poole-Wilson PA. 17-beta-Estradiol therapy lessens angina in postmenopausal women with syndrome X. *J. Am. Coll. Cardiol.* 1996; 28: 1500-1505.

[22] Levine R, Chen S-J, Durand J, Chen Y-F, Oparil S. Medroxyprogesterone attenuates estrogen-mediated inhibition of neointima formation after balloon injury of the rat carotid artery. *Circulation* 1996; 94: 2221-2227.

[23] Manson J, Hsia J, Johnson K, Rossouw J, Assaf A, Lasser N, Trevisan M, Black H, Heckbert S, Detrano R, Strickland O, Won N, Crouse J, Stein E, Cushman M, Women's Health Initiative Investigators. Estrogen plus progestin and the risk of coronary heart disease. *N. Engl. J. Med.* 2003; 349: 523-534.

[24] Manson J, Allison M, Rossouw J, Carr J, Langer R, Hsia J, Kuller L, Cochrane B, Hunt J, Ludlam S, Pettinger M, Gass M, Margolis K, Nathan L, Ockene J, Prentice R, Robbins J, Stefanick M, WHI, WHI-CACS Investigators. Estrogen therapy and coronary-artery calcification. *N. Engl. J. Med.* 2007; 358: 2591-2602.

[25] Hsia J, Langer R, Manson J, Kuller L, Johnson K, Hendrix S, Pettinger M, Heckbert S, Greep N, Crawford S, Eaton C, Kostis J, Caralis P, Prentice R, Women's Health Initiative Investigators. Conjugated equine estrogens and coronary heart disease: the Women's Health Initiative. *Arch. Intern. Med.* 2006; 166: 357-365.

[26] Committee WsHIS. Effects of conjugated equine estrogen in postmenopausal women with hysterectomy. *J. Am. Med .Assoc.* 2004; 291: 1701–1712.

[27] Kim-Schulze S, McGowan K, Hubchak S, Cid M, Martin M, Kleinman H, Greene G, Schnaper H. Expression of an estrogen receptor by human coronary artery and umbilical vein endothelial cells. *Circulation* 1996; 94: 1402-1407.

[28] Rubanyi G, Freay A, Kauser K, Sukovich D, Burton G, Lubahn D, Couse J, Curtis S, Korach K. Vascular estrogen receptors and endothelium-derived nitric oxide production in the mouse aorta. Gender difference and effect of estrogen receptor gene disruption. *J. Clin. Invest.* 1997; 99: 2429-2437.

[29] Kneale B, Chowienczyk P, Brett S, Coltart D, Ritter J. Gender differences in sensitivity to adrenergic agonists of forearm resistance vasculature. *J. Am. Coll. Cardiol.* 2000; 36: 1233–1238.

[30] Dechering K, Boersma C, Mosselman S. Estrogen receptors alpha and beta: two receptors of a kind? *Curr. Med. Chem.* 2000; 7: 561-576.

[31] Ma W, Tan J, Matsumoto H, Robert B, Abrahamson D, Das S, Dey S. Adult tissue angiogenesis: evidence for negative regulation by estrogen in the uterus. *Mol. Endocrinol.* 2001; 15: 1983-1992.

[32] Taylor A, Al-Azzawi F. Immunolocalisation of oestrogen receptor beta in human tissues. *J. Mol. Endocrinol.* 2000; 24: 145-155.

[33] Chen D, Bird I, Zheng J, Magness R. Membrane estrogen receptor-dependent extracellular signal-regulated kinase pathway mediates acute activation of endothelial nitric oxide synthase by estrogen in uterine artery endothelial cells. *Endocrinology* 2004; 145: 113-125.

[34] Gargett C, Zaitseva M, Bucak K, Chu S, Fuller P, Rogers P. 17Beta-estradiol up-regulates vascular endothelial growth factor receptor-2 expression in human myometrial microvascular endothelial cells: role of estrogen receptor-alpha and -beta. *J. Clin. Endocrinol. Metab.* 2002; 87: 4341-4349.

[35] Johns A, Freay A, Fraser W, Korach K, Rubanyi G. Disruption of estrogen receptor gene prevents 17beta estradiol-induced angiogenesis in transgenic mice. *Endocrinology* 1996; 137: 4511-4513.

[36] Brouchet L, Krust A, Dupont S, Chambon P, Bayard F, Arnal J. Estradiol accelerates reendothelialization in mouse carotid artery through estrogen receptor-a but not estrogen receptor-beta. *Circulation* 2001; 103: 423-428.

[37] Lindner V, Kim S, Karas R, Kuiper G, Gustafsson J, Mendelsohn M. Increased expression of estrogen receptor mRNA in male blood vessels after vascular injury. *Circ. Res.* 1998; 83: 224–229.

[38] Weston G, Rogers P. Endometrial angiogenesis. Bailliere's best practice and research clinical obstetrics and gynaecology 2000; 14: 919-936.

[39] Findlay J. Angiogenesis in reproductive tissues. *J. Endocrinol.* 1986; 111: 357-366.

[40] Harrison D. Cellular and molecular mechanisms of endothelial cell dysfunction. *J. Clin. Invest.* 1997; 100: 2153–2157.

[41] Krasinski K, Spyridopoulos I, Asahara T, van der Zee R, Isner J, Losordo D. Estradiol accelerates functional endothelial recovery after arterial injury. *Circulation* 1997; 95: 1768-1772.

[42] Karas R, Hodgin J, Kwoun M, KregeDagger J, Aronovitz M, Mackey W, Gustafsson J, Korach K, Smithies O, Mendelsohn M. Estrogen inhibits the vascular injury response in estrogen receptor ß-deficient mice. *Proc. Natl. Acad. Sci.* U S A 1999; 96: 15133–15136.

[43] Albrecht E, Babischkin J, Lidor Y, Anderson L, Udoff L, Pepe G. Effect of estrogen on angiogenesis in co-cultures of human endometrial cells and microvascular endothelial cells. *Hum. Reprod.* 2003; 18: 2039-2047.

[44] Fujimoto J, Hori M, Ichigo S, Tamaya T. Ovarian steroids regulate the expression of basic fibroblast growth factor and its mRNA in fibroblasts derived from uterine endometrium. *Ann. Clin. Biochem.* 1997; 34: 91-96.

[45] Kuiper G, Carlsson B, Grandien K, Enmark E, Häggblad J, Nilsson S, Gustafsson J. Comparison of the ligand binding specificity and transcript tissue distribution of estrogen receptors alpha and beta. *Endocrinology* 1997; 138: 863-870.

[46] Fotsis T, Pepper M, Adlercreutz H, Hase T, Montesano R, Schweigerer L. Genistein, a dietary ingested isoflavonoid, inhibits cell proliferation and in vitro angiogenesis. *J. Nutr.* 1995; 125: 790S-797S.

[47] Kuiper G, Lemmen J, Carlsson B, Corton J, Safe S, van der Saag P, van der Burg B, Gustafsson J. Interaction of estrogenic chemicals and phytoestrogens with estrogen receptor beta. *Endocrinology* 1998; 139: 4252-4263.

[48] Figtree G, McDonald D, Watkins H, Channon K. Truncated Estrogen Receptor *Circulation* 2003; 107: 120-126.

[49] Leung S, Teoh H, Keung W, Man R. Non-genomic vascular actions of female sex hormones: physiological implications and signaling pathways. *Clin. Exp. Pharmacol. Physiol.* 2007; 34: 822-826.

[50] Haynes M, Sinha D, Russell K, Collinge M, Fulton D, Morales-Ruiz M, Sessa W, Bender J. Membrane estrogen receptor engagement activates endothelial nitric oxide synthase via the PI3-kinase-Akt pathway in human endothelial cells. *Circ. Res.* 2000; 87: 677-682.

[51] Klinge C, Blankenship K, Risinger K, Bhatnagar S, Noisin E, Sumanasekera W, et al. Resveratrol and estradiol rapidly activate MAPK signaling through estrogen receptors alpha and beta in endothelial cells. *J. Biol. Chem.* 2005; 280: 7460-7468.

[52] Critchley H, Brenner R, Drudy T, Williams K, Nayak N, Millar M, Saunders P. Estrogen receptor β, but not estrogen receptor α, is present in the vascular endothelium of the human and nonhuman primate endothelium. *J. Clin. Endocrinol. Metab.* 2001; 86: 1370-1378.

[53] Krikun G, Schatz F, Taylor R, Critchley H, Rogers P, Huang J, Lockwood C. Endometrial endothelial cell steroid receptor expression and steroid effects on gene expression. *J. Clin. Endocrinol. Metab.* 2005; 90: 1812-1818.

[54] Kim H, Lee J, Jeong J, Bae S, Lee H, Jo I. Nongenomic stimulation of nitric oxide release by estrogen is mediated by estrogen receptor alpha localized in caveolae. *Biochem .Biophys. Res. Commun.* 1999; 263: 257-262.

[55] Castoria G, Migliaccio A, Bilancio A, Di Domenico M, de Falco A, Lombardi M, Fiorentino R, Varricchio L, Barone M, Auricchio F. PI3-kinase in concert with Src promotes the S-phase entry of oestradiol-stimulated MCF-7 cells. *EMBO J.* 2001; 20: 6050-6059.

[56] Marino M, Ascenzi P, Acconcia F. S-palmitoylation modulates estrogen receptor alpha localization and functions. *Steroids* 2006; 71: 298-303.

[57] Ing N, Tornesi M. Estradiol up-regulates estrogen receptor and progesterone receptor gene expression in specific ovine uterine cells. *Biol. Reprod.* 1997; 56: 1205-1215.

[58] Ing N, Spencer T, Bazer F. Estrogen enhances endometrial estrogen receptor gene expression by a posttranscriptional mechanism in the ovariectomized ewe. *Biol. Reprod.* 1996; 54: 591-599.

[59] Vázquez F, Rodríguez-Manzaneque J, Lydon J, Edwards D, O'Malley B, Iruela-Arispe M. Progesterone regulates proliferation of endothelial cells. *J. Biol. Chem.* 1999; 274: 2185-2192.

[60] Rao B. Pregnancy associated highly vascularized tumours negatively correlate with the levels of anti-angiogenic 17 alpha-hydroxy-progesterone. *Anticancer. Res.* 1997; 17: 1019-1021.

[61] Heryanto B, Rogers P. Regulation of endomtrial cell proliferation by oestrogen and progesterone in the ovariectomized mouse. *Reproduction* 2002; 123: 107-113.

[62] Pierro E, Minici F, Alesiani O, Miceli F, Proto C, Screpanti I, Mancuso S, Lanzone A. Stromal-epithelial interactions modulate estrogen responsiveness in normal human endometrium. *Biol. Reprod.* 2001; 64: 831-838.

[63] Tsutsumi O, Yano T, Taketani Y. Phosphofructokinase activity as a measure of maturation of rat oocytes developed in vivo and in vitro. *Horm. Res.* 1994; 41: Suppl 63-67.

[64] Vlodavsky I, Brown K, Gospodarowicz D. A comparison of the binding of epidermal growth factor to cultured granulosa and luteal cells. *J. Biol. Chem.* 1978; 253: 3744-3750.

[65] Giudice L. Genes associated with embryonic attachment and implantation and the role of progesterone. *J. Reprod. Med.* 1999; 44: 165-171.

[66] Einspanier R, Miyamoto A, Schams D, Müller M, Brem G. Tissue concentration, mRNA expression and stimulation of IGF-I in luteal tissue during the oestrous cycle and pregnancy of cows. *J. Reprod. Fertil.* 1990; 90: 439-445.

[67] Schams D, Kosmann M, Berisha B, Amselgruber W, Miyamoto A. Stimulatory and synergistic effects of luteinising hormone and insulin like growth factor 1 on the secretion of vascular endothelial growth factor and progesterone of cultured bovine granulosa cells. *Exp. Clin. Endocrinol. Diabetes* 2001; 109: 155-162.

[68] Hill D, Clemmons D, Riley S, Bassett N, Challis J. Immunohistochemical localization of insulin-like growth factors (IGFs) and IGF binding proteins -1, -2 and -3 in human placenta and fetal membranes. *Placenta* 1993; 14: 1-12.

[69] Price W, Moats-Staats B, D'Ercole A, Stiles A. Insulin-like growth factor binding protein production and regulation in fetal rat lung cells. *Am. J. Respir. Cell. Mol. Biol.* 1993; 8: 425-432.

[70] Kapur S, Tamada T, Dey S, Andrews G. Expression of insulin-like growth factor-I (IGF-I) and its receptor in the peri-implantation mouse uterus, and cell-specific regulation of IGF-I gene expression by estradiol and progesterone. *Biol. Reprod.* 1992; 46: 208-219.

[71] Murphy L, Friesen H. Differential effects of estrogen and growth hormone on uterine and hepatic insulin-like growth factor I gene expression in the ovariectomized hypophysectomized rat. *Endocrinology* 1988; 122: 325-332.

[72] Evans W, Phernetton T, Magness R. 17beta-estradiol effect on critical cardiac output with reduction of cardiac output in oophorectomized sheep. *Am. J. Physiol.* 1998; 275: H57-H64.

[73] Magness R, Parker CJ, Rosenfeld C. Systemic and uterine responses to chronic infusion of estradiol-17 beta. *Am. J. Physiol.* 1993; 265: E690-E698.

[74] Tamemoto H, Kadowaki T, Tobe K, Yagi T, Sakura H, Hayakawa T, Terauchi Y, Ueki K, Kaburagi Y, Satoh S. Insulin resistance and growth retardation in mice lacking insulin receptor substrate-1. *Nature* 1994; 372: 182-186.

[75] Boulton T, Nye S, Robbins D, Ip N, Radziejewska E, Morgenbesser S, DePinho R, Panayotatos N, Cobb M, Yancopoulos G. ERKs: a family of protein-serine/ threonine kinases that are activated and tyrosine phosphorylated in response to insulin and NGF. *Cell* 1991; 65: 663-675.

[76] Cardona-Gómez G, Mendez P, DonCarlos L, Azcoitia I, Garcia-Segura L. Interactions of estrogens and insulin-like growth factor-I in the brain: implications for neuroprotection. *Brain Res. Rev.* 2001; 37: 320-334.

[77] Cardona-Gómez G, Mendez P, DonCarlos L, Azcoitia I, Garcia-Segura L. Interactions of estrogen and insulin-like growth factor-I in the brain: molecular mechanisms and functional implications. *J. Steroid. Biochem. Mol. Biol.* 2002; 83: 211-217.

[78] Duenas M, Torres-Aleman I, Naftolin F, Garcia-Segura L. Interaction of insulin-like growth factor-I and estradiol signaling pathways on hypothalamic neuronal differentiation. *Neuroscience* 1996; 74: 531-539.

[79] Risau W, Flamme I. Vasculogenesis. Annu Rev Cell Dev Biol 1995; 11: 73-91.

[80] Flamme I, Risau W. Induction of vasculogenesis and hematopoiesis in vitro. *Development* 1992; 116: 435-459.

[81] Asahara T, Murohara T, Sullivan A, Silver M, van der Zee R, Li T, Witzenbichler B, Schatteman G, Isner J. Isolation of putative progenitor endothelial cells for angiogenesis. *Science* 1997; 275: 965-967.

[82] Peichev M, Naiyer A, Pereira D, Zhu Z, Lane W, Williams M, Oz M, Hicklin D, Witte L, Moore M, Raffi S. Expression of VEGFR-2 and AC133 by circulating human CD34+ cells identifies a population of functional endothelial precursors. *Blood* 2000; 95: 952-958.

[83] Asahara T, Masuda H, Takahashi T, Kalka C, Pastore C, Silver M, Kearney M, Magner M, Isner J. Bone marrow origin of endothelial progenitor cells responsible for postnatal vasculogenesis in physiological and pathological neovascularization. *Circ. Res.* 1999; 85: 221-228.

[84] Murayama T, Tepper O, Silver M, Ma H, Losordo D, Isner J, Asahara T, Kalka C. Determination of bone marrow-derived endothelial progenitor cell significance in angiogenic growth factor-induced neovascularization in vivo. *Exp. Hematol.* 2002; 30: 967-972.

[85] Gussin H, Bischoff F, Hoffman R, Elias S. Endothelial precursor cells in the peripheral blood of pregnant women. *J. Soc. Gynecol. Investi.* 2000; 9: 357-361.

[86] Ottersbach K, Dzierzak E. The murine placenta contains hematopoietic stem cells within the vascular labyrinth region. *Dev. Cell* 2005; 8: 377-387.

[87] Matsubara K, Abe E, Matsubara Y, Kameda K, Ito M. Circulating endothelial progenitor cells during normal pregnancy and pre-eclampsia. *ARJI* 2006; 56: 1-7.

[88] Albrecht E, Robb V, Pepe G. Regulation of placental vascular endothelial growth/permeability factor expression and angiogenesis by estrogen during early baboon pregnancy. *J. Clin. Endocrinol. Metab.* 2004; 89: 5803-5809.

[89] Kalka C, Masuda H, Takahashi T, Kalka-Moll W, Silver M, Kearney M, Li T, Isner J, Asahara T. Transplantation of ex vivo expanded endothelial progenitor cells for therapeutic neovascularization. *Proc. Natl. Acad. Sci.* USA 2000; 28: 3422-3427.

[90] Orlic D, Kajstura J, Chimenti S, Jakoniuk I, Anderson S, Li B, Pickel J, McKay R, Nadal-Ginard B, Bodine D, Leri A, Anversa P. Bone marrow cells regenerate infarcted myocardium. *Nature* 2001; 410: 701-705.

[91] Strehlow K, Werner N, Berweiler J, Link A, Dirnagl U, Priller J, Laufs K, Ghaeni L, Milosevic M, Böhm M, Nickenig G. Estrogen increases bone marrow-derived endothelial progenitor cell production and diminishes neointima formation. *Circulation* 2003; 107: 3059-3065.

[92] Hamada H, Kon Kim M, Iwakura A, Li M, Thorne T, Qin G, Asai J, Tsutsumi Y, Sekiguchi H, Silver M, Wecker A, Bord E, Zhu Y, Kishore R, Losordo D. Estrogen receptors α and β mediate contribution of bone marrow-derived endothelial progenitor cells to functional recovery after myocardial infarction. *Circulation* 2006; 114: 2261-2270.

[93] Iwakura A, Shastry S, Luedemann C, Hamada H, Kawamoto A, Kishore R, Zhu Y, Qin G, Silver M, Thorne T, Eaton L, Masuda H, Asahara T, Losordo D. Estradiol enhances recovery after myocardial infarction by augmenting incorporation of bone marrow-derived endothelial progenitor cells into sites of ischemia-induced neovascularization via endothelial nitric oxide synthase-mediated activation of matrix metalloproteinase-9. *Circulation* 2006; 113: 1605-1614.

[94] Iwakura A, Luedemann C, Shastry S, Hanley A, Kearney M, Aikawa R, Isner J, Asahara T. Estrogen-mediated, endothelial nitric oxide synthase-dependent mobilization of bone marrow-derived endothelial progenitor cells contributes to reendothelialization after arterial injury. *Circulation* 2003; 108: 3115-3121.

[95] Vasa M, Fichtlscherer S, Aicher A, Adler K, Urbich C, Martin H, Zeiher A, Dimmeler S. Number and migratory activity of circulating endothelial progenitor cells inversely correlate with risk factors for coronary artery disease. *Circ. Res.* 2001; 89: E1-E7.

[96] Albrecht E, Robb V, Pepe G. Regulation of placental vascular endothelial growth/ permeability factor expression and angiogenesis by estrogen during early baboon pregnancy. *J. Clin. Endocrinol. Metab.* 2004; 89: 5803-5809.

[97] Imanishi T, Kobayashi K, Hano T, Nishio I. Effect of estrogen on differentiation and senescence in endothelial progenitor cells derived from bone marrow in spontaneously hypertensive rats. *Hypertens. Res.* 2005; 28: 763-772.

In: Cell Growth Processes: New Research
Editor: Daiki Kimura, pp. 183-201
ISBN: 978-1-60456-132-6

Chapter VII

Influence of Rapid Prototyped Scaffolds' Pore Volume and Shape on Cartilage Regeneration

Lorenzo Moroni [a, *1], ***Joost R. de Wijn*** [a] ***and Clemens A. van Blitterswijk*** [a]

[a] Institute for Biomedical Technology, University of Twente,
P.O. Box 217, 7500 AE Enschede, The Netherlands

[*] currently at Johns Hopkins University – Biomedical Engineering Dept.,
3400 N Charles St., MD21218 Baltimore, USA

Abstract

Solid free form (SFF) technologies offer nowadays established means to manufacture 3D scaffolds for tissue engineering and regenerative medicine applications. The resulting structures are characterized by defined porosity, pore volume and shape. Their pore network can be periodically designed and is completely accessible and interconnected, which is beneficial for optimal cell nutrient supply and diffusion. For these reasons, SFF scaffolds are interesting 3D models to study not only tissue regeneration, but also cell growth processes. No systematic study has been done on the influence of different scaffold's porosities, pore volume and shape on cell growth and tissue formation. Therefore, the aim of this investigation was to evaluate whether SFF scaffolds with different pore architectures can influence cell seeding efficiency, attachment, and tissue regeneration. 3D fiber deposition (3DF) was used as a fabrication technique and cartilage regeneration was considered as a specific tissue engineering application. Results showed that scaffolds with lower porosity and pore volume supported a higher amount of attached cells both after 10 and 30 days in dynamic culture. Scaffolds with a decreased pore volume along the longitudinal direction, but equal overall porosity, supported a higher number of entrapped cells as well. This was accompanied by a correspondent higher production of extracellular matrix (ECM), as analyzed by quantitative GAG expression. No significant influence of the analyzed pore shapes could

[1] email: lm@jhu.edu.

be detected in terms of cell attachment and growth, or in terms of tissue formation, as also noticed by different microscopic analysis. Hence, when 3D scaffolds with completely accessible and interconnected pore networks are used for tissue engineering and regenerative medicine, pore volume may be selectively tailored for the overall structure, or regionally within the same structure, to appropriately modulate cell growth and tissue formation for specific applications.

Keywords: rapid prototyping, scaffold, pore volume, pore shape, tissue engineering.

1. Introduction

Tissue engineering is generally pursued by three different strategies: (i) cell therapy, (ii) scaffold-based regeneration, and (iii) drug delivery. Although cell therapy and drug delivery have given promising results for different tissues [1-6], a scaffold-based approach is necessary when mechanical properties, tissue guidance and architecture are critical to mimic for the success of the desired tissue regeneration [7-10]. 3D scaffolds can be considered as templates that facilitate cell attachment, proliferation, and differentiation into the proper phenotype to produce extracellular matrix (ECM) forming a specific tissue [11, 12]. To satisfy all these functions, scaffolds need to offer a completely accessible and interconnected pore network where cells can easily migrate into and can be still viable to nutrient supply [9, 12, 13]. A mainstay in scaffold designing is that porosity should be maximized to obtain cell migration and nutrient diffusion inside the constructs. However, this generally results in a loss of mechanical properties [14-16]. Furthermore, it has been shown that nutrient diffusion is still hampered in the center of large scaffolds manufactured by conventional foaming- or porogen-fabrication techniques and limited to 0.5-1 mm in thickness [12, 13, 17]. Pore tortuosity and the lack of their complete interconnection is thought to be responsible for nutrient limitations [13, 18, 19].

Another important parameter to be considered when designing scaffolds is pore network architecture. Scaffolds should initially replace tissue ECM while cells adhere and start producing their own. In this respect, their internal architecture could be beneficial for optimal tissue repair if it is designed to match the original ECM [20-23]. Yet, conventional scaffolds are not optimal since their pore architecture is difficult to be controlled. An alternative to conventional fabrication platforms that demonstrated to improve cell-scaffold interactions while maintaining adequate mechanical properties and pore architecture design flexibility is solid free form fabrication (SFF) [13, 24, 25]. SFF scaffolds are characterized by a uniform, accessible, and completely interconnected pore network that allows enhanced nutrient diffusion for larger constructs [9, 19]. These structures are designed and fabricated by computer aided design/computer aided machining (CAD/CAM) software, thus broadening the possibility to create different pore networks with specific architectures that can be maintained or varied through manufacturing [26-28]. Scaffolds with different degrees of complexity can be created, their pore network being either homogenous or hierarchical depending from the required tissue architecture [28-30]. Due to these characteristics, it may be possible that cell and tissue in-growth occur at reduced porosity. This could imply an enhanced cell entrapment capacity of the scaffolds because of the correspondent decrease in pore size. Analogously, scaffolds with different pore shape could alter the ability of the whole structure to "sieve" cells during attachment.

Therefore, the aim of this study was to assess the influence of pore size and shape on cell attachment and tissue formation. Articular cartilage was chosen as an application, since this tissue displays high compressive mechanical properties and a hierarchical internal architecture [7, 31, 32]. 3D scaffolds were fabricated with cubical or polygonal homogenous pores with different sizes by a rapid prototyping technique known as 3D fiber deposition (3DF) [27, 33]. This technique has shown a great versatility in fabricating scaffolds for tissue engineering with different pore volumes and architectures, resulting in modulating mechanical properties [16, 27]. Scaffolds are generated by depositing an extruded material in the form of a fiber, following a pattern in a CAD/CAM fashion. Fibers form a patterned layer and layers are deposited on top of each other forming the final 3D structure. Pore shape was varied by modifying the fiber deposition orientation. Pore size was changed by either varying the width while maintaining the height constant or *vice versa*. This resulted in scaffolds with similar porosity but different pore shape and volume, or in scaffolds with similar pore shape but different porosity and pore volume. Combining cell response to different scaffold architectures with their intrinsic different mechanical properties may lead to create a predictive tool to choose the best performing scaffold design for specific tissue engineering and regenerative medicine applications.

2. Materials and Methods

2.1. Scaffold Fabrication

Poly(ethylene oxide terephthalate)-poly(butylene terephthalate) (PEOT/PBT) copolymers (IsoTis S.A., Irvine, USA) were used to fabricate 3DF scaffolds. Their chemical composition is typically represented by the notation aPEOTbPBTc (a/b/c), where a is the molecular weight of the starting poly(ethylene glycol) (PEG) segments employed in the copolymerization reaction, while b and c refer to the weight ratio between PEOT and PBT blocks, respectively. In this study, 300/55/45 was used as a biomaterial that found interest for cartilage regeneration [18, 19, 27].

3D scaffolds were produced with a Bioplotter machine (Envisiontec GmbH, Germany), which is an XYZ plotter unit driven by computer aided design/computer aided manufacturing (CAD/CAM) software, as previously described [27]. Few modifications were done to extrude highly viscous fibers from hot melt polymers. The polymer was loaded in a stainless steel syringe and heated at T = 190 °C through a thermo-set cartridge unit, fixed on the "X"-mobile arm of the device. When the polymer reached its molten phase, a nitrogen pressure of 5 Bars was applied to the syringe through a pressurized cap. Rectangular block models were plotted layer by layer, through the extrusion of the polymer on a stage as a fiber from a nozzle attached to the syringe.

3DF fabricated scaffolds are typically characterized by varying the fiber diameter d1 (through the nozzle diameter or the deposition speed), the spacing between fibers in the same layer d2, the layer thickness d3, and the configuration of the deposited fibers within the whole architecture. The nozzle used to extrude polymer fibers were stainless steel Luer Lock needles with internal diameter (ID) of 400 μm, shortened to a length of approximately 16.2 mm. The corresponded plotted fibers are expected to have a diameter d1 ≤ 400 μm, depending on the deposition speed. The fiber spacing d2 was set to 600 μm, 800 μm, and 1000 μm, while the layer thickness d3 was set to 150 μm and 250 μm. Scaffolds' pore shape and volume were changed by plotting fibers with 45° and 90° angle steps between two successive layers (called 0-45 and 0-90 configurations, respectively) and by modifying the fiber orientation after one or after two printed layers (referred to as 00-4545 and 00-9090 architectures). The deposition speed was fixed at 300 mm/min.

2.2. Scaffold Characterization

Cylindrical plugs of 6 mm in diameter by 4 mm in height were cored out in the "Z-direction" from the rectangular 3DF plotted blocks, and either analyzed by microscopical analysis or used for culture. Scaffolds geometry and architecture was characterized by scanning electron microscopy (SEM, Philips XL 30 ESEM-FEG) analysis, after gold sputtering (Cressingdon). The porosity of 3DF scaffolds was calculated as [34]:

$$P = 1 - \frac{V_{scaffold}}{V_{cube}} = 1 - \frac{\pi}{4} \cdot \frac{1}{\frac{d_2}{d_1}} \cdot \frac{1}{\frac{d_3}{d_1}} \qquad (1)$$

where P is the scaffold porosity, d1 the fiber diameter, d2 the fiber spacing and d3 the layer thickness, within each different structure.

2.3. Cell Culture

Chondrocytes were isolated via collagenase (collagenase type II, Worthington) digestion from articular cartilage harvested from 6-month old New Zealand white rabbits. Primary cells were expanded for two passages, dynamically seeded in spinner flasks at a density of 5 millions/scaffold and cultured on 3DF scaffolds for 10, and 30 days (n=4). Seeding was performed in a working volume of 80 ml at 40 rpm, for 24 hours. During culturing the medium volume was kept constant, while its revolution speed increased to 60 rpm. 300/55/45 scaffolds of 6 mm in diameter and 4 mm in height fabricated with 00-9090, 00-4545, and 0-45 architectures were considered. These architectures were chosen to assess any influence of pore volume and shape on tissue formation. Chondrocyte culture medium contained HEPES (Invitrogen)-buffered DMEM (Invitrogen) supplemented with 10% fetal bovine serum (FBS, Sigma-Aldrich), 0.2 mM ascorbic acid 2-phosphate (Invitrogen), 0.1 mM non-essential amino acids (Sigma-Aldrich), 0.4 mM proline (Sigma-Aldrich), 100 units/ml penicillin (Invitrogen), and 100 μg/ml streptomycin (Invitrogen). Constructs were cultured at 37°C in a humid atmosphere with 5% CO2. Medium was refreshed twice a week.

2.4. Microscopical Analysis of Tissue Constructs

Cell viability was assessed by a live-dead assay (Molecular Probe). Ethidium-bromide homodimer was used as a marker for dead cells due to its specific binding to fragmented DNA in cells that no longer have an intact plasma membrane, and fluoresces red. Calcein AM permeates the plasma membrane of viable cells, where it is cleaved by intracellular esterases and fluoresces green [35]. After harvesting tissue constructs, 1mm thick sections were incubated in a PBS solution containing 6 mmol/l ethidium-bromide homodimer and 2 mmol/l calcein AM for 30 min at 37 °C. Sections were examined in an inverted fluorescent microscope (Nikon Eclipse E400) using an FITC Texas Red filter. Representative images were captured using a digital camera (Sony Corporation, Japan) and Matrix Vision software (Matrix Vision GmbH, Germany).

Cell attachment and 3D distribution in the scaffolds was analyzed by confocal and optical microscopy. Samples for confocal laser microscopy were stained as above described for live-dead assay. For optical microscopy, construct sections were stained with methylene blue. Samples were fixed overnight in 0.14 M cacodylate buffer (pH = 7.2 – 7.4) containing 1.5% glutaraldehyde (Merck) and incubated in a 1% methylene blue solution for 1 minute.

To analyze tissue formation and distribution, constructs were characterized by SEM. Scaffolds were fixed with the cacodylate-glutaraldehyde buffer above mentioned, subsequently dehydrated in sequential ethanol series, and critical point dried from liquid carbon dioxide using a Balzers CPD 030 machine. Samples were then gold sputtered and studied under the SEM.

2.5. Biochemical Analysis

Deoxyribonucleic acid (DNA) and glycosaminoglycans (GAG) assays were performed at 10 and 30 days of culture. Constructs were digested overnight at 56 °C in a Tris-EDTA buffered solution containing 1 mg/ml proteinase K, 18.5 μg/ml pepstatin A, and 1 μg/ml iodoacetamide (Sigma-Aldrich). Quantification of total DNA was done with Cyquant dye kit according to the manufacturers description (Molecular Probes) using a spectrofluorometer (LS 50B, Perkin Elmer). GAG amount was determined spectrophotometrically (EL 312e Bio-TEK Instruments) after reaction with dimethylmethylene blue dye (DMMB, Sigma-Aldrich) by measuring absorbance at 520 nm. The final amount was calculated using a standard of chondroitin sulphate B (Sigma-Aldrich).

2.6. Statistical Analysis

Statistical Analysis was performed using a Student's t-test, where the confidence level was set to 0.05 for statistical significance. Values in this study are reported as mean and standard deviation.

3. Results

3.1. Scaffolds Characterization

3DF scaffolds with different pore volume and shape were successfully generated and demonstrated to have characteristic dimensions (d1, d2, and d3) close to the theoretical set values (figure 1). SEM analysis revealed a fiber diameter d1 of 268 ± 32 μm (nozzle ID = 400 μm; deposition speed V = 300 mm/min). The fiber spacing d2 was 605 ± 12 μm, 807 ± 28 μm, and 1001 ± 0.00 μm. The layer thickness d3 was 148 ± 10 μm, and 257 ± 11 μm. All the scaffolds produced had completely interconnected pores and no layer delamination phenomenon occurred. Scaffolds with a 00-9090 architecture possessed quadrangular pores, while scaffolds with a 0-45 or a 00-4545 architecture had triangular and polygonal pores. The pore geometry was defined by fiber diameter and spacing (d1 and d2), and layer thickness (d3). The deposition of two successive layers with the same fiber orientation resulted in doubling the pore height. Analogously, an increase of the fiber spacing from 600 μm to 1000 μm implied an increase in pore width. The pore height was mainly defined by the layer thickness, while the pore width was defined by the difference between fiber spacing and fiber diameter. This resulted in pore volumes (xyz) varying from 337x337x296 ± 16x16x20 μm^3 to 733x733x296 ± 26x26x20 μm^3 for quadrangular pores. In case of 0-45 and 00-4545 scaffold architecture, triangular pores varied from 182x302x257 ± 8x38x11 μm^3 to 182x302x514 ±.8x38x22 μm^3, while polygonal pores varied from 355x329x257 ± 55x16x11 μm^3 to 355x329x514 ± 55x16x22 μm^3, respectively. This corresponded in a variation of scaffold porosity from 41 ± 3% to 75 ± 1%. Table 1 summarizes scaffolds characteristics.

Table 1 – Characteristic fiber and pore dimensions of 3 DF scaffolds with different pore volume and shape

Scaffold Architecture	Fiber Diameter d1 (μm)	Fiber Spacing d2 (μm)	Layer Thickness d3 (μm)	Pore Size x-y-z (μmxμmxμm)	Porosity (%)
00-9090	268 ± 32	605 ± 12	148 ± 10	337x337x296 ± 16x16x20	41 ± 3
	268 ± 32	807 ± 28	148 ± 10	539x539x296 ± 3x3x20	56 ± 3
	268 ± 32	1001 ± 0.00	148 ± 10	733x733x296 ± 26x26x20	64 ± 2
0-45	268 ± 32	807 ± 28	257 ± 11	182x302x257 ± 8x38x11 182x302x514 ±.8x38x22	75 ± 1
00-4545	268 ± 32	807 ± 28	257 ± 11	355x329x257 ± 55x16x11 355x329x514 ± 55x16x22	75 ± 1

3.2. Cartilage Tissue Formation

Rabbit chondrocytes attached on all 3DF scaffolds and cell viability was maintained over a month of culture (figure 2). Confocal microscopy revealed a relatively homogenous cell distribution after 10 days on the outer as well as on the inner surface and pores of the scaffolds (figure 3). Optical microscopy analysis confirmed the homogenous cell distribution over the scaffolds cylindrical constructs after a month of culture (figures 4). Tissue formation was uniform in 00-9090, 00-4545, and 0-45 scaffold architecture as shown by SEM (figure 5). Qualitatively, enhanced tissue formation seemed to occur with decreasing pore volume both in pore width (00-9090) and in pore height (00-4545 vs 0-45). These findings were confirmed by biochemical analysis (figures 6 and 7). DNA and GAG increased with decreasing pore volume after 10 and 30 days. In particular, it seemed that with decreasing pore width (00-9090 architectures) DNA amount was optimal on scaffolds with 56% porosity (figure 6b) after 30 days, while GAG accumulation seemed to be optimal at the same porosity after 10 days to continuously increase with decreasing porosity after 30 days of culture. Analogously, with decreasing pore height (00-4545 vs 0-45 architectures) DNA and GAG contents significantly increased. Cell seeding efficiency was enhanced in scaffolds with a 00-9090 architecture as compared to a 00-4545 one with similar porosity. However, no influence of pores shape could be detected at 30 days. Interestingly, cells displayed a higher GAG/DNA ratio with increasing porosity and with increasing pore height after 10 days of culture (figure 8a). 00-4545 scaffold architecture maintained the highest GAG/DNA ratio over one month of culture as compared to 0-45 scaffolds with equal porosity but different pore height and to 00-9090 scaffolds with similar porosity but different pore shape (figure 8b). No statistically significant difference was found when compared to 00-9090 scaffolds with lower porosity.

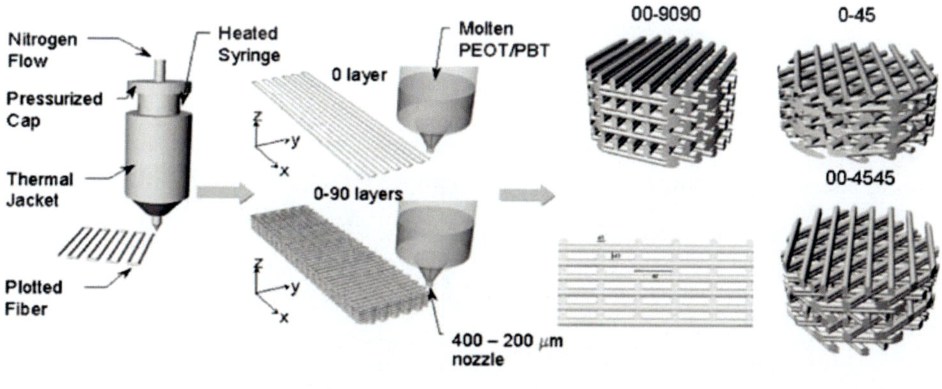

Figure 1. Schematic drawing of 3DF fabrication process. The parameters characterizing a scaffold (d1, d2, and d3) and the scaffold architectures considered in this study are shown. Adapted from [27].

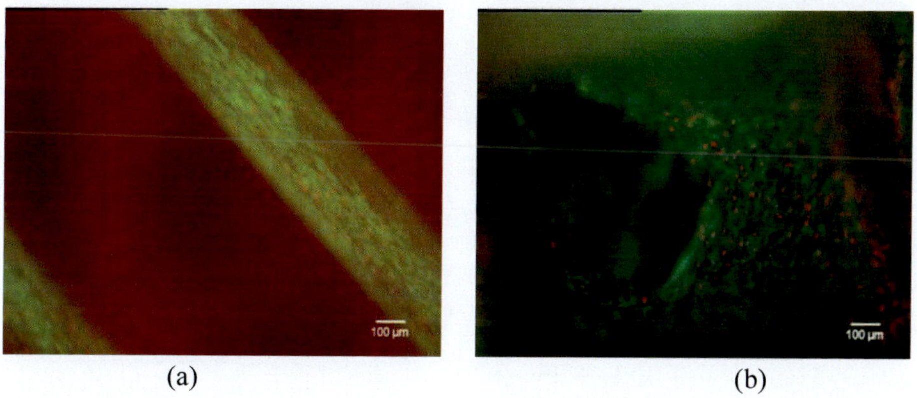

(a) (b)

Figure 2. Live-Dead assay demonstrates that most of the cells were alive at 10 (a) and 30 (b) days of culture. (a) cells attached to the scaffolds' fiber; (b) cells entrapped in a scaffold pore. Scale bar: 100 μm.

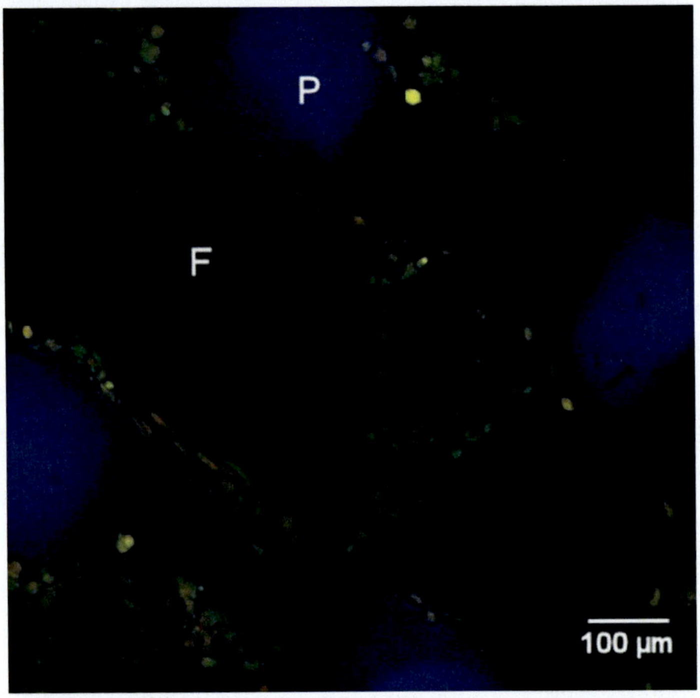

a

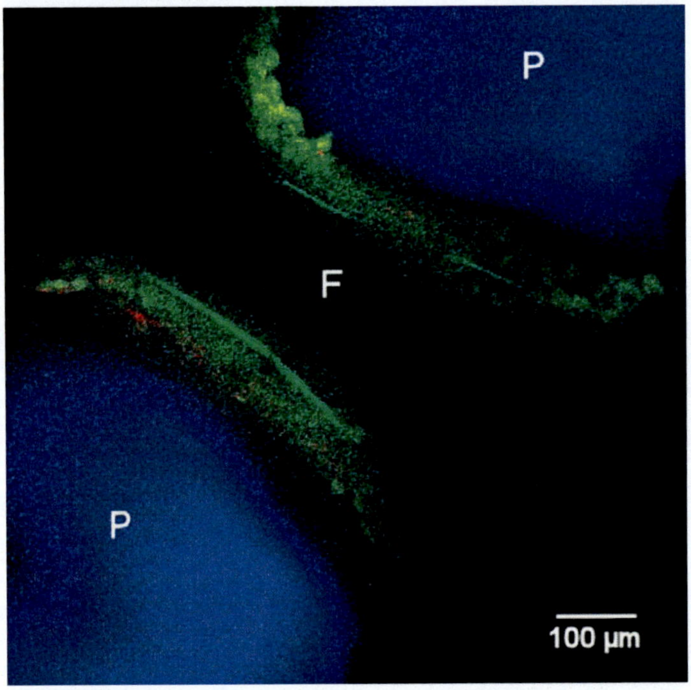

b

Figure 3. Continued on next page

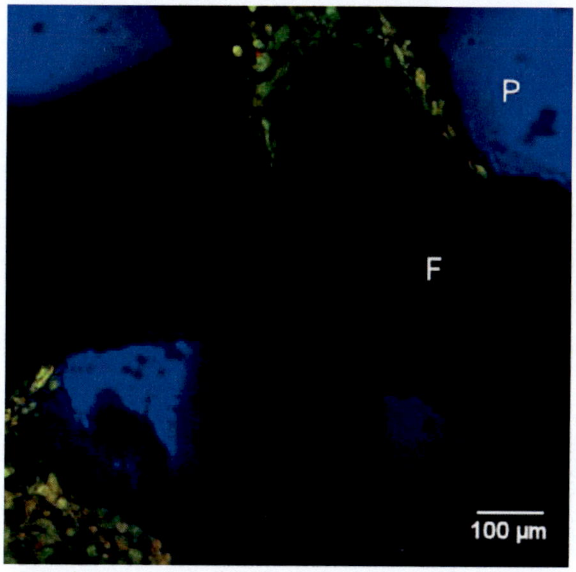

c

Figure 3. Confocal light microscopy detecting alive cells attached on the fibers or entrapped in the pores in the middle of 00-9090 (a, b) and 0-45 or 00-4545 (c) scaffolds. F = fiber; P = pore. Scale bar; 100 µm.

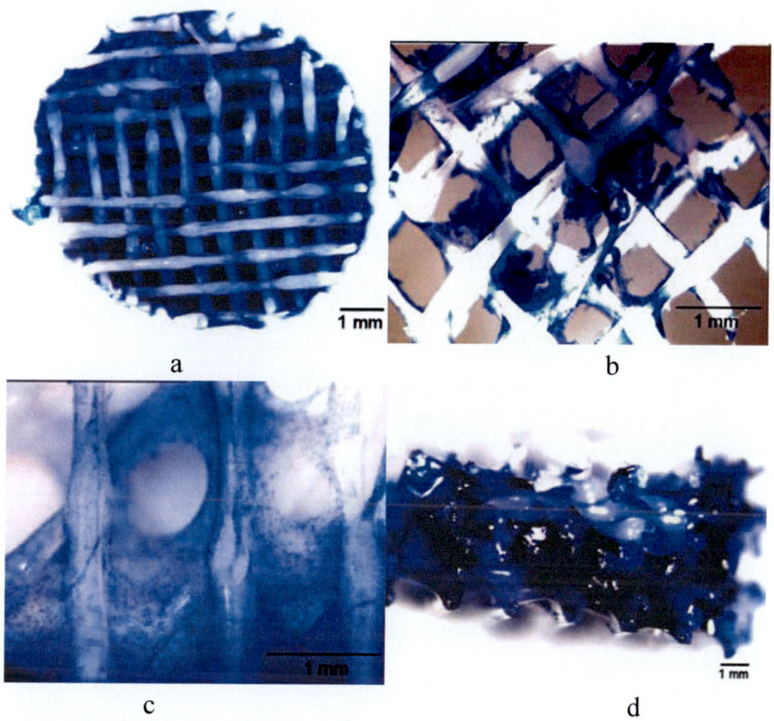

Figure 4 .Optical microscopy depicts cell distributed over the entire scaffolds volume after a month of culture. (a) view of a whole 00-9090 scaffold architecture; (b, c) detailed view of a 00-9090 and a 0-45 scaffold, respectively, showing cells bridging scaffold's pores; (d) cross section view of a 0-45 scaffold indicate cell cover all the scaffold thickness. Scale bar: 1 mm.

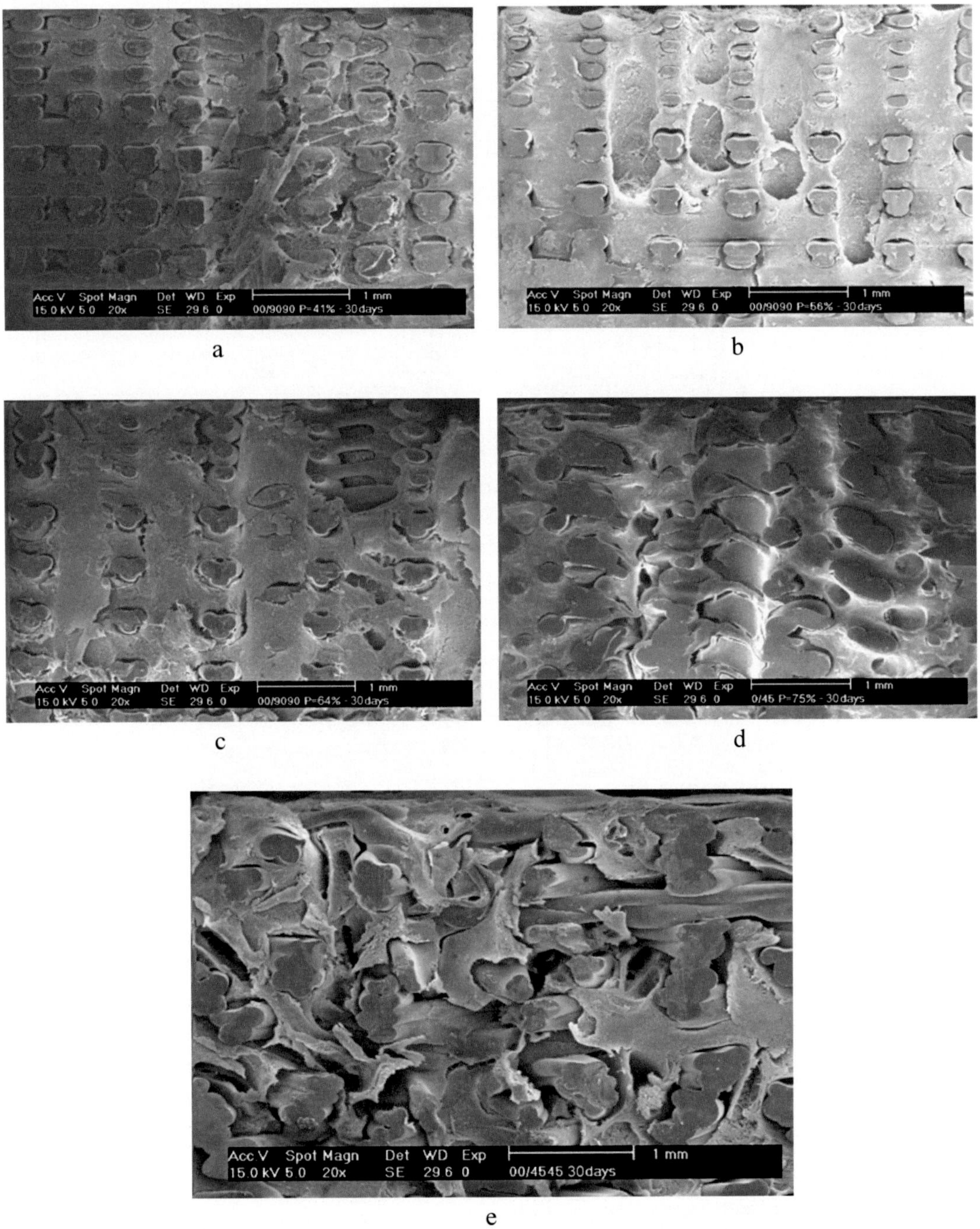

Figure 5. SEM micrographs illustrating tissue formation in the middle of the scaffolds. (a) 00-9090 scaffold with 41% porosity; (b) 00-9090 scaffolds with 56% porosity; (c) 00-9090 scaffold with 64% porosity; (d) 0-45 and (e) 00-4545 scaffolds with 75% porosity. Scale bar: 1 mm.

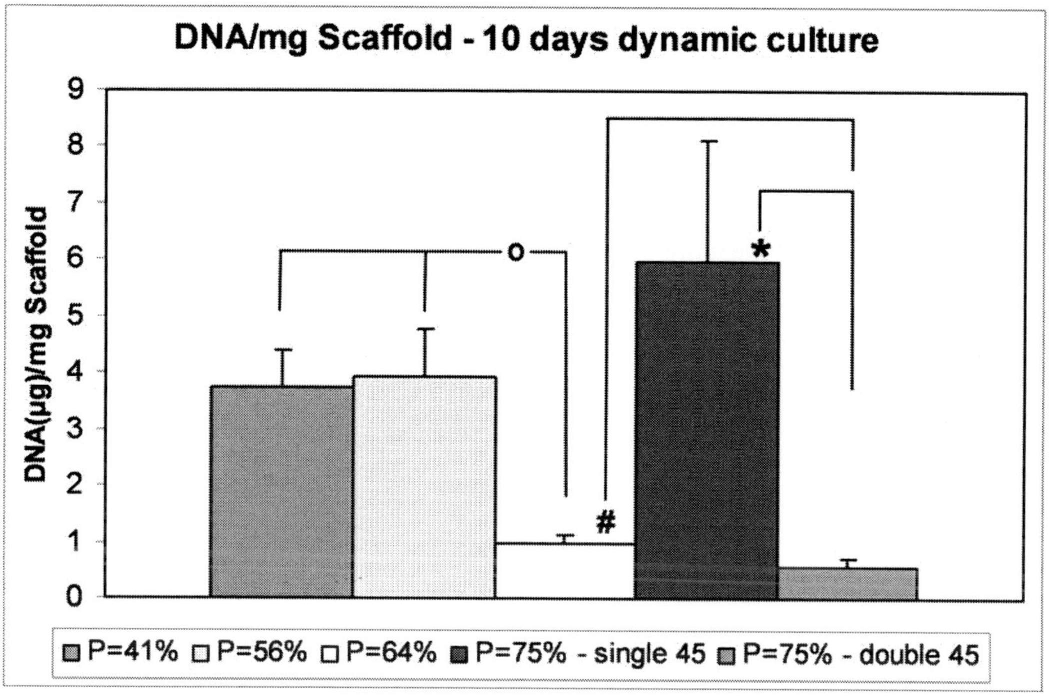

a

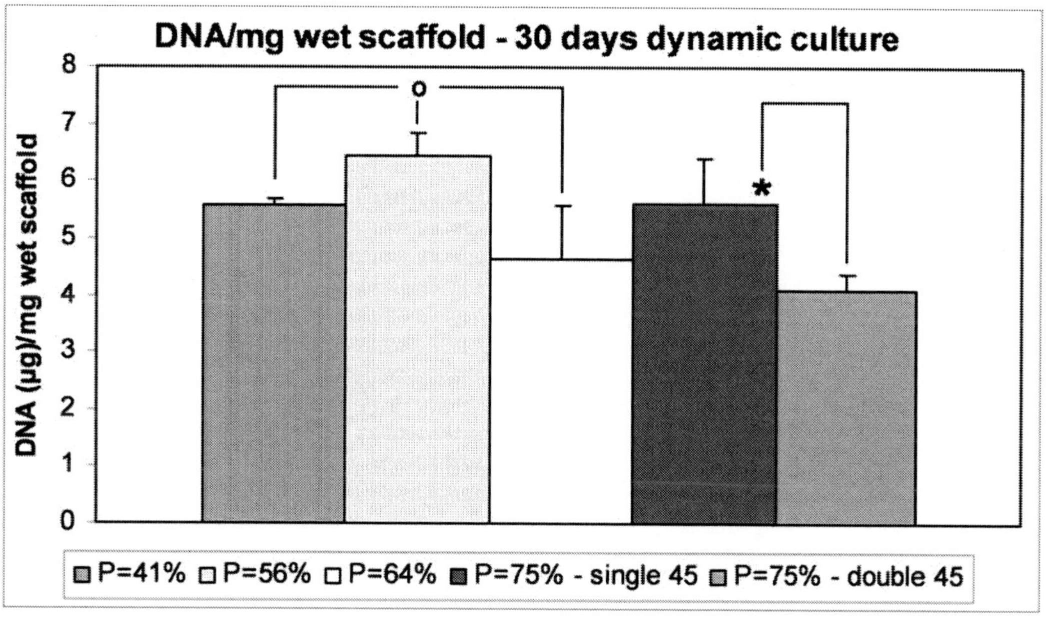

b

Figure 6. DNA assay on scaffolds with different pore volume and shape after (a) 10 and (b) 30 days of culture. (*) shows significant differences between 0-45 and 00-4545 architectures; (#) indicates significant differences between 00-9090 and 00-4545 architectures with similar porosity. (o) points out significant differences between 00-9090 scaffolds with different porosities.

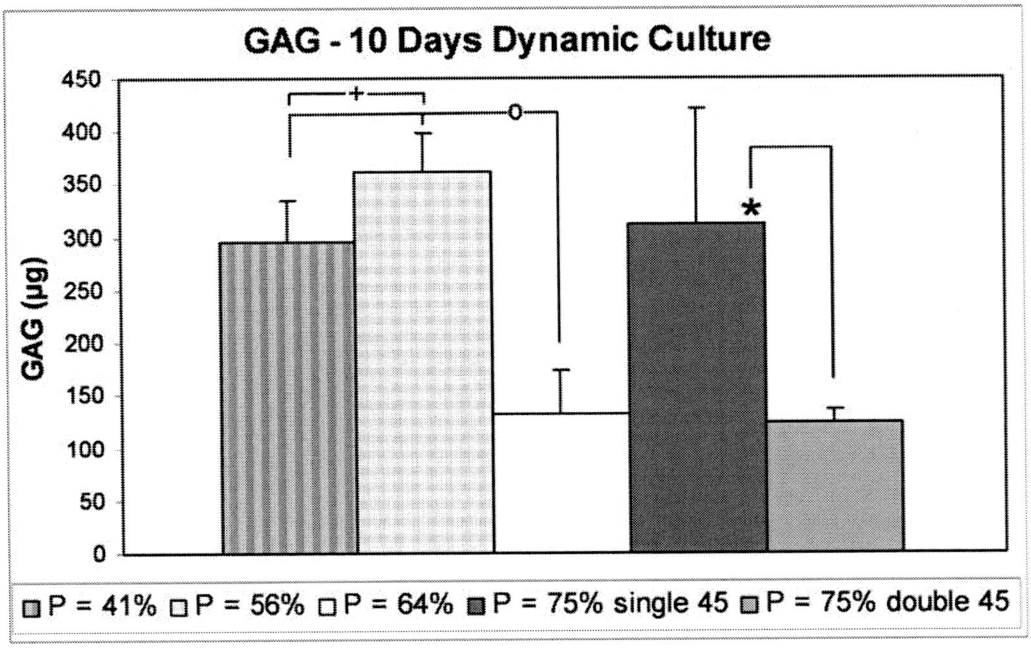

a

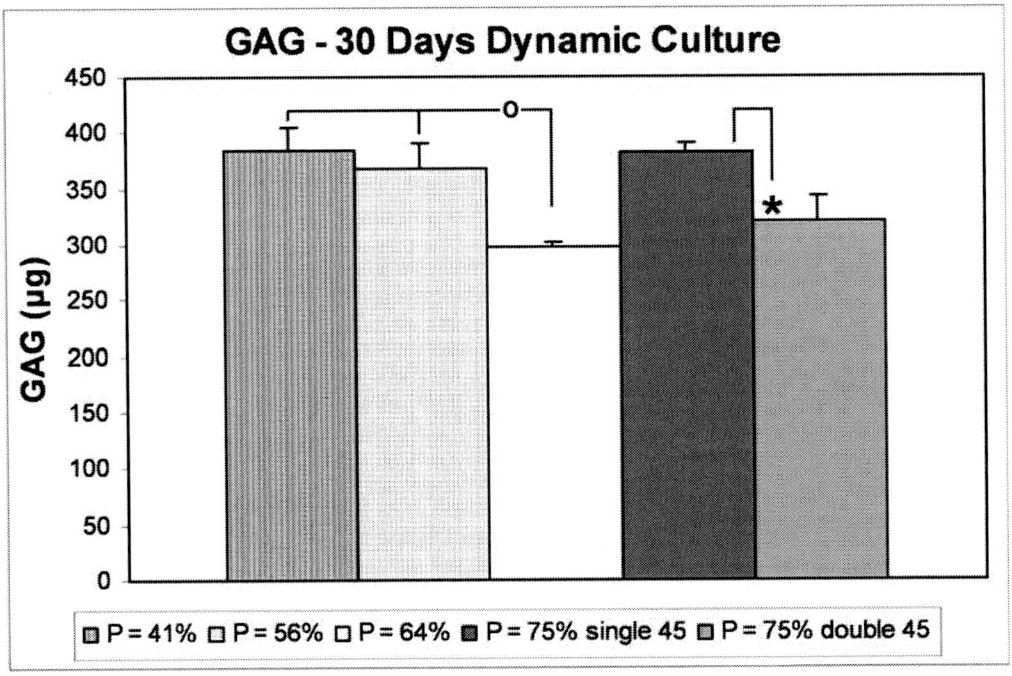

b

Figure 7. GAG assay on scaffolds with different pore volume and shape after (a) 10 and (b) 30 days of culture. (*) illustrates significant differences between 0-45 and 00-4545 architectures; (o) and (+) depict significant differences between 00-9090 scaffolds with different porosities.

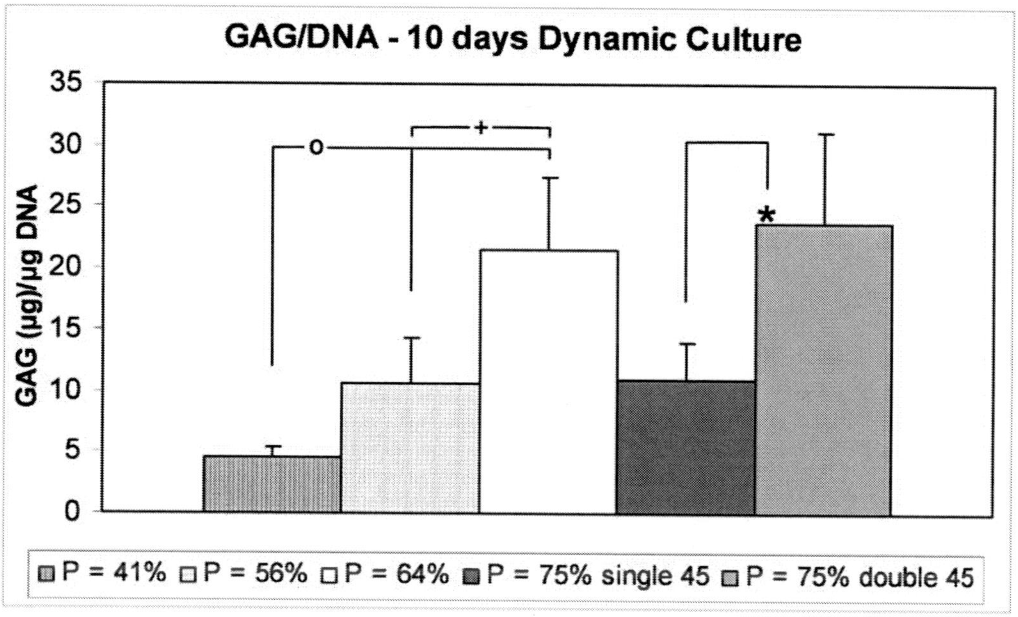

a

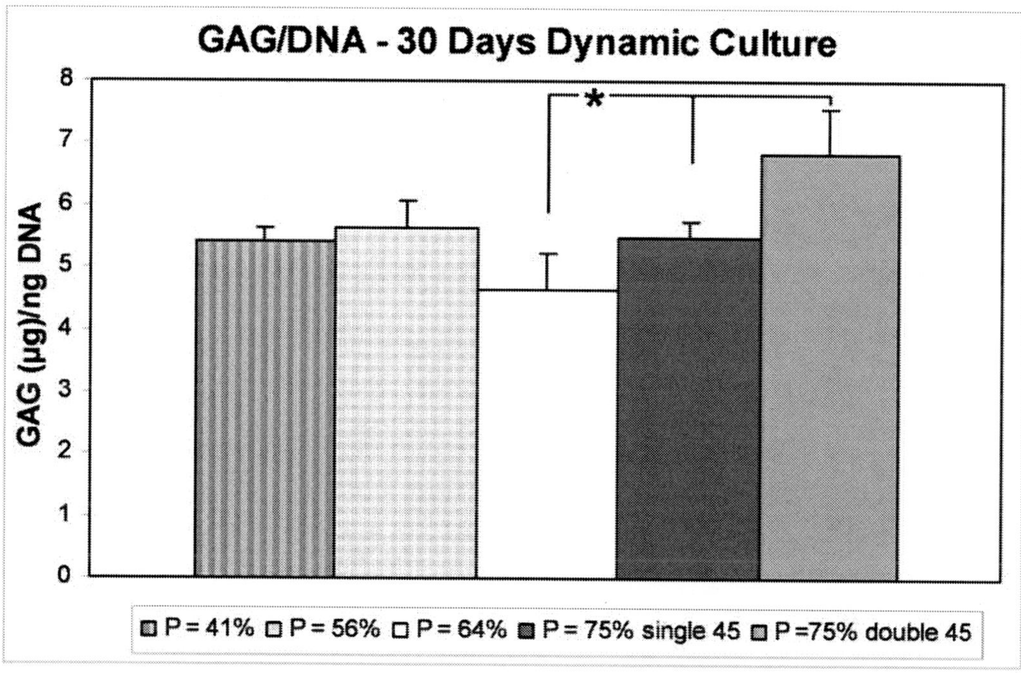

b

Figure 8. GAG/DNA ratio of chondroctyes cultured on scaffolds with different pore volume and shape after (a) 10 and (b) 30 days. (*) indicates significant differences between 0-45 and 00-4545 architectures; (o) and (+) show significant differences between 00-9090 scaffolds with different porosities.

4. Discussion

Rapid prototyped (RP) scaffolds with different pore size and shape have been fabricated by 3D fiber deposition and studied to assess the influence of these design parameters on tissue formation. Solid free-form fabricated (SFF) scaffolds offer the possibility of designing scaffolds with a wide range of pore geometry and architecture. This typically results in periodical and often hierarchical structures with a completed interconnected pore network that allow a more efficient diffusion of nutrients within the constructs [9, 13, 19]. Consequently, a more homogenous distribution of cells throughout the scaffolds is observed when compared to scaffolds manufactured with porogen and textile conventional techniques [17, 18, 36]. Natural tissues possess an intrinsic architecture in the way the extra cellular matrix (ECM) is deposited by the cells to give them their unique and specific properties. In this respect, SFF scaffolds offer again a higher flexibility, since they could be designed so that their porous network would resemble that of the tissue to be regenerated. Here, a preliminary investigation on simple parameters governing scaffolds architecture such as pore size and shape was conducted, in the attempt to better understand whether these entities could be useful for functional tissue engineering. In particular, cartilage was chosen as an application. Cartilage displays a different ECM architecture depending on its location and, within it, depending on the region where the matrix is deposited [7, 31, 37, 38]. Conversely to what is generally expected for scaffolds fabricated with conventional techniques, a decrease in the porosity of RP scaffolds resulted in a more efficient cell entrapment and enhanced tissue formation (figures 6 and 7). This was likely due to the periodical and completely interconnected pore architecture of SFF scaffolds, which compensated for the expected reduction in cell and nutrient permeability with decreasing porosity [18]. When cell and nutrient diffusion were maintained, a decrease in pore size implied a better "sieving" capacity of the scaffold and a consequent enhanced cell entrapment in its pores. This might have resulted in a correspondent increase in cell-cell communication and in a correspondent enhancement of ECM production as shown by pellet and mass-culture studies [39-41]. Augmented cell entrapment and matrix formation in low pore volume constructs can be of further advantage when regenerating hard tissues, as scaffolds with decreased porosity display higher stiffness and strength [16]. When characterized by high porosity these scaffolds could result in poor cell seeding efficiency, due to the intrinsic full accessibility of their pore network. Yet, such structures are desirable in some tissue engineering applications to approach mechanical properties and architecture of soft tissues. In this case, poor cell seeding efficiency might be circumvented by integrating other scaffolding techniques like electrospinning and sol-gel formation. This would allow creating multifunctional scaffolds where solid or gel micro/nano fibrillar networks are interspersed in the macro porous network of RP matrices, which would result in enhanced cell entrapment and cell-material interactions.

A higher amount of cells was also measured on 00-9090 scaffolds with respect to 00-4545 architecture with similar overall porosity, after 10 days of culture. However, no effect of scaffold architecture on DNA content was observed after one month. It seems that the more regular cubical pores of a 00-9090 architecture resulted in a more efficient early cell attachment and entrapment in its pore network as compared to the more irregular polygonal pores of a 00-4545 architecture. Hutmacher *et al.* also observed a higher cell seeding efficiency of more regular SFF scaffold architectures after 1 and 2 weeks of culture [42]. This might be related to preferential nutrient flow pathways during dynamic culture when the scaffold's pore network displays a more regular configuration [43, 44]. When GAG/DNA was considered, scaffolds with higher porosity and pore volume supported a higher ratio after 10 days of culture (figure 8). No difference could be measured over a month. Similarly, 3DF scaffolds with a polygonal shaped pore network (00-4545) displayed a higher index as compared to cubical pores at 10 days and after one month of culture. Although GAG/DNA is often related to a better chondrocyte redifferentiation capacity on scaffolds, here the differences in this ratio could be more likely attributed to the initial different cell seeding efficiency. In fact, no significant change in cell phenotype was observed over time and chondrocytes maintained a spindle-like spread morphology independently from the scaffold internal architecture (figure 6). With lower DNA content, attached cells might have stimulated each other to increase ECM production [37, 45], resulting in the measured GAG/DNA increase.

Cartilage regeneration has been studied in a various number of different scaffolds [17, 37, 39, 46-49]. Although conventional scaffolds showed to support chondrocyte attachment and homogenous tissue formation [31, 50, 51], scaffold thickness was limited to small size (1-2 mm). When larger defect sizes need to be regenerated, SFF scaffolds offer a promising alternative as they improve cell nutrition and avoid cell apoptosis and lack of tissue formation in the center of the constructs [17, 18, 36]. Different pore size and pore shape have been singularly evaluated on tissue formation in RP scaffolds [23, 26, 42, 52, 53]. However, to our knowledge no comparative study on the influence of these parameters has been reported before. Woodfield *et al.* [28] investigated cartilage tissue formation on RP scaffolds with pore gradients designed to mimic the zonal porosity of articular cartilage in the deep, middle, and surface regions. Similarly to what we have found, chondrocyte and ECM contents were higher in the superficial zone of constructs characterized by lower pore size compared to the middle and deep regions of the scaffolds. Here, it was shown that small pore size is not as critical as in scaffolds fabricated with conventional technologies. Its reduction caused an increase in cell seeding efficiency and in ECM formation. It is expected that pore size cannot be infinitely decreased, as this would ultimately hinder cell migration and survival. It seems from this study that optimal scaffold porosity for cartilage formation would range between 40 and 60%. Pore network anisotropy along the longitudinal axis of RP scaffolds is also desired to increment cell and matrix contents and should be considered during design, unless a homogenous pore volume is required. No significant influence of pore shape on tissue regeneration could be drawn from this preliminary investigation. Further studies should focus on different pore shapes to shed lights on the effective importance of this scaffold design parameter. Furthermore, cells with a fairly high metabolic activity were used here. Future investigations should consider less active cell source, which may help in expanding our understanding on how critical the influence of these parameters in tissue engineering is. Yet, pore volume is a fundamental scaffold design parameter and can influence the dynamics of tissue regeneration by modulating cell seeding efficiency and extra cellular matrix deposition.

5. Conclusions

Rapid prototyping scaffolds showed to modulate cell attachment and tissue formation dynamics by varying their pore volume either in width or in height. No significant influence of cubical or polygonal pore shape was observed on cartilage tissue regeneration. Therefore, scaffold's pore volume should be considered as a primary parameter when designing 3D scaffolds for specific tissue engineering and regenerative medicine applications.

Acknowledgements

We are grateful to Dr. T.B.F. Woodfield for his insights on rapid prototyping fabrication. The help of Dr. J. Tibbe and Dr. H.J. van Manen with fluorescence, optical, and confocal microscopy was also very much appreciated.

References

[1] Uhrich KE, Cannizzaro SM, Langer RS, Shakesheff KM. Polymeric systems for controlled drug release. *Chem. Rev.* 1999;99(11):3181-98.

[2] Brittberg M, Peterson L, Sjogren-Jansson E, Tallheden T, Lindahl A. Articular cartilage engineering with autologous chondrocyte transplantation. A review of recent developments. *J. Bone Joint Surg .Am.* 2003;85-A Suppl 3:109-15.

[3] Takahashi J. Stem cell therapy for Parkinson's disease. *Expert Rev. Neurother.* 2007;7(6):667-75.

[4] *Arnesen H, Lunde K, Aakhus S, Forfang K. Cell therapy in myocardial infarction. Lancet* 2007;369(9580):2142-3.

[5] Babensee JE, McIntire LV, Mikos AG. Growth factor delivery for tissue engineering. *Pharm. Res.* 2000;17(5):497-504.

[6] Hunziker E, Spector M, Libera J, Gertzman A, Woo SL, Ratcliffe A, Lysaght M, Coury A, Kaplan D, Vunjak-Novakovic G. Translation from research to applications. *Tissue Eng.* 2006;12(12):3341-64.

[7] Mow VC, Ratcliffe A, Poole AR. Cartilage and diarthrodial joints as paradigms for hierarchical materials and structures. *Biomaterials* 1992;13(2):67-97.

[8] Moutos FT, Freed LE, Guilak F. A biomimetic three-dimensional woven composite scaffold for functional tissue engineering of cartilage. *Nat. Mater.* 2007;doi: 10.1038/nmat1822.

[9] Hollister SJ. Porous scaffold design for tissue engineering. *Nat. Mater.* 2005;4(7):518-24.

[10] Wang XH, Yan YN, Pan YQ, Xiong Z, Liu HX, Cheng B, Liu F, Lin F, Wu RD, Zhang RJ, Lu QP. Generation of three-dimensional hepatocyte/gelatin structures with rapid prototyping system. *Tissue Engineering* 2006;12(1):83-90.

[11] Langer R, Vacanti JP. Tissue engineering. *Science* 1993;260(5110):920-6.

[12] Muschler GF, Nakamoto C, Griffith LG. Engineering principles of clinical cell-based tissue engineering. *J. Bone Joint Surg. Am.* 2004;86-A(7):1541-58.

[13] Sachlos E, Czernuszka JT. Making tissue engineering scaffolds work. Review: the application of solid freeform fabrication technology to the production of tissue engineering scaffolds. *Eur. Cell Mater.* 2003;5:29-39; discussion 39-40.

[14] Hutmacher DW. Scaffold design and fabrication technologies for engineering tissues--state of the art and future perspectives. *J. Biomater. Sci. Polym. Ed.* 2001;12(1):107-24.

[15] Yang S, Leong KF, Du Z, Chua CK. The design of scaffolds for use in tissue engineering. Part II. Rapid prototyping techniques. *Tissue Eng.* 2002;8(1):1-11.

[16] Moroni L, Poort G, Van Keulen F, de Wijn JR, van Blitterswijk CA. Dynamic mechanical properties of 3D fiber-deposited PEOT/PBT scaffolds: An experimental and numerical analysis. *J. Biomed .Mater. Res.* A 2006(78):605-614.

[17] Demarteau O, Wendt D, Braccini A, Jakob M, Schafer D, Heberer M, Martin I. Dynamic compression of cartilage constructs engineered from expanded human articular chondrocytes. *Biochem. Biophys. Res. Commun.* 2003;310(2):580-8.

[18] Malda J, Woodfield TB, van der Vloodt F, Wilson C, Martens DE, Tramper J, van Blitterswijk CA, Riesle J. The effect of PEGT/PBT scaffold architecture on the composition of tissue engineered cartilage. *Biomaterials* 2005;26(1):63-72.

[19] Malda J, Woodfield TB, van der Vloodt F, Kooy FK, Martens DE, Tramper J, van Blitterswijk CA, Riesle J. The effect of PEGT/PBT scaffold architecture on oxygen gradients in tissue engineered cartilaginous constructs. *Biomaterials* 2004;25(26):5773-80.

[20] Stevens MM, George JH. Exploring and engineering the cell surface interface. *Science* 2005;310(5751):1135-8.

[21] Lin CY, Kikuchi N, Hollister SJ. A novel method for biomaterial scaffold internal architecture design to match bone elastic properties with desired porosity. *J. Biomech.* 2004;37(5):623-36.

[22] Dagalakis N, Flink J, Stasikelis P, Burke JF, Yannas IV. Design of an artificial skin. Part III. Control of pore structure. *J. Biomed. Mater. Res.* 1980;14(4):511-28.

[23] Kim SS, Utsunomiya H, Koski JA, Wu BM, Cima MJ, Sohn J, Mukai K, Griffith LG, Vacanti JP. Survival and function of hepatocytes on a novel three-dimensional synthetic biodegradable polymer scaffold with an intrinsic network of channels. *Ann. Surg.* 1998;228(1):8-13.

[24] Hutmacher DW, Sittinger M, Risbud MV. Scaffold-based tissue engineering: rationale for computer-aided design and solid free-form fabrication systems. *Trends Biotechnol.* 2004;22(7):354-62.

[25] Yeong WY, Chua CK, Leong KF, Chandrasekaran M. Rapid prototyping in tissue engineering: challenges and potential. *Trends Biotechnol.* 2004;22(12):643-52.

[26] Zein I, Hutmacher DW, Tan KC, Teoh SH. Fused deposition modeling of novel scaffold architectures for tissue engineering applications. *Biomaterials* 2002;23(4):1169-85.

[27] Moroni L, de Wijn JR, van Blitterswijk CA. 3D fiber-deposited scaffolds for tissue engineering: influence of pores geometry and architecture on dynamic mechanical properties. *Biomaterials* 2006;27(7):974-85.

[28] Woodfield TB, Van Blitterswijk CA, De Wijn J, Sims TJ, Hollander AP, Riesle J. Polymer scaffolds fabricated with pore-size gradients as a model for studying the zonal organization within tissue-engineered cartilage constructs. *Tissue Eng.* 2005;11(9-10):1297-311.

[29] Sherwood JK, Riley SL, Palazzolo R, Brown SC, Monkhouse DC, Coates M, Griffith LG, Landeen LK, Ratcliffe A. A three-dimensional osteochondral composite scaffold for articular cartilage repair. *Biomaterials* 2002;23(24):4739-51.

[30] Lin AS, Barrows TH, Cartmell SH, Guldberg RE. Microarchitectural and mechanical characterization of oriented porous polymer scaffolds. *Biomaterials* 2003;24(3):481-9.

[31] Woodfield TB, Bezemer JM, Pieper JS, van Blitterswijk CA, Riesle J. Scaffolds for tissue engineering of cartilage. *Crit. Rev. Eukaryot. Gene Expr.* 2002;12(3):209-36.

[32] Hunzinker EB. Articular cartilage repair: basic science and clinical progress. A review of the current status and prospects. *Osteoarthritis and Cartilage* 2001;10:432-463.

[33] Woodfield TB, Malda J, de Wijn J, Peters F, Riesle J, van Blitterswijk CA. Design of porous scaffolds for cartilage tissue engineering using a three-dimensional fiber-deposition technique. *Biomaterials* 2004;25(18):4149-61.

[34] Landers R, Pfister A, Hubner U, John H, Schmelzeisen R, Mullhaupt R. Fabrication of soft tissue engineering scaffolds by means of rapid prototyping techniques. *Journal of materials science* 2002;37:3107-3116.

[35] Haugland RP. Handbook of Fluorescent Probes and Research Chemicals. 7th ed., Eugene, OR: Molecular Probes 1999.

[36] Miot S, Woodfield T, Daniels AU, Suetterlin R, Peterschmitt I, Heberer M, van Blitterswijk CA, Riesle J, Martin I. Effects of scaffold composition and architecture on human nasal chondrocyte redifferentiation and cartilaginous matrix deposition. *Biomaterials* 2005;26(15):2479-89.

[37] Kim TK, Sharma B, Williams CG, Ruffner MA, Malik A, McFarland EG, Elisseeff JH. Experimental model for cartilage tissue engineering to regenerate the zonal organization of articular cartilage. *Osteoarthritis Cartilage* 2003;11(9):653-64.

[38] Klein TJ, Schumacher BL, Schmidt TA, Li KW, Voegtline MS, Masuda K, Thonar EJ, Sah RL. Tissue engineering of stratified articular cartilage from chondrocyte subpopulations. *Osteoarthritis Cartilag*e 2003;11(8):595-602.

[39] Hendriks JAA, de Bruijn E, van Blitterswijk CA, Riesle J. A powerful tool in cartilage tissue engineering; coculturing primary chondrocytes with expanded chondrocytes enhances chondrogenesis. ORS Transactions of 51st annual meeting 2006.

[40] Solursh M, Meier S. Effects of cell density on the expression of differentiation by chick embryo chondrocytes. J Exp Zool 1974;187(3):311-22.

[41] Schulze-Tanzil G, de Souza P, Villegas Castrejon H, John T, Merker HJ, Scheid A, Shakibaei M. Redifferentiation of dedifferentiated human chondrocytes in high-density cultures. Cell Tissue Res 2002;308(3):371-9.

[42] Hutmacher DW, Schantz T, Zein I, Ng KW, Teoh SH, Tan KC. Mechanical properties and cell cultural response of polycaprolactone scaffolds designed and fabricated via fused deposition modeling. J Biomed Mater Res 2001;55(2):203-16.

[43] Wang H, Pieper J, Peters F, van Blitterswijk CA, Lamme EN. Synthetic scaffold morphology controls human dermal connective tissue formation. J Biomed Mater Res A 2005;74(4):523-32.

[44] Singh H, Teoh SH, Low HT, Hutmacher DW. Flow modelling within a scaffold under the influence of uni-axial and bi-axial bioreactor rotation. J Biotechnol 2005;119(2):181-96.

[45] Tsuchiya K, Chen GP, Ushida T, Matsuno T, Tateishi T. The effect of coculture of chondrocytes with mesenchymal stem cells on their cartilaginous phenotype in vitro. Materials Science and Engineering C-Biomimetic and Supramolecular Systems 2004;24(3):391-396.

[46] Barry JJ, Gidda HS, Scotchford CA, Howdle SM. Porous methacrylate scaffolds: supercritical fluid fabrication and in vitro chondrocyte responses. Biomaterials 2004;25(17):3559-68.

[47] Freed LE, Marquis JC, Nohria A, Emmanual J, Mikos AG, Langer R. Neocartilage formation in vitro and in vivo using cells cultured on synthetic biodegradable polymers. J Biomed Mater Res 1993;27(1):11-23.

[48] VunjakNovakovic G, Freed LE, Biron RJ, Langer R. Effects of mixing on the composition and morphology of tissue-engineered cartilage. Aiche Journal 1996;42(3):850-860.

[49] Martin I, Shastri VP, Padera RF, Yang J, Mackay AJ, Langer R, Vunjak-Novakovic G, Freed LE. Selective differentiation of mammalian bone marrow stromal cells cultured on three-dimensional polymer foams. J Biomed Mater Res 2001;55(2):229-35.

[50] VunjakNovakovic G, Obradovic B, Martin I, Bursac PM, Langer R, Freed LE. Dynamic cell seeding of polymer scaffolds for cartilage tissue engineering. Biotechnol Prog 1998;14(2):193-202.

[51] Martin I, Vunjak-Novakovic G, Yang J, Langer R, Freed LE. Mammalian chondrocytes expanded in the presence of fibroblast growth factor 2 maintain the ability to differentiate and regenerate three-dimensional cartilaginous tissue. Exp Cell Res 1999;253(2):681-8.

[52] Lin CY, Schek RM, Mistry AS, Shi X, Mikos AG, Krebsbach PH, Hollister SJ. Functional bone engineering using ex vivo gene therapy and topology-optimized, biodegradable polymer composite scaffolds. Tissue Eng 2005;11(9-10):1589-98.

[53] Hutmacher DW, Ng KW, Kaps C, Sittinger M, Klaring S. Elastic cartilage engineering using novel scaffold architectures in combination with a biomimetic cell carrier. *Biomaterials* 2003;24(24):4445-58.

In: Cell Growth Processes: New Research
Editor: Daiki Kimura, pp. 203-217

ISBN: 978-1-60456-132-6

Chapter VIII

Transcriptional Regulation of Neurotrophic Factor in the Process of Neurite Growth (Review)

Seyeon Park[1]
Department of Applied Chemistry, Dongduk Women's University Seoul 136-714, Korea

Abstract

Glial-cell-line-derived neurotrophic factor (GDNF) was known as a survival factor for midbrain dopaminergic neurons. GDNF and related ligands, neurturin, persephin and artemin maintain several neuronal populations in the central nervous systems, including midbrain dopamine neurons and motoneurons. In addition, GDNF, neurturin and artemin support the survival and regulate the differentiation of many peripheral neurons, including sympathetic, parasympathetic, sensory and enteric neurons. The GDNF family ligands signal through a receptor complex consisting of a ligand binding component, GDNF family receptor α (GFRα) and RET receptor tyrosine kinase. A recent study showed that systemic GDNF family ligands prevents and reverses neuropathic pain and normalizes spinal nerve ligation (SNL)-induced morphological changes. It was suggested that GDNF family ligands plays a role in inducing neurite outgrowth in the early stages of neuron development. Furthermore, the mature DRG neurons showed some response to artemin, suggesting that they exhibited a developmental shift to the ligand as prenatal DRG neurons.

Upon ligand binding, GDNF in complex with GFR may interact with heparan sulphate glycosaminoglycans to activate the Met receptor tyrosine kinase through cytoplasmic Src-family kinases. GDNF family ligands also signal through the neural cell adhesion molecule NCAM. In cells lacking RET, GDNF binds with high affinity to the NCAM and GFRα complex, which activates Fyn and FAK. At the final stage of intracellular signaling, a set of genes involved in regulation of actin dynamics including

1 Correspondence should be addressed: Seyeon Park, Ph. D. Department of Applied Chemistry, Dongduk Women's University, 23-1 Wolgok-dong, Sungbuk-ku, Seoul 136-714, Korea,Tel 82-2-940-4514 Fax 82-2-940-4193,E-mail sypark21@dongduk.ac.kr, seypark21@hotmail.com .

Wiskott-Aldrich syndrome protein (WASP) interacting protein, cofilin and dynamin were transcriptionally regulated by artemin, suggesting a previously undefined role in the regulation of neurite growth by artemin. DNA methylation provides a mechanism for artemin-dependent genetic regulation responsible for axonal growth.

Neurotrophic Factor and Cell Biology

Artemin is a member of the glial cell line-derived neurotrophic factor (GDNF) family of ligands that helps to ensure the survival of sensory neurons. Neurotrophic factors are a heterogenous group of signaling molecules that regulate multiple aspects of the development and maintenance of the central and peripheral nervous systems. The GDNF family is a subfamily of the transforming growth factor β (TGFβ) superfamily and includes artemin [28,34], persephin [32], and neurturin [50]. The GDNF family receptors comprise the GDNF receptor accessory proteins (GFR-α1, α2, α3 and α4), each of which binds preferentially to a different ligand, and the tyrosine kinase receptor, RET, which mediate signaling [74]. In recent studies of the impact of GDNF on posttraumatic nerve regeneration and the treatment of neuropathic pain, GDNF was found to reverse the posttraumatic changes [4,5,48,54]. Artemin expression is observed in the adult brain and spinal cord [3], although not at a very high level, suggesting that it may play a role in the post-developmental stage.

Artemin, GDNF, and neurturin were all capable of influencing neurite outgrowth from the dorsal root and sympathetic ganglia, but the responses of each type of neuron to the different ligands varied during development. Neurites from the dorsal root ganglia responded to artemin at P0 and P7, to GDNF at E15.5 and P0, and to neurturin at E15.5, P0, and P6/7 [78]. Neurites from the superior cervical sympathetic ganglia responded significantly to artemin at E15.5, to GDNF at E15.5 and P0, and to neurturin at E15.5. Neurites from the lumbar sympathetic ganglia responded to artemin at all stages from E11.5 to P7, to GDNF at P0 and P7 and to neurturin at E11.5 to P6/7. Artemin plays a role in inducing neurite outgrowth from young sympathetic neurons in the early stages of sympathetic axon pathfinding, whereas GDNF and neurturin are likely to be important at later stages of sympathetic neuron development, by inducing axons to enter particular target tissues once they are in their vicinity or by inducing branching within target tissues [78]. However, artemin expression is observed in the adult brain and spinal cord [10]. Moreover, artemin promoted the axonal growth and affected the transcriptional patterns in DRG neurons of the P7 stage [45]. GDNF promotes the survival of postnatal—but not embryonic—mouse dorsal root ganglia cells in vitro, despite the fact that its receptors are expressed at both ages [39]. Thus, although receptors for GDNF were present at both ages, embryonic cells did not respond to GDNF treatment by exhibiting any significant level of survival.

The differential effects of GDNF at E14 and P14 could be caused by the changing phenotype of DRG cells during development [29]. Indeed, over 300 genes changed between the different ages. Among those confirmed by real-time PCR as being down-regulated by age were NCAM (a neuron-specific cell-adhesion molecule) and cerebroglycan, a cell-surface heparan sulfate proteoglycan known to be transiently expressed by immature neurons and that typically disappears after axon outgrowth and after cell migration has been completed [60].

Intracellular Signal Transduction

Like other GDNF family ligands artemin contains seven conserved, similarly spaced cysteine residues, which are referred to as a cysteine knot motif, and show an approximately 40% amino acid identity to each other [62]. The GDNF family ligands signal through a unique multicomponent receptor complex consisting of the glycosyl-phosphatidylinositol (GPI)-anchored coreceptor (GFRα1–4) which acts as a ligand binding component and RET receptor tyrosine kinase which acts as a signaling component [3,16,12,22,65,28,6,2,40,52,23,61,9,1,37, 39,74,67,14,62,64]. GDNF, neurturin, artemin and persephin use GFRα1, GFRα2, GFRα3 and GFRα4 as the preferred ligand-binding receptors, respectively, although alternative ligand-coreceptor interactions also appear to occur in culture. That is, GDNF can bind to GFRα2 and GFRα3, and neurturin and artemin can bind to GFRα1 [3,62,2,52,23].

Like other receptor tyrosine kinases, RET is known to activate various signaling pathways including the RAS/extracellular signal-regulated kinase (ERK), phosphatidylinositol 3-kinase (PI3K)/AKT, p38 mitogen activated protein kinase (MAPK) and c-Jun N-terminal kinase (JNK) pathways [71,75,70,42,64,7,35,36, 33,68,59, 15,55, 16]. It is also known that these signaling events are performed via the phosphotyrosine present in the intracellular domains of activated receptors [64]. A recent study demonstrated that the RAS/ERK, PI3K/AKT, p38MAPK and JNK pathways are all activated mainly through tyrosine 1062 of RET, which acts as a binding site for SHC adaptor proteins [16]. In addition, the p38MAPK, JNK and MEK5/big mitogen-activated protein kinase 1 (BMK1) pathways are also activated through tyrosine 1062 [16]. Because it is possible that PI3K signaling regulates the functions of Rho/Rac-related small G proteins following GDNF stimulation, the JNK pathway may be situated downstream of PI3K. In addition, the tyrosine phosphorylation of several other signaling molecules including FAK, paxillin, p130CAS and p62Dok was found be induced by PI3K signaling [35,36].

GDNF also promoted the survival of cultured neuronal cells expressing GFRα1, but lacking RET, and induced CREB and Fos activation in them [68,47]. Thus, it was suggested that RET-independent GDNF signaling through GFRα1 may play a role postnatally in neuronal functions rather than in neuronal differentiation during embryogenesis.

More recently, the presence of RET-independent signals triggered by GDNF was suggested [73,51]. GDNF was able to activate Src-family kinases (SFKs) and PLCγ in sensory neurons from RET−/− mice and Src kinases were associated with GFRα1 [62]. However, whether or not artemin can transmit signals in RET-dependent or independent ways has not yet been fully demonstrated. SFK is known to be related to neurite outgrowth. Src-dependent tyrosine phosphorylation at the tips of growth cone filopodia promotes their extension [49]. In our studies, it was found that artemin induced neurite growth in mature DRG neurons [45]. Because SFKs are activated by GDNF in several cell lines and in some neurons [13,46,47], we investigated the role of SFKs and ERKs in artemin-mediated axonal growth in mature DRG neurons [In press: Cell Biochemistry and Function]. Various signaling inhibitors (ERK1/2 inhibitor; PD 98059 50μM, PI3K inhibitor; LY294002 20μM, SFK inhibitor; SU6656 5μM) were used to investigate the effect of artemin on each intracellular signaling pathway during neurite outgrowth. SU6656, a specific inhibitor of

SFKs suppressed the artemin-dependent axonal growth in a statistically significant manner at a concentration of 5 μM. In addition, the treatment of the DRG neurons with the ERK-specific inhibitor, PD98059, diminished the artemin-dependent axonal growth, indicating the downstream engagement of ERK in an artemin-mediated signal transduction pathway [In press: Cell Biochemistry and Function]. Although LY294002, a specific inhibitor of PI3 kinases, at a concentration of 20 μM partially reduced the growth of the neurites, the statistical analysis did not show any significant differences. This result for PI3 kinases is inconsistent with the previous data [19] suggesting that PI3K is related to neurite extension. The tightly regulated and localized activation of PI3K at the growth cone is essential for the rapid axon growth induced by NGF [19].

Activation of RET stimulates kinase domain, and RET has docking site for various cytoplasmic adaptor proteins such as Grb 7/10, PLCγ, Shc/ENIGMA and Grb2. This is consistent with the data indicating that several second messenger pathways are activated by RET including PLCγ, JNK, PI3K and Ras-MAPK pathways [31]. First of all, the ERK1/2 pathway is known to be related to neurite growth. The ability of axons to regenerate their growth cones is regulated by calcium, cAMP and ERK [8]. MEKK1 was found to control neurite regrowth after experimental injury by balancing ERK1/2 and JNK2 signaling [72]. An analysis of platelet-derived growth factor receptor mutants showed that sustained ERK activation alone is insufficient to control neuritogenesis, and that additional signals, possibly involving Src are also required [69].

Transcriptional Change

Artemin, GDNF, and neurturin were all capable of influencing neurite outgrowth from the dorsal root and sympathetic ganglia, but the responses of each type of neuron to the different ligands varied during development [5]. GDNF promotes the survival of postnatal—but not embryonic—mouse dorsal root ganglia cells in vitro, despite the fact that its receptors are expressed at both ages [29]. Over 300 genes changed between the different ages [60]. Among those confirmed by real-time PCR were shown in Table 1.

It has been suggested that several hundred genes are regulated between the embryonic and postnatal stages, and that several important classes of genes are differentially regulated by GDNF treatment, including genes related to translation and to phenotypic specification and maturation [29]. A set of genes related to cell adhesion, cytoskeleton and cellular morphology was found to be consistently down-regulated by GDNF, suggesting that GDNF plays a role in repressing neurite growth and/or branching. This nuclear program initiated by GDNF was functionally confirmed in cultures of embryonic wild-type neurons sustained with nerve growth factor and in bax−/− neurons that survive in the absence of trophic support [29].

We investigated the effect of artemin on the activation of axonal growth in mature DRG neurons. To demonstrate the correlation between the transcriptional programming of artemin and neurite growth or its systemic role in the neuronal system, gene profiling was performed to identify those genes that are differentially regulated in response to artemin, using microarray analysis in the post-developmental rat DRG. In our study, the transcriptional program initiated by artemin was functionally relevant to the axonal growth in DRG neurons [shown in 45]. Among the genes that were differentially regulated by artemin, coronin, Myr5, Wiskott-Aldrich syndrome protein interacting protein (WASPIP), cofilin, drebrin and Eps8 are known to interact with actin protein (Table 2). Functionally, artemin resulted in an evident increase in the total neurite length of the mature DRG neurons [45]. Out of GDNF-regulated genes (Table 1), actin, neuronatin and tubulinβ are also regulated by artemin (our unpublished data). These proteins are commonly related to the cytoskeleton formation.

Table 1. GDNF-regulated genes confirmed by real-time PCR (shown in Ref 29)

Accession No.	Gene	E14	P14
Mm.297	A-X actin	-5.2	3.2
Mm.4968	Translation elongation factor 2	1.0	-19.1
Mm.16323	Translation initiation factor 4A2	2.1	-5.1
Mm.1391	GATA-binding protein 2	1.2	-12.0
Mm.2344	Guanine nucleotide binding protein _-1	-5.8	6.9
Mm.5079	Hydroxysteroid 11-_ dehydrogenase 2	-8.4	-1.7
Mm.20354	Kinesin light chain 1	-1.5	-7.9
Mm.7362	Lamin B2	-9.9	-17.8
Mm.4071	Laminin receptor 1 (LBP67)	-9.3	3.1
TC35356	Neuritin	-1.6	-6.6
Mm.7414	Neuron specific gene family member 1	3.4	(-38)
Mm.3210	Neuronatin	1.6	(-165)
Mm.66	Ribosomal protein S4, X-linked	-4.1	5.6
Mm.1493	SRY-box containing gene 10 (Sox10)	-6.3	-1.4
TC38163	Tubulin _-4	1.1	-13.6
Mm.18538	ESTs	-4.4	(403)
Mm.20417	ESTs	(2362)	15.0
Mm.2498	ESTs	-5.0	6.6
Mm.5063	ESTs	-5.8	2.8
Mm.30134	60S Ribosomal protein L44	7.8	2.5

EGF stimulation of serum-starved cells induced invadopodium formation. RNA interference and dominantnegative mutant expression analyses revealed that neural WASP (N-WASP), Arp2/3 complex, and their upstream regulators, Nck1, Cdc42, and WIP, are necessary for invadopodium formation [77]. Suppression of cofilin expression by RNA interference inhibited the formation of long-lived invadopodia, resulting in formation of only short-lived nvadopodia with less matrix degradation activity [77]. These results indicate that the N-WASP–Arp2/3 pathway and cofilin is necessary for the stabilization and maturation of invadopodia not only through EGF receptor signaling pathway but through neurotrophic factor signaling. In this case, cofilin is one of actin depolymerization factors.

Drebrin located in dendritic spines regulates their morphological changes and plays a role in the synaptic plasticity via spine function [27]. Drebrins are actin-binding proteins found in the central nervous system of various species [reviewed in 58]. Drebrin forms a complex with gelsolin, myosin, and F-actin in vivo [18]. Previous in vitro studies [53,21] have revealed that drebrin binds to filamentous actin (F-actin) competitively with F-actin-stabilizing proteins such as fascin, α-actinin, and tropomyosin. Overexpression of drebrin in fibroblasts results in actin filament remodeling [57,17]. In addition to its ability to bind to F-actin, drebrin has been shown to interact with profilin, an actin monomer-binding protein that stimulates actin to polymerize, probably via the proline-rich domain of drebrin [30]. These observations indicate a general involvement of drebrin in actin cytoskeleton dynamics. Antisense knockdown of drebrin A, a dendritic spine protein, causes stronger preference, impaired pre-pulse inhibition, and an increased sensitivity to psychostimulant [27].

Table 2. Artemin-regulated genes confirmed by real-time PCR (shown in Ref 45)

Accession No.	Gene	DRG Culture			Isolated neuron culture
		Microarray	Real-time PCR		
		Normalized ratio (3 h)	Normalized ratio (3 h)	+CdR, Normalized ratio (3 h)	Normalized ratio (3 h)
AI071582	Dynamin 1	0.478*	0.463±0.281*^	0.894±0.186^	0.587±0.112*
AA924428	U1-R-A1-dy-b-10-0-U1.s1 U1-R-A1 Coronin	0.480*	0.497±0.108*^	0.967±0.361^	0.336±0.094*
AA874917	Biglycan	0.517	-	-	-
AA818082	U1-R-A0-ag-d-02-0-U1.s1 U1-R-A0 Myr5	0.567*	0.572±0.136*	0.364±0.261	0.466±0.188*
AA964363	Plectin	0.576	-	-	-
AI030685	Nestin	0.606	0.639±0.241	-	0.911±0.160
AA899166	WASP interacting protein	0.636*	0.449±0.122*^	0.976±0.106^	0.472±0.097*
AA899395	Cofilin	0.637*	0.614±0.101*^	1.010±0.245^	0.787±0.069*
AA858569	Neuronatin	0.656	0.861±0.245	-	0.723±0.296
AA924797	Parvin, alpha	0.663	-	-	-
AA924727	Collagen, type 1, alpha 1	0.676	-	-	-
AI044670	Ninjurin 1	0.760*	0.720±0.078*	0.676±0.310	0.370±0.180*
AA899901	Ataxin-2	0.767	0.745±0.301	-	0.764±0.211
AA998487	Dynamin 2	0.780	-	-	-
AI144588	Neuron-glia-CAM-related cell adhesion molecule	0.789*	0.776±0.068*	0.421±0.215	0.680±0.138*
AA900039	Drebrin 1	0.790*	0.546±0.211*	0.689±0.133	0.461±0.224*
AA818315	Seipin	0.794	-	-	-
AA900660	U1-R-E0-di-f-09-0-U1.s1 U1-R-E0 CofactorA	1.57	-	-	-
AA875641	U1-R-E0-ct-f-12-0-U1.s1 U1-R-E0 Eps 8	1.88**	1.86±0.108**^	0.944±0.459^	1.92±0.211**

* The values were consistently down-regulated between microarray and real-time PCR with statistical significance ($p<0.05$)

** The values were consistently up-regulated between microarray and real-time PCR with statistical significance ($p<0.05$)

- Single peak data were not obtained using designed probes in real-time PCR experiment.

^ The values represented statistically significant difference between the control (without CdR) and CdR treatment ($P<0.05$).

All coronin proteins have a variable N-terminal domain, followed by WD repeats and a C-terminal coiled-coil dimerization domain. Transfection of coronin-1-GFP into RAW 264.7 cells revealed that coronin rapidly and transiently associates with the phagosome. Dominantnegative approach by expressing only the central WD domains caused cell rounding and dissociation from the substratum, hampering analysis of their phenotype [79]. These data demonstrate that coronin-1 is required for an early step in phagosome formation, consistent with a role in actin polymerization [79].

The Eps8 family of proteins links growth factor stimulation to actin reorganization generating functional redundancy in the Ras/Rac pathway [41]. The Eps8 shows barbed-end capping activity, which resides in their conserved carboxy-terminal effector domain. The isolated effector domain of Eps8 caps barbed ends with an affinity in the nanomolar range [11].

Biglycan belongs to the family of small leucine-rich proteoglycans and is functionally involved in matrix assembly, cellular migration, and the regulation of growth factor such as TGF-β activity [20,26].

Plectin is a versatile cytolinker protein [43]. It can bind to all major cytoskeletal filament networks and it plays an important role as mechanical linker and anchoring protein of intermediate filaments. Several of a series of specific binding partners of plectin identified are constituents of the subplasma membrane protein skeleton or of junctional membrane complexes, such as integrin β4, spectrin/fodrin, and desmoplakin, suggesting that the protein provides a linkage between the cytoskeleton and the cell periphery [76]. Plectin has actin binding domain [56].

We also found differences in the vesicle distribution, such that the total clusters of the vesicles were dispersed along the neurites in the artemin-regulated neurons [shown in 45]. Some enlarged parts were observed along the neurites. The number of vesicles was increased by the treatment with artemin. Synaptic varicosities are known to contain clusters of vesicles as well as other presynaptic elements such as machineries for exocytosis and endocytosis [50]. These results suggest that artemin induces an enhancement of synaptic vesicle clustering and accumulation.

DNA methylation is known to play a key role in regulating gene expression during cell development. Many genes contain CpG islands in their promoter regions and the methylation of these regions leads to transcriptional silencing in cells [54,63]. Also, the methylation of the promoter region of some genes is thought to be an important mechanism in modulating the expression of genes [73,80]. In the previous study, we investigated the effect of DNA methyltransferase inhibition on normal rat DRG neurons treated with artemin. The DNA methyltransferase inhibitor, 5-aza-2-deoxycytidine (CdR), suppressed the artemin-dependent axonal growth [45]. Our results support the hypothesis that the DNA or promoter methylation levels could act as an artemin-induced axonal growth mechanism in mature DRG neurons.

It was suggested that artemin plays a role in inducing neurite outgrowth in the early stages of neuron development [80]. However, the mature DRG neurons showed some response to artemin, suggesting that they exhibited a developmental shift to the ligand as prenatal DRG neurons. Our data suggest that DNA methylation provides a mechanism for artemin-dependent genetic regulation responsible for axonal growth. It would be of therapeutical relevance to determine if artemin, a GDNF family member, promotes axonal regeneration after nerve injury.

Proteome Change

To understand the molecular basis involved in neurotrophic factor-induced cell biology, it is important to identify proteins that are regulated by GDNF or artemin. We analyzed proteome changes induced by GDNF or artemin in SH-SY-5y human neuroblastoma cells using 2-dimentional gel approach. This methodology provides important qualitative information on post-translational modifications to each protein and quantitative data on protein expressions in response to a particular stimulus. This information is particularly important because it provides data on early cellular events, such as, on the stimulus and signaling cascades triggered independently of protein neosynthesis.

The immediate effects of cell stimuli are associated with protein post-translational modifications, such as, phosphorylation, glutathionylation and cysteine oxidation. To study these early modifications, SH-SY-5y human neuroblastoma cells were treated with 500 ng/ml GDNF or artemin and then collected for two-dimensional electrophoresis analysis. Approximate 120 different spots that were focused in the 4-7 pH range were analyzed per sample. After exposing cells to GDNF or artemin, we observed; intensified spots, and attenuated spots. Each of these spots were excised, digested with trypsin and analyzed by MALDI–TOF. Peptide mass fingerprint analysis and nonredundant sequence database matching allowed the unambiguous identification of all of the analyzed species. The proteins were shown in Table 3 and 4.

Table 3. GDNF-regulated proteins

Spot number	Fold-change	MW (Dalton)	pI	Identification(MS/MS)
G7314	12304.9	42120	6.05	CDC_Septin
G8307	26011.1	47481	7.01	Enolase 1
G1204	14932.1	31888	4.84	Laminin-binding protein
G1106	14332.5	28517	4.89	Tropomyosin isoform
G7313	9360.2	42108	5.31	Actin, gamma 1 propeptide
G6408	0.15	57146	5.98	Protein disulfide isomerase-associated 3 precursor
G1001	11.99	27899	4.73	Tyrosine 3/tryptophan 5 -monooxygenase activation protein, zeta polypeptide

Table 4. Artemin-regulated proteins

Spot number	Fold-change	MW (Dalton)	pI	Identification(MS/MS)
A3704	196.56	36,202	8.26	Glyceraldehyde-3-phosphate dehydrogenase
A8002	0.27	36,824	6.85	CDC_Septin
A8802	0.35	95,078	5.87	Matrin 3
A3201	0.42	25,151	5.33	Ubiquitin carboxyl-terminal esterase L1 (ubiquitin thiolesterase)
A401	2.5	48,283	4.29	Calreticulin precursor
A8804	0.3	107158	5.71	Glycosyl hydrolases family 31

From the result, tropomyosin was identified as a marker of GDNF signaling in SH-SY-5y cells. The spot corresponding to tropomyosin was positioned at a pI value of approximately 5.0. This spot appeared in GDNF-treated cells. The significance of this is unclear but may be related to differences in the regulation of the actomyosin contractile system in neuronal cells as compared with that present in muscle cells. Although the function of these proteins in neouronal differentiation needs further investigation, tropomyosin may have value as a cytoskeleton rearrangement-associated protein and as a molecular target in neurotrophic factor-based cell change. From this point of view, it is important that GDNF causes a tropomyosin isoform to be modified during the immediate early response.

Another protein identified from the proteomic analysis is a thiol/disulfide exchange catalyst as a marker of the effect of GDNF in SH-SY-5y cells. Protein disulfide isomerase (PDI) belongs to the superfamily of thiol/disulfide exchange catalyst which acts as protein-thiol oxidoreductase enzymes and shares sequence similarities with thioredoxin [38]. PDI is composed of four domains which have similarities with thioredoxin folds (a-b-b'-a') [24].

Interestingly, of the proteins with expressions changed by artemin treatment, septin protein comprises a eukaryotic guanine nucleotide binding protein subfamily which forms filamentous heteropolymer complexes [25].

Taken together, these results suggest that neurotrophic factors reorganize intracellular cytoskeleton through transcriptional and translational regulation.

References

[1] Baloh, R. H.; Gorodinsky, A.; Golden, J. P.; Tansey, M. G.; Keck, C. L.; Popescu, N. C. Johnson Jr., E. M.; Milbrandt J. GFR-alpha3 is an orphan member of the GDNF/neurturin/persephin receptor family. *Proc. Natl. Acad.* Sci. U.S.A. 95:5801–5806; 1998.

[2] Baloh, R. H.; Tansey, M. G.; Golden, J. P.; Creedon, D. J.; Heuckeroth, R. O.; Keck, C. L.; Zimonjic, D. B.; Popescu, N. C.; Johnson Jr., E. M.; Milbrandt, J. TrnR2, a novel receptor that mediates neurturin and GDNF signaling through Ret. *Neuron*. 18:793–802; 1997.

[3] Baloh, R. H.; Tansey, M. G.; Lampe, P. A.; Fahrner, T. J.; Enomoto, H.; Simburger K. S.; Leitner, M. L.; Araki, T.; Johnson Jr., E. M.; Milbrandt, J. Artemin, a novel member of the GDNF ligand family, supports peripheral and central neurons and signals through the GFRα3-RET receptor complex. *Neuron.* 21:1290–1302; 1998.

[4] Bennett, D. L.; Boucher, T. J.; Armanini, M. P.; Poulsen, K. T.; Michael, G. J.; Priestley, J. V.; Philips, H. S.; McMahon, S. B.; Shelton, D. L. The glial cell line-derived neurotrophic factor family receptor components are differentially regulated within sensory neurons after nerve injury. *J. Neurosci.* 20:427–437; 2000.

[5] [5] Boucher, T. J.; Okuse, K.; Bennett, D. L.; Munson, J. B.; Wood, J. N.; McMahon, S. B. Potent analgesic effects of GDNF in neuropathic pain states. *Science*. 290:124–127; 2000.

[6] Buj-Bello, A.; Adu, J.; Pinon, L. G.; Horton, A.; Thompson, J.; Rosenthal, A.; Chinchetru, M.; Buchman, V. L.; Davies, A. M. Neurturin responsiveness requires a GPI-linked receptor and the Ret receptor tyrosine kinase. *Nature*. 387;721–724; 1997.

[7] Chiariello, M.; Visconti, R.; Carlomagno, F.; Melillo, R. M.; Bucci, C.; de Franciscis, V.; Fox, G. M.; Jing, S.; Coso, O. A.; Gutkind, J. S.; Fusco, A.; Santoro, M. Signalling of the Ret receptor tyrosine kinase through the c-Jun NH2-terminal protein kinases (JNKs): evidence for a divergence of the ERKs and JNKs pathways induced by Ret. *Oncogene*. 16:2435–2445; 1998.

[8] Chierzi, S.; Ratto, G.. M.; Verma, P.; Fawcett, J. W. The ability of axons to regenerate their growth cones depends on axonal type and age, and is regulated by calcium, cAMP and ERK. *Eur. J. Neurosci.* 21:2051-2062; 2005.

[9] Creedon, D. J.; Tansey, M. G.; Baloh, R. H.; Osborne, P. A.; Lampe, P. A.; Fahrner, T. J.; Heuckeroth, R. O.; Milbrandt, J.; Johnson Jr., E. M. Neurturin shares receptors and signal transduction pathways with glial cell line-derived neurotrophic factor in sympathetic neurons. *Proc. Natl. Acad. Sci.* U.S.A. 94:7018–7023; 1997.

[10] Davies, A. M. Neurotrophin switching: where does it stand? Curr. Opin. *Neurobiol.* 7:110-118; 1997.

[11] Disanza, A.; Carlier, M. F.; Stradal, T. E.; Didry, D.; Frittoli, E.; Confalonieri, S.; Croce, A.; Wehland, J.; Di Fiore, P. P.; Scita, G. Eps8 controls actin-based motility by capping the barbed ends of actin filaments. *Nat. Cell. Biol.* 6:1180-1188; 2004.

[12] Durbec, P.; Marcos-Gutierrez, C. V.; Kilkenny, C.; Grigoriou, M.; Wartiowaara, K.; Suvanto, P.; Smith, D.; Ponder, B.; Costantini, F.; Saarma, M.; Sariola, H.; Pachnis, V. GDNF signalling through the Ret receptor tyrosine kinase. *Nature.* 381:789–793; 1996.

[13] Encinas, M.; Tansey, M. G.; Tsui-Pierchala, B. A.; Comella, J. X.; Milbrandt, J.; Johnson Jr., E. M. C-Src is required for glial cell line-derived neurotrophic factor(GDNF) family ligand-mediated neuronal survival via a phosphatidylinositol-3 kinase (PI-3K)-dependent pathway. *J Neurosci.* 21:1464-1472; 2001.

[14] Enokido, Y.; de Sauvage, F.; Hongo, J. A.; Ninkina, N.; Rosenthal, A.; Buchman, V. L.; Davies, A. M. GFRα4 and the tyrosine kinase Ret form a functional receptor complex for persephin. *Curr. Biol.* 8:1019–1022; 1998.

[15] [15] Feng, L.; Wang, C. Y.; Jiang, H.; Oho, C.; Dugich-Djordjevic, M.; Mei, L.; Lu, B. Differential signaling of glial cell line-derived neurotrophic factor and brain-derived neurotrophic factor in cultured ventral mesencephalic neurons. *Neuroscience*. 93:265–273; 1999.

[16] Hayashi, H.; Ichihara, M.; Iwashita, T.; Murakami, H.; Shimono, Y.; Kawai, K.; Kurokawa, K.; Murakumo, Y.; Imai, T.; Funahashi, H.; Nakao, A.; Takahashi, M. Characterization of intracellular signals via tyrosine 1062 in RET activated by glial cell line-derived neurotrophic factor. *Oncogene.* 19:4469–4475; 2000.

[17] Hayashi, K.; Ishikawa, R.; Kawai-Hirai, R.; Takagi, T.; Taketomi, A.; Shirao, T. Domain analysis of the actin-binding and actin-remodeling activities of drebrin. *Exp. Cell. Res*. 253:673–680; 1999.

[18] [18] Hayashi, K.; Ishikawa, R.; Ye, L. H.; He, X. L.; Takata, K.; Kohama, K.; Shirao, T. Modulatory role of drebrin on the cytoskeleton within dendritic spines in the rat cerebral cortex. *J. Neurosci.* 15:7161-7170; 1996.

[19] Higuchi M, Onishi K, Masuyama N, Gotoh Y. The phosphatidylinositol-3 kinase (PI3K)-Akt pathway suppresses neurite branch formation in NGF-treated PC12 cells. *Genes Cells* 2003 8: 657-669.

[20] Iozzo, R. V. The biology of the small leucine-rich proteoglycans. Functional network of interactive proteins. *J. Biol. Chem.* 274:18843-18846; 1999.

[21] Ishikawa, R.; Hayashi, K.; Shirao, T.; Xue, Y.; Takagi, T.; Sasaki, Y.; Kohama, K. Drebrin, a development-associated brain protein from rat embryo, causes the dissociation of tropomyosin from actin filaments. *J. Biol. Chem.* 269:29928–29933; 1994

[22] Jing, S.; Wen, D.; Yu, Y.; Holst, P. L.; Luo, Y.; Fang, M.; Tamir, R.; Antonio, L.; Hu, Z.; Cupples, R.; Louis, J. C.; Hu, S.; Altrock, B. W.; Fox, G. M. GDNF-induced activation of the Ret protein tyrosine kinase is mediated by GDNFR-α, a novel receptor for GDNF. *Cell.* 85:1113–1124; 1996.

[23] Jing, S.; Yu, Y.; Fang, M.; Hu, Z.; Holst, P. L.; Boone, T.; Delaney, J.; Schultz, H.; Zhou, R.; Fox, G. M. GFRα-2 and GFRα-3 are two new receptors for ligands of the GDNF family. *J. Biol. Chem.* 272:33111–33117; 1997.

[24] Kemmink, J.; Darby, N. J.; Dijkstra, K.; Nilges, M.; Creighton, T. E. The folding catalyst protein disulfide isomerase is constructed of active and inactive thioredoxin modules. *Curr. Biol.* 7:239-245; 1997.

[25] Kinoshita, M.; Noda, M. Roles of septins in the mammalian cytokinesis machinery. *Cell. Struct. Funct.* 26:667-670; 2001.

[26] Kinsella, M. G.; Bressler, S. L.; Wight, T. N. The regulated synthesis of versican, decorin, and biglycan: extracellular matrix proteoglycans that influence cellular phenotype. Crit. Rev. Eukaryot. *Gene. Expr.* 14:203-234; 2004.

[27] Kobayashi, R.; Sekino, Y.; Shirao, T.; Tanaka, S.; Ogura, T.; Inada, K.; Saji, M. Antisense knockdown of drebrin A, a dendritic spine protein, causes stronger preference, impaired pre-pulse inhibition, and an increased sensitivity to psychostimulant. *Neurosci. Res.* 49:205-217; 2004.

[28] Kotzbauer, P. T.; Lampe, P. A.; Heuckeroth, R. O.; Golden, J. P.; Creedon, D. J.; Johnson Jr, E. M.; Milbrandt, J. Neuturin, a relative of glial-cell-line-derived neurotrophic factor. Nature. 384:467–470; 1996.

[29] Linnarsson, S.; Mikaels, A.; Baudet, C.; Ernfors, P. Activation by GDNF of a transcriptional program repressing neurite growth in dorsal root ganglia. *Proc. Natl. Acad. Sci.* U.S.A. 98:14681-14686; 2001.

[30] Mammoto, A.; Sasaki, T.; Asakura, T.; Hotta, I.; Imamura, H.; Takahashi, K.; Matsuura, Y.; Shirao, T.; Takai, Y. Interactions of drebrin and gephyrin with profilin. *Biochem. Biophys. Res. Commun.* 243:86–89; 1998.

[31] Manson, I. The RET receptor tyrosine kinase: activation, signalling and significance in neural development and disease. *Pharm Acta Helv.* 74:261-264; 2000.

[32] Masure, S.; Geerts, H.; Cik, M.; Hoefnagel, E.; Van Den Kieboom, G.; Tuytelaars, A.; Harris, S.; Lesage, A. S.; Leysen, J. E.; Van Der Helm, L.; Verhasselt, P.; Yon, J.; Gordon, R. D. Enovin, a member of the glial-cell-line-derived neurotrophic factor (GDNF) family with growth promoting activity on neuronal cells. Existence and tissue-specific expression of different splice variants. *Eur. J. Biochem.* 266:902–982; 1999.

[33] McQuay, H. J.; Jadad, A. R.; Carroll, D.; Faura, C.; Glynn, C. J.; Moore, R. A.; Liu, Y. Opioid sensitivity of chronic pain: a patient-controlled analgesia method. *Anaesthesia.* 47:757–767; 1992.

[34] Milbrandt, J.; de Sauvage, F. J.; Fahrner, T. J.; Baloh, R. H.; Leitner, M. L.; Tansey, M. G.; Lampe, P. A.; Heuckeroth, R. O.; Kotzbauer, P. T.; Simburger, K. S.; Golden, J. P.; Davies, J. A.; Vejsada, R.; Kato, A. C.; Hynes, M.; Sherman, D.; Nishimura, M.; Wang, L. C.; Vandlen, R.; Moffat, B.; Klein, R. D.; Poulsen, K.; Gray, C.; Garces, A.; Johnson Jr., E. M. Persephin, a novel neurotrophic factor related to GDNF and neurturin. *Neuron.* 20:245–253.1: 1998.

[35] Murakami, H.; Iwashita, T.; Asai, N.; Iwata, Y.; Narumiya, S.; Takahashi, M. Rho-dependent and -independent tyrosine phosphorylation of focal adhesion kinase, paxillin and p130Cas mediated by Ret kinase. *Oncogene.* 18:1975–1982; 1999.

[36] Murakami, H.; Iwashita, T.; Asai, N.; Shimono, Y.; Iwata, Y.; Kawai, K.; Takahashi, M. Enhanced phosphatidylinositol 3-kinase activity and high phosphorylation state of its downstream signalling molecules mediated by Ret with the MEN 2B mutation. Biochem. Biophys. R*es. Commun.* 262:68–75; 1999.

[37] Naveilhan, P.; Baudet, C.; Mikaels, A.; Shen, L.; Westphal, H.; Ernfors, P. Expression and regulation of GFRα3, a glial cell line-derived neurotrophic factor family receptor. *Proc. Natl. Acad. Sci.* U.S.A. 95:1295–1300; 1998.

[38] Noiva, R. Protein disulfide isomerase: the multifunctional redox chaperone of the endoplasmic reticulum. Semin. *Cell. Dev. Biol.* 10:481-493; 1999.

[39] Nomoto, S.; Ito, S.; Yang, L. X.; Kiuchi, K. Molecular cloning and expression analysis of GFRα-3, a novel cDNA related to GDNFRα and NTNRα. *Biochem. Biophys. Res. Commun.* 244:849–853; 1998.

[40] Nozaki, C.; Asai, N.; Murakami, H.; Iwashita, T.; Iwata, Y.; Horibe, K.; Klein, R. D.; Rosenthal, A.; Takahashi, M. Calcium-dependent Ret activation by GDNF and neurturin. *Oncogene.* 16:293–299; 1998.

[41] Offenhäuser, N.; Borgonovo, A.; Disanza, A.; Romano, P.; Ponzanelli, I.; Iannolo, G.; Di Fiore, P. P.; Scita, G. The eps8 family of proteins links growth factor stimulation to actin reorganization generating functional redundancy in the Ras/Rac pathway. *Mol. Biol. Cell.* 15:91-98; 2004.

[42] Ohiwa, M.; Murakami, H.; Iwashita, T.; Asai, N.; Iwata, Y.; Imai, T.; Funahashi, H.; Takagi, H.; Takahashi, M. Characterization of Ret-Shc-Grb2 complex induced by GDNF, MEN 2A and MEN 2B mutations. Biochem. Biophys. Res. *Commun.* 237:747–751; 1997.

[43] Osmanagic-Myers, S.; Wiche, G. Plectin-RACK1 (receptor for activated C kinase 1) scaffolding: a novel mechanism to regulate protein kinase C activity. *J. Biol. Chem.* 279:18701-18710; 2004.

[44] Ossipov, M. H.; Zhang, E. T.; Carvajal, C.; Gardell, L.; Quirion, R.; Dumont, Y.; Lai, J.; Porreca, F. Selective mediation of nerve injury-induced tactile hypersensitivity by neuropeptide Y. *J. Neurosci.* 22:9858–9867; 2002.

[45] Park, S.; Hong, Y. W. Transcriptional regulation of artemin is related to neurite outgrowth and actin polymerization in mature DRG neurons. *Neurosci. Lett.* 404:61-66; 2006.

[46] Paveliev, M.; Airaksinen, M. S.; Saarma, M. GDNF family ligands activate multiple events during axonal growth in mature sensory neurons. *Mol. Cell. Neurosci.* 25:453-459; 2004.

[47] Poteryaev, D.; Titievsky, A.; Sun, Y. F.; Thomas-Crusells, J.; Lindahl, M.; Billaud, M.; Arumae, U.; Saarma, M. GDNF triggers a novel Ret-independent Src kinase family-coupled signaling via a GPI-linked GDNF receptor α1. *FEBS Lett.* 463:63–66; 1999.

[48] Ramer, M. S.; Priestley, J. V.; McMahon, S. B. Functional regeneration of sensory axons into the adult spinal cord. *Nature.* 403:312–316; 2000.

[49] Robles, E.; Woo, S.; Gomez, T. M. Src-dependent tyrosine phosphorylation at the tips of growth cone filopodia promotes extension. *J Neurosci.* 25:7669-7681; 2005.

[50] Rosenblad, C.; Gronborg, M.; Hansen, C.; Blom, N.; Meyer, M.; Johansen, J.; Dago, L.; Kirik, D.; Patel, U. A.; Lundberg, C.; Trono, D.; Bjorklund A.; Johansen, T. E. In vivo protection of nigral dopamine neurons by lentiviral gene transfer of the novel GDNF-family member neublastin/artemin. *Mol. Cell. Neurosci.* 15:199–214; 2000.

[51] Rusanescu, G.; Qi, H.; Thomas, S. M.; Brugge, J. S.; Halegoua, S. Calcium influx induces neurite growth through a Src-Ras signaling cassette. *Neuron.* 15:1415–1425; 1995.

[52] Sanicola, M.; Hession, C.; Worley, D.; Carmillo, P.; Ehrenfels, C.; Walus, L.; Robinson, S.; Jaworski, G.; Wei, H.; Tizard, R.; Whitty, A.; Pepinsky, R. B.; Cate, R. L. Glial cell line-derived neurotrophic factor-dependent RET activation can be mediated by two different cell-surface accessory proteins. *Proc. Natl. Acad. Sci.* U.S.A. 94:6238–6243; 1997.

[53] Sasaki, Y.; Hayashi, K.; Shirao, T.; Ishikawa. R.; Kohama, K. Inhibition by drebrin of the actin-bundling activity of brain fascin, a protein localized in filopodia of growth cones. *J. Neurochem.* 66:980–988; 1996.

[54] Schwab, M. E. Neurobiology. Finding the lost target. Nature. 403:257–260; 2000.

[55] Segouffin-Cariou, C.; Billaud, M. Transforming ability of MEN2A-RET requires activation of the phosphatidylinositol 3-kinase/AKT signaling pathway. *J. Biol. Chem.* 275:3568–3576; 2000.

[56] Sevcík, J.; Urbániková, L.; Kost'an, J.; Janda, L.; Wiche, G. Actin-binding domain of mouse plectin. Crystal structure and binding to vimentin. *Eur. J Biochem.* 271:1873-1884; 2004.

[57] Shirao, T.; Hayashi, K.; Ishikawa, R.; Isa, K.; Asada, H.; Ikeda, K.; Uyemura, K. Formation of thick, curving bundles of actin by drebrin A expressed in fibroblasts. *Exp. Cell. Res.* 215:145–153; 1994.

[58] Shirao, T. The roles of microfilament-associated proteins, drebrins, in brain morphogenesis: a review. *J. Biochem.* 117:231–236; 1995.

[59] Soler, R. M.; Dolcet, X.; Encinas, M.; Egea, J.; Bayascas, J. R.; Comella, J. X. Receptors of the glial cell line-derived neurotrophic factor family of neurotrophic factors signal cell survival through the phosphatidylinositol 3-kinase pathway in spinal cord motoneurons. *J. Neurosci.* 19:9160–9169; 1999.

[60] Stipp, C. S.; Litwack, E. D.; Lander, A. D. Cerebroglycan: an integral membrane heparan sulfate proteoglycan that is unique to the developing nervous system and expressed specifically during neuronal differentiation. *J. Cell Biol.* 124:149-160; 1994.

[61] Suvanto, P.; Wartiovaara, K.; Lindahl, M.; Arumae, U.; Moshnyakov, M.; Horelli-Kuitunen, N.; Airaksinen, M. S.; Palotie, A.; Sariola, H.; Saarma, M. Cloning, mRNA distribution and chromosomal localisation of the gene for glial cell line-derived neurotrophic factor receptor β, a homologue to GDNFR-α. *Hum. Mol. Genet.* 6:1267–1273; 1997.

[62] Takahashi, M. The GDNF/RET signaling pathway and human diseases. *Cytokine Growth Factor Rev.* 12:361-373; 2001.

[63] Tan, L. W.; Bianco T.; Dobrovic, A. Variable promoter region CpG island methylation of the putative tumor suppressor gene Connexin 26 in breast cancer. *Carcinogenesis.* 23:231–236; 2002.

[64] Thompson, J.; Doxakis, E.; Pinon, L. G.; Strachan, P.; Buj-Bello, A.; Wyatt, S.; Buchman, V. L.; Davies, A. M. GFRα-4, a new GDNF family receptor. *Mol. Cell. Neurosci.* 11:117–126; 1998.

[65] Treanor, J. J.; Goodman, L.; de Sauvage, F.; Stone, D. M.; Poulsen, K. T.; Beck, C. D.; Gray, C.; Armanini, M. P.; Pollock, R. A.; Hefti, F.; Phillips, H. S.; Goddard, A.; Moore, M. W.; Buj-Bello, A.; Davies, A. M.; Asai, N.; Takahashi, M.; Vandlen, R.; Henderson, C. E.; Rosenthal, A. Characterization of a multicomponent receptor for GDNF. *Nature.* 382:80–83; 1996.

[66] Trupp, M.; Arenas, E.; Fainzilber, M.; Nilsson, A. S.; Sieber, B. A.; Grigoriou, M.; Kilkenny, C.; Salazar-Grueso, E.; Pachnis, V.; Arumae, U. Functional receptor for GDNF encoded by the c-ret proto-oncogene. *Nature.* 381:785–789; 1996.

[67] Trupp, M.; Raynoschek, C.; Belluardo, N.; Ibanez, C. F.; Multiple GPI-anchored receptors control GDNF-dependent and independent activation of the c-Ret receptor tyrosine kinase. *Mol. Cell. Neurosci.* 11:47–63: 1998.

[68] Trupp, M.; Scott, R.; Whittemore, S. R.; Ibanez, C. F. Ret-dependent and -independent mechanisms of glial cell line-derived neurotrophic factor signaling in neuronal cells. *J. Biol. Chem.* 274:20885–20894; 1999.

[69] Vaillancourt, R. R.; Heasley, L. E.; Zamarripa, J.; Storey, B.; Valius, M.; Kazlauskas, A.; Johnson, G. L. Mitogen-activated protein kinase activation is insufficient for growth factor receptor-mediated PC12 cell differentiation. *Mol. Cell. Biol.* 15:3644–3653; 1995.

[70] van Weering, D. H.; Bos, J. L. Glial cell line-derived neurotrophic factor induces Ret-mediated lamellipodia formation. *J. Biol. Chem.* 272:249–254; 1997.

[71] van Weering, D. H.; Medema, J. P.; van Puijenbroek, A.; Burgering, B. M.; Baas, P. D.; Bos, J. L. Ret receptor tyrosine kinase activates extracellular signal-regulated kinase 2 in SK-N-MC cells. *Oncogene.* 11:2207–2214; 1995.

[72] Waetzig, V.; Herdegen, T. MEKK1 controls neurite regrowth after experimental injury by balancing ERK1/2 and JNK2 signaling. Mol. Cell. *Neurosci.* 30:67-78; 2005.

[73] Wang, C. Y.; Yang, F.; He, X. P.; Je, H. S.; Zhou, J. Z.; Eckermann, K.; Kawamura, D.; Feng, L.; Shen, L.; Lu, B. Regulation of neuromuscular synapse development by glial cell line-derived neurotrophic factor and neurturin. *J. Biol. Chem.* 277:10614–10625; 2002.

[74] Worby, C. A.; Vega, Q. C.; Chao, H. H. J.; Seasholtz, A. F.; Thompson R. C.; Dixon, J. E. Identification and characterization of GFRα-3, a novel co-receptor belonging to the

glial cell line-derived neurotrophic receptor family. *J. Biol. Chem.* 273:3502–3508; 1998.

[75] Worby, C. A.; Vega, Q. C.; Zhao, Y.; Chao, H. H.; Seasholtz, A. F.; Dixon, J .E. Glial cell line-derived neurotrophic factor signals through the RET receptor and activates mitogen-activated protein kinase. *J. Biol. Chem.* 271:23619–23622; 1996.

[76] Wiche, G. Role of plectin in cytoskeleton organization and dynamics. *J. Cell. Sci.* 111:2477-2486; 1998.

[77] Yamaguchi, H.; Lorenz, M.; Kempiak, S.; Sarmiento, C.; Coniglio, S.; Symons, M.; Segall, J.; Eddy, R.; Miki, H.; Takenawa, T.; Condeelis, J. Molecular mechanisms of invadopodium formation: the role of the N-WASP-Arp2/3 complex pathway and cofilin. *J. Cell. Biol.* 31:441-452; 2005.

[78] Yan, H,; Newgreen, D. F.; Young, H. M. Developmental changes in neurite outgrowth responses of dorsal root and sympathetic ganglia to GDNF, neurturin, and artemin. *Dev. Dyn.* 2003 227:395-401; 2003.

[79] Yan, M.; Collins R. F.; Grinstein, S.; Trimble, W. S. Coronin-1 function is required for phagosome formation. *Mol. Biol. Cell.* 16:3077-308; 2005.

[80] Zöchbauer-Müller, S.; Fong, K. M.; Virmani, A. K.; Geradts, J.; Gazdar, A. F.; Minna, J. D. Aberrant promoter methylation of multiple genes in non-small cell lung cancers. *Cancer. Res.* 61:249–255; 2001.

Index

A

B

E

F

G

H

I

J

K

L

M

N

Q

R

S

T

U

V

W

X

Y

Z